Energy Balance and Cancer

Series Editor:
Nathan A. Berger
Case Western Reserve University
Cleveland, OH, USA

For further volumes:
http://www.springer.com/series/8282

Deborah J. Bowen • Gerald V. Denis •
Nathan A. Berger

Editors

Impact of Energy Balance on Cancer Disparities

 Springer

Editors
Deborah J. Bowen
Department of Community
  Health Sciences
Boston University
Boston, MA, USA

Gerald V. Denis
Boston University
Boston, MA, USA

Nathan A. Berger
Center for Science, Health & Society
Case Western Reserve University
  School of Medicine
Cleveland, OH, USA

ISBN 978-3-319-06102-3   ISBN 978-3-319-06103-0 (eBook)
DOI 10.1007/978-3-319-06103-0
Springer Cham Heidelberg New York Dordrecht London

Library of Congress Control Number: 2014942691

Printed on acid-free paper

Springer is part of Springer Science+Business Media (www.springer.com)

Preface

Energy Balance and Cancer Disparities

While great progress has been made across the spectrum of cancer research, extending from prevention, diagnosis, and therapy to survivorship, the benefits of these advances have not been realized by all groups. Significant disparities exist due to a variety of factors including age, gender, ethnicity, socioeconomic status, geography, built environment, and others. Since energy balance impacts the entire continuum of cancer care, from prevention through survivorship, groups affected by disparities in energy balance including the complex issues influencing obesity, exercise, sedentary behavior, sleep, insulin resistance, and more may show profound differences in cancer outcomes. Moreover, these disparities may have diverse contributors and consequences in different regions throughout the world.

The goal of this volume is to identify cancer disparities in different groups in the USA and around the world and compare similarities and variations in energy balance to identify commonalities in order to inform further opportunities for transdisciplinary research and interventions. Specific chapters have been included to provide information regarding application of current state-of-the-art strategies to analyze and alter biologic, behavioral, community, and policy effects on energy balance and the disparities that result from barriers that restrict their generalized implementation.

In Chap. 1, Rory Weier, James Fisher, and Electra Paskett (Ohio State University) along with Jesse Plascak (University Washington) discuss the distinctive features of Appalachia and its unique contribution to the burden of obesity, cancer incidence, and mortality in the USA. In Chap. 2, Donald Nicolson and Una Macleod (Hull York Medical School) and David Weller (University Edinburgh) examine socioeconomic factors that determine disparities in lifestyle factors, cancer incidence and outcomes in the United Kingdom.

In Chap. 3, Donna Spruijt-Metz, Lauren Cook, CK Freddy Wen, Robert Garcia, Gillian A. O'Reilly, Jennifer B. Unger, (University Southern California Keck School of Medicine), Selena T. Nguyen-Rodriguez (California State University,

Long Beach), and Ya-Wen Hsu (Chia Nan University of Pharmacy and Science, Taiwan) discuss behavioral influences on racial/ethnic and socioeconomic disparities versus incidence and mortality by cancer sites. Chapter 4, by Kathryn Schmitz (University Pennsylvania), Tanya Agurs-Collins (National Cancer Institute, Marian Neuhouser (Fred Hutchinson Cancer Research Center), Lisa Pollack and Sarah Gehlert (Washington University in St. Louis), reviews the impact of obesity, race, and ethnicity on cancer survivorship, which is particularly important in view of the projected increase in this group of patients. In Chap. 5, Nathan LeBrasseur (Mayo Clinic), Derek Huffman (Albert Einstein College of Medicine), and Gerald Denis (Boston University College of Medicine) discuss the impact of aging on obesity, inflammation, and cancer. They raise the possibility that healthy aging may maintain fitness or protect against these chronic disorders and examine the social determinants of healthy and unhealthy aging. Focusing on specific malignancies with established disparities, in Chap. 6, Graham Colditz, Kari Bohlke, Su-Hsin Chang, and Kenneth Carson (Washington University School of Medicine) review the evidence that obesity, more common in African Americans, and other factors such as lower serum levels of 25-hydroxy vitamin D, may contribute to the significantly higher incidence of Multiple Myeloma. In Chap. 7, Melissa Kang and Temitope Keku (University of North Carolina) discuss single nucleotide polymorphisms (SNPs) that occur in a racially oriented manner, resulting in differences in obesity and inflammatory genes that may contribute to racial disparities in colorectal cancer incidence and survival. Rebecca Hasson (University Michigan) and Michael Goran (University Southern California) in Chap. 8 and Sarah Cohen (EpidStat Institute) and Loren Lipworth (Vanderbilt-Ingram Cancer Center) in Chap. 9 provide comprehensive assessments of racial differences in biological mechanisms linking obesity to cancer with particular focus on insulin resistance, sex steroids, inflammatory mediators, and adipokines. In Chap. 10, Melinda Stolley (University of Illinois at Chicago) analyzes behavioral factors contributing to disparities in breast cancer survival and describes community-based strategies to alter energy balance and decrease disparities. In Chap. 11, Deborah Bowen (University of Washington) and Stacey Zawacki (Boston University) examine differential responsibilities and potential contributions to change neighborhood-based policies that impact the obesogenic environment or, as uniquely defined in this chapter, the inflammatory environment. In the last section, Chap. 12, Debra Haire-Joshu (Washington University) focuses on the important issue of how public and social policy has been and can be used to prevent obesity-related disparities in young children thereby reducing their predisposition to cancer at later stages of life.

This current volume in the series on Energy Balance and Cancer provides a unique transdisciplinary approach to analyze problems associated with disparities in energy balance and cancer in diverse geographic areas and among different ethnicities from a biological, behavioral, socioeconomic, environmental, and policy basis as well as to suggest where and how potential interventions may be helpful. This volume should provide a valuable resource to all investigators, practitioners, and policy makers dealing with problems of obesity, energy balance, and cancer. It is the first major book dealing with biology, behavior, and policy that contributes to and results from disparities in energy balance and cancer. It should provide a valuable resource to disparity-focused investigators at the molecular, psychosocial, community, and policy levels and serve as an important guide to the broad range of professionals who regularly deal with these issues.

Boston, MA   Deborah J. Bowen
Boston, MA   Gerald V. Denis
Cleveland, OH   Nathan A. Berger

Contents

1 Obesity and Cancer in Appalachia 1
Rory C. Weier, Jesse J. Plascak, James L. Fisher,
and Electra D. Paskett

2 Disparities in Cancer Outcomes: A UK Perspective 19
Donald J. Nicolson, Una Macleod, and David Weller

**3 Behavioral Differences Leading to Disparities in Energy Balance
and Cancer** 37
Donna Spruijt-Metz, Lauren Cook, C.K. Freddy Wen, Robert Garcia,
Gillian A. O'Reilly, Ya-Wen Hsu, Jennifer B. Unger,
and Selena T. Nguyen-Rodriguez

**4 Impact of Obesity, Race, and Ethnicity on Cancer
Survivorship** 63
Kathryn H. Schmitz, Tanya Agurs-Collins, Marian L. Neuhouser,
Lisa Pollack, and Sarah Gehlert

**5 The Biology of Aging: Role in Cancer, Metabolic Dysfunction,
and Health Disparities** 91
Nathan K. LeBrasseur, Derek M. Huffman, and Gerald V. Denis

6 Energy Balance and Multiple Myeloma in African Americans 119
Graham A. Colditz, Kari Bohlke, Su-Hsin Chang,
and Kenneth Carson

**7 Single Nucleotide Polymorphisms in Obesity and Inflammatory
Genes in African Americans with Colorectal Cancer** 131
Melissa Kang and Temitope O. Keku

**8 Ethnic Differences in Insulin Resistance as a Mediator
of Cancer Disparities** 165
Rebecca E. Hasson and Michael I. Goran

9   Role of Ethnic Differences in Mediators of Energy Balance 201
      Sarah S. Cohen and Loren Lipworth

**10   Community-Based Strategies to Alter Energy Balance
      in Underserved Breast Cancer Survivors** 233
      Melinda Stolley

11   The Role of Policy in Reducing Inflammation 259
      Deborah J. Bowen and Stacey Zawacki

**12   Cancer Prevention Through Policy Interventions That Alter
      Childhood Disparities in Energy Balance** 283
      Debra Haire-Joshu

Index 305

List of Contributors

Tanya Agurs-Collins Health Behaviors Research Branch (HBRB), Behavioral Research Program (BRP), Division of Cancer Control and Population Sciences, National Cancer Institute, Rockville, MD, USA

Nathan A. Berger Center for Science, Health and Society, Case Western Reserve University School of Medicine, Cleveland, OH, USA

Kari Bohlke Alvin J. Siteman Cancer Center, Barnes-Jewish Hospital, Washington University School of Medicine, St. Louis, MO, USA

Deborah J Bowen Department of Bioethics and Humanities, University of Washington, Seattle, WA, USA

Kenneth Carson Alvin J. Siteman Cancer Center, Barnes-Jewish Hospital, Washington University School of Medicine, St. Louis, MO, USA

Su-Hsln Chang Alvin J. Siteman Cancer Center, Barnes-Jewish Hospital, Washington University School of Medicine, St. Louis, MO, USA

Sarah S. Cohen EpidStat Institute, Ann Arbor, MI, USA

Graham A. Colditz Department of Epidemiology, Barnes-Jewish Hospital, Washington University School of Medicine, St. Louis, MO, USA

Lauren Cook Department of Preventive Medicine, Keck School of Medicine, University of Southern California, Los Angeles, CA, USA

Gerald V. Denis Cancer Research Center and Department of Pharmacology and Experimental Therapeutics, Boston University School of Medicine, Boston, MA, USA

James L. Fisher Comprehensive Cancer Center, The Ohio State University, Columbus, OH, USA

Robert Garcia Department of Preventive Medicine, USC Institute for Prevention Research, Keck School of Medicine, University of Southern California, Claremont, CA, USA

Sarah Gehlert George Warren Brown School of Social Work, Washington University, St. Louis, MO, USA

Michael I. Goran University of Southern California, Los Angeles, CA, USA

Debra Haire-Joshu Brown School of Social Work, Washington University, St. Louis, MO, USA

Rebecca E. Hasson Schools of Kinesiology and Public Health, University of Michigan, Ann Arbor, MI, USA

Ya-Wen Hsu Department of Hospital and Health Care Administration, Chia Nan University of Pharmacy & Science, Jen-Te, Tainan, Taiwan

Derek M. Huffman Institute for Aging Research, Albert Einstein College of Medicine, Bronx, NY, USA

Melissa Kang Division of Gastroenterology and Hepatology, University of North Carolina, Chapel Hill, NC, USA

Temitope O. Keku Division of Gastroenterology and Hepatology, Department of Medicine, University of North Carolina CB # 7032, Chapel Hill, NC, USA

Nathan K. LeBrasseur Physical Medicine and Rehabilitation, Physiology and Biomedical Engineering, Robert and Arlene Kogod Center for Aging, Mayo Clinic, Rochester, MN, USA

Loren Lipworth Division of Epidemiology, Department of Medicine, Vanderbilt-Ingram Cancer Center, Nashville, TN, USA

Una Macleod Centre for Health and Population Sciences, Hull York Medical School, University of Hull, Hull, UK

Marian L. Neuhouser Division of Public Health Sciences, Fred Hutchinson Cancer Research Center, Seattle, WA, USA

Selena T. Nguyen-Rodriguez Department of Health Science, California State University, Long Beach, CA, USA

Donald J Nicolson Centre for Health and Population Sciences, Hull York Medical School, University of Hull, Hull, UK

Gillian A. O'Reilly Department of Preventive Medicine, Keck School of Medicine, University of Southern California, Los Angeles, CA, USA

Electra D. Paskett College of Public Health—Comprehensive Cancer Center, Columbus, OH, USA

Jesse J. Plascak Biobehavioral Cancer Prevention and Control, University of Washington, Seattle, WA, USA

Lisa Pollack George Warren Brown School of Social Work, Washington University, St. Louis, MO, USA

Kathryn H. Schmitz Division of Clinical Epidemiology, Center for Clinical Epidemiology and Biostatistics, Perelman School of Medicine, University of Pennsylvania, Philadelphia, PA, USA

Donna Spruijt-Metz Mobile and Connected Health Program, Center for Economic and Social Research, University of Southern California, Playa Vista, CA, USA

Melinda Stolley Institute for Health Research and Policy, University of Illinois at Chicago (MC 275), Chicago, IL, USA

Jennifer B. Unger Department of Preventive Medicine, Institute for Health Promotion and Disease Prevention Research, Keck School of Medicine, University of Southern California, Los Angeles, CA, USA

Rory Cusack Weier Comprehensive Cancer Center, The Ohio State University, Columbus, OH, USA

David Weller Edinburgh Cancer Research UK Centre, The University of Edinburgh, Edinburgh, UK

CK Freddy Wen Department of Preventive Medicine, Keck School of Medicine, University of Southern California, Los Angeles, CA, USA

Stacey A. Zawacki Department of Nutritional Sciences, Boston University, Boston, MA, USA

Chapter 1
Obesity and Cancer in Appalachia

Rory C. Weier, Jesse J. Plascak, James L. Fisher, and Electra D. Paskett

Abstract Appalachia, a diverse, federally designated region that spans 13 states, is home to nearly 25 million residents. It is also an area in which the leading cause of death is cancer and financial and physical access to healthcare are known barriers to regular medical care. Obesity and its risk factors contribute to the region's burden of cancer incidence and mortality. Disparate prevalences of overweight and obesity have been found in Appalachia as early as preschool, and, compared to the rest of the country, parts of Appalachia have higher rates of physical inactivity and lower prevalence of fruit and vegetable consumption. Obesity is related to at least eight types of cancer, of which colorectal cancer and female breast cancer have been the most heavily examined in Appalachia. This report reviews what is known about obesity and cancer in the Appalachian region and provides suggestions for future intervention and research to address Appalachia's cancer and obesity burdens.

R.C. Weier
Comprehensive Cancer Center, The Ohio State University, 1590 North High Street, Suite 525, Columbus, OH 43201, USA
e-mail: Rory.Weier@osumc.edu

J.J. Plascak
Biobehavioral Cancer Prevention and Control Training Program, University of Washington, Box 359455, Seattle, WA 98195-9455, USA
e-mail: plascak@uw.edu

J.L. Fisher
James Cancer Hospital and Solove Research Institute, The Ohio State University, 1590 North High Street, Suite 525, Columbus, OH 43201, USA
e-mail: Jay.Fisher@osumc.edu

E.D. Paskett (✉)
Division of Cancer Prevention and Control, Department of Internal Medicine, College of Medicine, The Ohio State University, 1590 North High Street, Suite 525, Columbus, OH 43201, USA
e-mail: Electra.Paskett@osumc.edu

D.J. Bowen et al. (eds.), *Impact of Energy Balance on Cancer Disparities*,
Energy Balance and Cancer 9, DOI 10.1007/978-3-319-06103-0_1,
© Springer International Publishing Switzerland 2014

Keywords Obesity • Diet • Exercise • Disparities • Cancer • Rural health • Appalachia

The Relationship Between Overweight and Obesity and Cancer

Body mass index (BMI), a ratio of weight in kilograms (kg) to height in meters (m) squared, is commonly used to categorize body weight. Among adults, overweight is defined as a BMI between 25.0 and 29.9 and obesity is defined as BMI greater than or equal to 30.0 [1]. Among children, overweight is defined as BMI between the 85th and 95th percentiles for children of the same sex and age, and obesity is defined as equal to or greater than the 95th percentile for children of the same sex and age [1].

Overweight and obesity have greatly increased over the past three decades in the USA [2]. Results from the 2009 to 2010 National Health and Nutrition Examination Survey (NHANES) show that 68.7 % of US adults aged 20 years and older are estimated to be overweight or obese, and among children and teens, aged 2–19 years, 17.0 % are estimated to be obese [3]. According to the American Cancer Society, diet, physical activity, and weight status are associated with cancer risk. One-third of the more than 500,000 cancer deaths that occur each year in the USA can be attributed to poor diet and physical inactivity, which are also risk factors for overweight and obesity. Additionally, it is estimated that overweight and obesity are responsible for approximately 14 % of all cancer deaths among men and 20 % of all cancer deaths among women [4].

Epidemiologic and molecular studies in various countries and different settings have provided supporting evidence of a causal relationship between excess adiposity (fat storage) and cancer risk [5, 6]. Adipose tissues are highly metabolically active and produce an array of hormones, growth factors, and signaling molecules fueling inflammation and cellular proliferation that may lead to cancer [5, 6]. According to the National Cancer Institute (NCI) overweight and obesity increases the risk of at least the eight following cancers: esophageal, pancreatic, colorectal, female breast (after menopause), endometrial, kidney, thyroid, and gallbladder cancers [7]. Obesity and physical inactivity may account for approximately 25–30 % of new cases of colon, female breast (postmenopausal), endometrial, kidney, and esophageal cancers [8]. In addition, overweight and obesity *may* increase the risk of several other sites/types of cancer (e.g., prostate, other male genitals, ovary, non-Hodgkin's lymphoma, leukemia, liver, and hemangioma) [7].

A population group suffering from a cancer burden that is significantly higher than that of the general population is defined as a cancer disparity population [9]. The Appalachian population suffers a disproportionate burden of cancer, and cancer risk factors such as obesity. The purpose of this report is to review what is known about obesity and cancer in Appalachia.

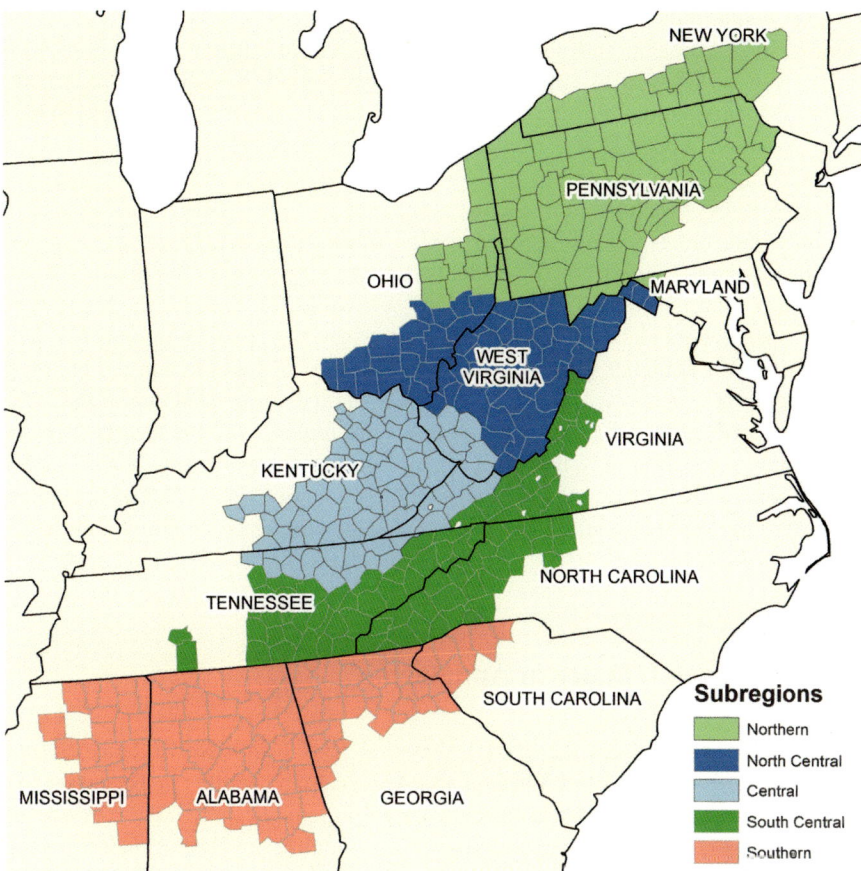

Fig. 1.1 The current Appalachian region of the USA (Adapted from the Appalachia Regional Commission) [12]

The Appalachian Region

Appalachia is a federally designated region of the USA that includes 24.8 million residents living within 420 contiguous counties across 13 states that include some portion of the Appalachian mountains or foothills [10]. In most cases, Appalachian status was given to a county because of lagging socioeconomic indicators [10, 11]. Once designated as "Appalachian" a county is qualified for special government funding and subsidies [10, 11]. The Appalachian Regional Commission (ARC)—a regional economic development agency—has been charged with coordinating programs in the region [10]. The ARC has categorized Appalachian counties into sub regions based on similar demographic and socioeconomic characteristics (Fig. 1.1) [12]. Despite this, subregions are not consistently defined throughout the Appalachian health literature.

The demographic and socioeconomic characteristics vary greatly between regions of Appalachia and non-Appalachia and within Appalachia. According to an analysis of 2007–2011 American Community Survey data, 83.9 % of Appalachian residents, compared to 64.2 % of US residents, were non-Hispanic Whites [13]. Between Appalachian subregions, the percentage of non-Hispanic Whites varied from 70.4 % (Southern) to 95.5 % (Central). The same analysis indicated that 16.5 % of Appalachian residents, compared to 14.6 % of US residents, aged 25 years and older had not attained a high school diploma. However, Northern Appalachia had a percentage of residents not attaining a high school diploma that was lower than that of US residents (11.8 %), while Central Appalachia had the highest percentage (27.2 %). Similarly, the poverty rate among Appalachian residents (16.1 %) is higher than that of the USA (14.3 %) [13]. Again, the Northern Appalachia poverty prevalence of 13.8 % was slightly lower than that of the USA, while all other regions had prevalences that were higher, with the Central Appalachian prevalence of 23.5 % being the highest among the subregions. These regionally dependent demographic and socioeconomic characteristics of Appalachia are important factors that could affect various health outcomes including cancer and risk factors such as obesity.

The Burden of Obesity in Appalachia

An analysis of data from the 2007 Behavioral Risk Factor Surveillance System (BRFSS) indicates that West Virginia and Appalachian counties in Tennessee and Kentucky were among the areas of the USA with the highest prevalence of obesity ($\geq$30.9 %) [14]. Estimates indicate that 81 % of counties in Kentucky, Tennessee, and West Virginia and 77 % in Alabama, Georgia, Louisiana, Mississippi, and South Carolina had obesity prevalences greater than 60 % of all US counties [14]. Many of these counties are located within Appalachia (Fig. 1.2). According to 2004–2007 state BRFSS data, Appalachian regions of Kentucky, Ohio, Pennsylvania, Virginia, and West Virginia had obesity prevalences ranging between 27 % (Pennsylvania) and 34.7 % (Kentucky) among men and 26.0 % (Pennsylvania, Virginia) and 31.7 % (Kentucky) among women [15]. Moreover, in each of these states, except Pennsylvania, the obesity prevalence was greater in the Appalachian region, compared to the non-Appalachian region.

A number of studies have estimated the burden of overweight and obesity among children and adults in Appalachia. There is evidence of disparate rates of overweight and obesity among low-income Appalachian children as early as preschool [16]. The rates of high BMI among nearly 500 preschool-aged children participating in Southeastern Ohio Head Start programs exceeded national estimates [16]. Among a sample of 2,000 children aged 6–11 years participating in a school-based screening program in Southeastern Ohio, the overweight (BMI $\geq$ 85th percentile) and obesity (BMI > 95th percentile) prevalences of (17 %) and (20.9 %), respectively, both significantly exceeded national estimates

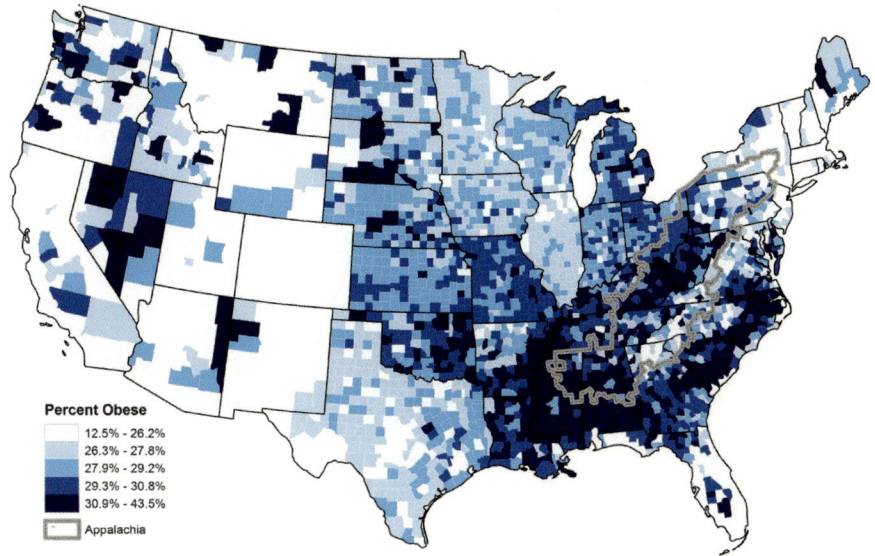

Fig. 1.2 County-level estimates of adult (≥20 years) obesity prevalence, 2007 (Adapted from the Centers for Disease Control and Prevention) [29]

[17]. Additionally, among a sample of over 1,500 Appalachian adults living in West Virginia, 39 % were overweight and 34 % were obese [18]. Despite the high BMI status of this population, 74 % viewed themselves as healthy [18]. Discrepancies between subjective and objective measures of health were also observed among over 200 adults in Appalachian Kentucky. Over 60 % of the sample considered their health to be good, while 75 % were overweight or obese [19]. The observed disconnect between self-reported health and BMI demonstrates the importance of understanding obesity from an Appalachian perspective in order to make progress.

The Burden of Cancer in Appalachia

Nationally and worldwide, the leading cause of death is heart disease [20]. However, within the Appalachian region, the leading cause of death is cancer [21]. Compared to the rest of the country, the Appalachian region had higher incidence rates of cancers diagnosed between 2001 and 2003 [22]. There were also differences between the Northern, Central, and Southern regions of Appalachia. Overall, cancer incidence rates were lowest in the Southern region of Appalachia and were highest in the Central and Northern regions [22]. Similar differences in cancer mortality rates were found using National Center for Health Statistics (NCHS) cancer mortality data from 2003 to 2007 [21]. Compared to the rest of the USA, the mortality rate for all cancers combined was 7 % higher in the 13 states that compromise the

Appalachian region [21]. Within these 13 states, the mortality rate for all cancers combined was 5 % higher in Appalachian counties than in non-Appalachian counties [21].

According to 2003–2007 state BRFSS data, Appalachian regions of Kentucky, Ohio, Pennsylvania, Virginia, and West Virginia had prevalences of tobacco smoking that ranged between 25.9 % (Virginia) and 33.6 % (Kentucky) among men and 25.9 % (Virginia) and 29.0 % (Kentucky) among women [15]. For both genders, the prevalence of smoking was higher in Appalachian regions of the five states [15]. And, for each state, the prevalence was greater than that for the non-Appalachian region. The high prevalence of tobacco smoking in Appalachia undoubtedly underlies some of the cancer differences observed between the Appalachian region and the rest of the USA. However, tobacco use and associated effects do not fully explain these disparities [23, 22]. Obesity also contributes to the region's disparate cancer incidence and mortality. Of the eight cancers related to obesity, colorectal cancer and female breast cancer have been the most heavily studied in Appalachia.

Colorectal Cancer

As the third leading cause of cancer incidence and mortality in the USA, colorectal cancer is estimated to account for 9 % of new cancer cases (73,680) and 9 % of deaths (26,300) among males and 9 % of new cancer cases (69,140) and 9 % of deaths (24,530) among females in 2013 [24]. According to the NCI, high BMI and waist circumference, a measure of abdominal obesity, are more strongly associated with increased colorectal cancer risk among men than women. A potential mechanism that may explain the relationship between obesity and increased risk of colorectal cancer is insulin and insulin-related growth factor levels, which tend to be higher in people who are obese [7].

Across Appalachian regions of Kentucky, Ohio, Pennsylvania, Virginia, and West Virginia, 2001–2006 state cancer registry data indicate that the average annual, age-adjusted incidence rate of colorectal cancer was 56.8 per 100,000 (Virginia) to 70.7 per 100,000 (West Virginia) among men and 39.9 per 100,000 (Virginia) to 52.3 per 100,000 (Kentucky) among women [15]. For both genders, the colorectal cancer incidence rate was higher in the Appalachian region of the states, except for Virginia [15].

An analysis of 2003–2007 NCHS colorectal cancer mortality data indicates that Appalachian regions of 9 of the 13 states experienced colorectal cancer mortality rates that were higher than the national rate [21]. Appalachia Kentucky had the highest colorectal cancer mortality rate (21.6 per 100,000), Appalachia Georgia had the lowest colorectal cancer mortality rate (17 per 100,000), and Ohio was the only state in which the Appalachian region had a significantly higher colorectal cancer mortality rate than the non-Appalachian region (9 %) [21]. Across Appalachian regions of Kentucky, Ohio, Pennsylvania, Virginia, and West Virginia, 2001–2006 state cancer registry data indicate that the average annual age-adjusted rate of colorectal cancer mortality ranged between 21.1 per 100,000 (Virginia) up to 26.06 per 100,000 (West Virginia) among men and 13.2 per 100,000 (Virginia) to 19.0 per 100,000 (Kentucky) among women [15]. For both genders, the colorectal cancer mortality rate was higher in the Appalachian region of the state, except for Virginia [15].

Unlike many other cancers, it is possible to prevent colorectal cancer using screening methods to detect precancerous polyps and prompt their surgical removal [24]. The 3 % annual decline in the rate of colorectal cancer morality nationally between 2000 and 2009 has primarily been attributed to an increase in the prevalence of screening. However, there is evidence of geographic variability in declining colorectal cancer mortality rates and uptake of screening recommendations [25]. In the early 1990s, states in the Northeast and North central region of the USA experienced the highest rates of colorectal cancer. However, in the mid-2000s, the highest rates of colorectal cancer were concentrated along the southern Appalachian region, which may be indicative of low screening rates and late stage diagnoses [25]. According to 2002–2006 state BRFSS data, Appalachian regions of Kentucky, Ohio, Pennsylvania, Virginia, and West Virginia had prevalences of adult colonoscopy or sigmoidoscopy use within the past 5 years ranging from 40.0 % (Ohio) to 66.6 % (Kentucky) [15]. The prevalence of adults having undergone a colonoscopy or sigmoidoscopy within the past 5 years was higher in the non-Appalachian region of the state, except for Kentucky [15]. Furthermore, 2002–2006 state cancer registry data indicate that the proportion of colorectal cancer cases diagnosed late stage ranged from 46.5 % (Ohio) to 54.9 % (Virginia) among men and 46.7 % (Ohio) to 55.2 % (Virginia) among women [15]. Only in Kentucky was the prevalence of late stage colorectal cancer higher among men and women in the Appalachian region of the state [15].

Studies also suggest screening uptake may explain Appalachian and non-Appalachian differences. For example, a 2008 telephone survey of over 1,000 Kentucky adults found that those living in Appalachia were less likely to have received a colonoscopy or sigmoidoscopy within the past 10 years compared with those living outside Appalachia [26].

Female Breast Cancer

As the leading cause of cancer incidence and second leading cause of cancer mortality among women in the USA, breast cancer is estimated to account for 29 % of new cancer cases (232,340) and 14 % of deaths (39,620) among females in 2013 [24]. Although overweight and obesity have been linked to a reduced risk of premenopausal breast cancer, they have also been tied to an increased risk of postmenopausal breast cancer, albeit modest [7]. Increased postmenopausal breast cancer risk is associated with weight gain in adulthood and is most common among women who have never used hormone therapy and whose tumors express estrogen and progesterone receptors, particularly among white women [7]. A potential mechanism that may explain the relationship between obesity and a modest increase in risk of postmenopausal breast cancer is estrogen levels, which tend to be higher among women who are obese [7]. State cancer registry data from 2002 to 2006 indicate that the average annual age-adjusted female breast cancer incidence rates among Appalachian regions of Kentucky, New York, Ohio, Pennsylvania, Virginia, and West Virginia ranged from 112.2 per 100,000 (Kentucky) to 126.5 per 100,000 (New York) [15]. Only in New York was the breast cancer incidence rate higher among women in the Appalachian region of the state [15].

Breast cancer mortality rates have decreased over the past two decades, a change which has been attributed to a combination of screening and adjuvant treatment [27, 24]. An analysis of 1969–2007 Surveillance, Epidemiology and End Results (SEER) mortality data indicates that the decrease in breast cancer mortality rates occurred at a slower rate in Appalachian counties (17.5 %) compared to non-Appalachian counties in the 13 states (30.5 %) and non-Appalachia US counties across the country (28.3 %), which may suggest a lower prevalence of screening and differences in treatment [28]. Analysis of NCHS breast cancer mortality data from 2003 to 2007 indicates that Appalachian regions of nine of the 13 states experienced breast cancer mortality rates that were higher than the national rate [21]. Unlike colorectal cancer mortality rates, there were no in-state breast cancer mortality rate differences between Appalachian and non-Appalachian regions. The highest breast cancer mortality rate was in Appalachia Virginia (27.0 per 100,000) and the lowest was in Appalachia Georgia (22.3 per 100,000) [21]. State cancer registry data from 2001 to 2006 indicate that the average annual, age-adjusted female breast cancer mortality rates among Appalachian regions of Kentucky, New York, Ohio, Pennsylvania, Virginia, and West Virginia ranged from 23.4 per 100,000 (New York) to 26.6 per 100,000 (Ohio) [15]. Only in Kentucky was the breast cancer mortality rate higher among women in the Appalachian region of the state [15].

According to 2002–2006 state BRFSS data, Appalachian regions of Kentucky, Ohio, Pennsylvania, Virginia, and West Virginia had prevalences of women having underwent a mammogram within the past 3 years of 68.1 % (Kentucky) to 75.0 % (Pennsylvania) [15]. The prevalences were higher in the non-Appalachian regions of the states [15]. Furthermore, 2002–2006 state cancer registry data indicate that the proportion of female breast cancer cases diagnosed late stage ranged from 27.0 % (Ohio) to 36.2 % (Virginia) [15]. Only in Kentucky and Virginia were the prevalences of late stage breast cancer diagnoses higher among women in the Appalachian region of the state [15]. Studies also suggest that screening uptake may explain Appalachian and non-Appalachian differences. For example, a 2008 telephone survey of nearly 700 Kentucky adult women found that those living in Appalachia were less likely to receive regular mammograms than those living outside Appalachia [26].

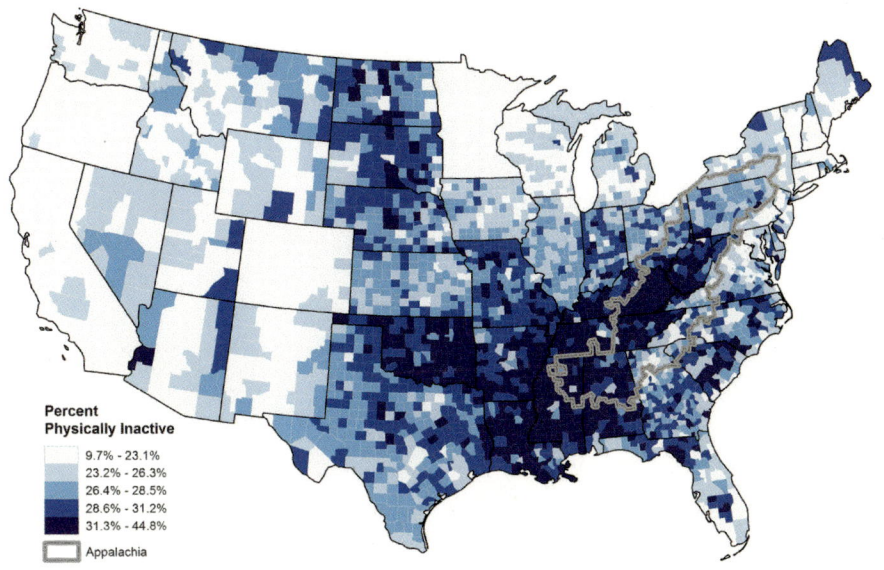

Fig. 1.3 County-level estimates of adult (≥20 years) physical inactivity prevalence, 2008 (Adapted from the Centers for Disease Control and Prevention) [29]

Risk Factors for Overweight and Obesity and Cancer: Physical Inactivity

An analysis of data from the 2008 BRFSS indicates that prevalences of leisure-time physical inactivity are highest in counties of the Southern and Appalachian regions of the USA [29]. Among the six states in which 70 % of counties had physical inactivity prevalences greater than or equal to 29.2 %, four are part of the Appalachian region (i.e., Alabama, Kentucky, Mississippi, and Tennessee) [29]. State BRFSS data from 2003 to 2007 indicate that the prevalence of no physical activity in the past month among Appalachian regions of Kentucky, Ohio, Pennsylvania, Virginia, and West Virginia ranged from 24 % (Pennsylvania) to 36.8 % (Kentucky) among men and 29.0 % (Pennsylvania) to 41.1 %(Kentucky) among women [15]. For both genders, the prevalences of no physical activity in the past month were higher in the Appalachian regions of the states [15] (Fig. 1.3). A study of 1,000 high school students in Southern Ohio found that only 5 % met the Centers for Disease Control and Prevention (CDC) recommendation of 60 min of moderate physical activity per day whereas 28 % and 78 % reported zero days of moderate and vigorous activity, respectively [30]. Among a church-based sample of over 1,200 adults in the Ohio Valley region of West Virginia, 48 % did not exercise on a regular basis each week whereas 42 % exercised 5 or more days per week for a total of 150 or more minutes [31]. Among over 200 adults in Appalachian Kentucky, 60 % viewed their health as good, just as 60 % reported no physical activity in the previous week [19].

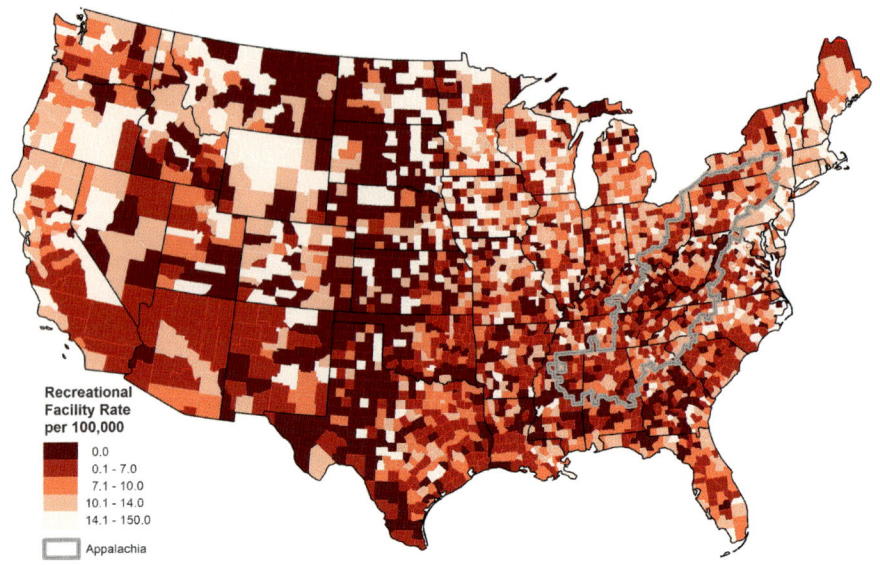

Fig. 1.4 County-level recreational facility rate per 100,000 population, 2008 (Adapted from the University of Wisconsin Population Health Institute and Robert Wood Johnson Foundation) [59]

Non-Appalachia focused research has suggested a number of determinants of physical activity behavior at both the individual level (e.g., limited time and financial resources, competing priorities and lack of knowledge) [32] and the environmental level (e.g., limited access to recreational facilities and sparse programming and activities) [33]. Many of these factors have been echoed across studies in Appalachia [34, 35, 31]. Among samples of Appalachia residents, participation in physical activity is influenced by transportation realities, such as long travel distances, poor road conditions, and limited transportation options [34]. A lack of local physical activity opportunities, sometimes due to low attendance to previously held programming and closure of recreational facilities within a reasonable distance, may also negatively affect physical activity participation [34]. As demonstrated in Fig. 1.4, access to recreational facilities is varied across the Appalachian region.

Perceptions of exercise and physical activity influence behavior among the Appalachian population. Focus groups conducted among Appalachian youth aged 8–17 years in Kentucky found that "physical activity" is viewed more positively than "exercise" among this population [35]. Reasons included that "exercise" is an activity that is planned for a specific duration and purpose and is often a requirement in school, whereas "physical activity" is less formal and more enjoyable because it is conducted at one's leisure [35]. Although exercise was viewed more positively across focus groups of over 110 adults in Appalachia Kentucky, the adults also viewed physical activity as a less structured activity that could be translated into exercise [34]. Common forms of physical activity among a church-based sample of over 1,200 adults in the Ohio Valley region of West Virginia were low intensity activities conducted in or around the home (e.g., work around the home, gardening, and leisure and brisk walks), whereas, participation in more formal activities (e.g., yoga, aerobics, swimming, sports) was much more infrequent [31]. Therefore, informal opportunities for physical activity may facilitate participation among this population [34, 35].

Risk Factors for Overweight and Obesity and Cancer: Poor Diet

National BRFSS data from 2000 to 2006 indicate that the prevalence of fruit and vegetable consumption was lower in the Mississippi Delta, Appalachian Mountains, and Great Plains compared to the West Coast, Northeast, and parts of the South [36]. State BRFSS data from 2002 to 2007 indicate that the prevalence of inadequate fruit and vegetable consumption among Appalachian regions of Kentucky, Ohio, Pennsylvania, Virginia, and West Virginia ranged from 52 % (Pennsylvania) to 88.9 % (Ohio) among men and 36.0 % (Pennsylvania) to 80.2 % (Kentucky) among women [15]. For both genders, the prevalence of inadequate fruit and vegetable consumption were higher in the Appalachian regions of the states [15]. Access to healthy food is not distributed equally across the country, or in the Appalachian region. A common indicator of access to healthy foods is the US Department of Agriculture Food Measurement Atlas's "low income and low access to store" variable. This variable represents the percentage of people in a given county who are of low income and, in urban areas, live more than one mile from the nearest supermarket or large grocery store or, in rural areas, live more than 10 miles from the nearest supermarket or large grocery store [37]. As demonstrated by Fig. 1.5, this measure shows that limited access to healthy food varies across the Appalachian region.

There is also variation in access to fast food in Appalachia (Fig. 1.6). Although there is mixed evidence about the relationship between access to fast food and health outcomes, fast-food establishments are known to offer items than are more calorically dense and nutritionally poor than meals prepared within the home [38, 39].

The relationship between diet quality and physical access to grocery stores, supermarkets and fast-food retailers is complicated, and physical access to healthy food does not necessarily result in consumption of a healthy diet. Among a sample of over 1,500 Appalachian adults living in West Virginia, of whom 74 % viewed themselves as healthy, 22 % reported consuming fast food three or more times a week [18]. However, 67 % reported drinking one or more cans of regular soda on a daily basis, which is concerning given that, on average, a 355 ml can of regular soda contains 40 g of sugar and 150 kcal [18]. Similarly, of over 200 adults in Appalachia Kentucky, 60 % viewed their health as good when just over a quarter reported

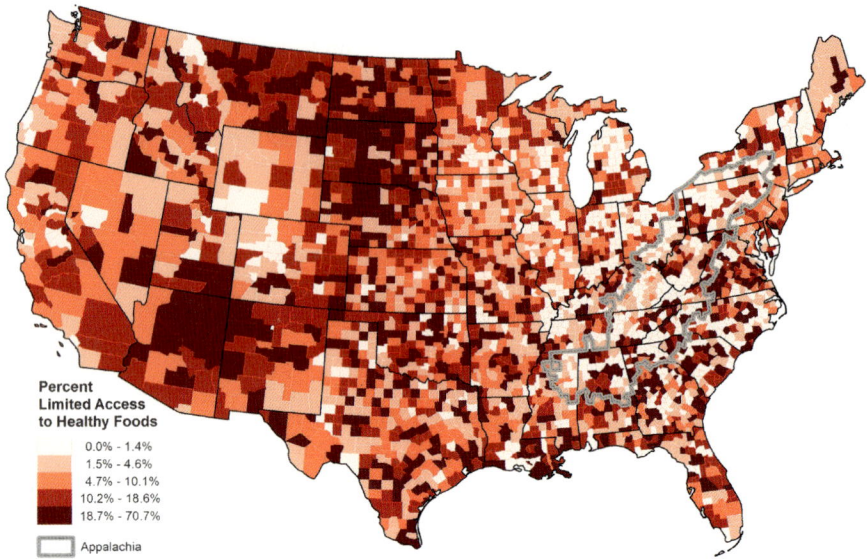

Fig. 1.5 County-level percent of the population that has limited access to healthy foods, 2006 (Adapted from the University of Wisconsin Population Health Institute and Robert Wood Johnson Foundation) [59]

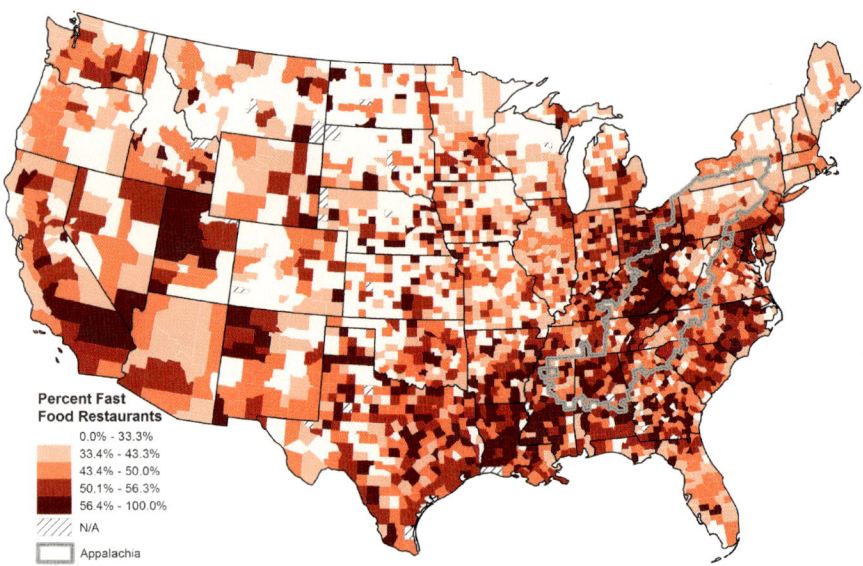

Fig. 1.6 County-level percent of all restaurants that are fast-food establishments, 2009 (Adapted from the University of Wisconsin Population Health Institute and Robert Wood Johnson Foundation) [59]

eating one or fewer daily servings of fruits and vegetables over the previous week [19]. Additionally, over 220 third graders across three Appalachian Ohio counties participating in a school-based dietary screening program did not achieve the recommended dietary intakes of grains, vegetables, fruits, milk, and meats/beans per the MyPyramid for Kids recommendations, the dietary recommendations for children that were current at the time of the study [40, 41]. Alarmingly, almost a fifth of the calories consumed by these children came from sweets, and this proportion was even higher among those of lower socioeconomic status [40].

As in other areas of the country, individual taste preferences and time constraints have been shown to influence diet quality in the Appalachian health literature [35]. However, other factors seem to be more specific to the region [35]. For example, social and familial norms and perceived support appear to greatly influence the food served at social gatherings and in the home as well as the dietary choices made by individuals [42–44]. Given the history of poverty in Appalachia, it is not surprising that cost is a salient determinant of what is purchased and consumed [42]. As with physical activity, diet is also affected by transportation realities in Appalachia [42]. It is also important to note that the region is not as isolated as it once was. As a result, historically common food practices like gardening and preservation are not as necessary as they once were given physical access to modern food retail outlets [45]. For example, compared to Amish adults living in Ohio Appalachia, non-Amish adults were more likely to purchase food outside of the home from grocery stores and restaurants and less likely to grow their own fruits and vegetables and use methods of preservation like canning and pickling [46].

Interventions

Disparate prevalences of overweight and obesity in Appalachia have been found as early as preschool [16, 47, 17, 48]. Consequently, many overweight and obesity interventions implemented in Appalachia have been focused on children, particularly in the context of school [49, 50]. For example, in Northern Tennessee, a school-based physical activity and healthy eating initiative based on the CDC's coordinated school health model was designed by a community coalition and piloted in a rural Appalachian elementary school [49]. Four-year follow-up data demonstrated significant improvements in the daily pedometer steps of students and their selection of healthy cafeteria food items [49]. Although commonly employed to improve academic outcomes, teen mentoring in the school setting is not commonly used in childhood overweight and obesity interventions [50]. A randomized control trial conducted among third and fourth graders in Appalachia participating in an afterschool health promotion program found that children whose program was mentored by teens experienced greater increases in physical activity than those whose program was led by adult instructors [50]. The need to integrate the support of family, primary care, and school-based efforts to foster long-term behavior changes among children has been noted in the literature [47, 40].

Methods currently being used to overcome barriers to improve access to health care and promote cancer screening across Appalachia offer additional insight into what can be done to address overweight and obesity in the region across age groups. For example, physical access and financial access have been noted as barriers to health care in Appalachia [51, 52]. Mobile health units, such as the Health Wagon, have had success reaching high risk individuals who lack health insurance and/or means of transportation [51]. The Health Wagon provides primary health care to Southwest Virginia's rural Appalachian population through clinics, cancer screening, case management services and telemedicine consultations with specialists that are free of charge [51]. By collaborating with local academic, medical, religious, and nonprofit partners and utilizing existing assistance programs and volunteers, the Health Wagon is able to maximize its limited resources [51]. However, staffing, financial support and coordination of care are just a few of the logistical challenges faced by this and likely other mobile clinics [51]. Despite these challenges, mobile health units may also be useful in the fight against overweight and obesity through the provision of biometric and dietary screening and case management services to individuals facing issues related to financial and physical access.

It is also possible to tailor evidence-based programs to address the region's burden of both overweight and obesity and cancer. Evidence-based programming is at the core of the "Cancer Control Plan, Link, Act, Network with Evidence-based Tools (PLANET)," an effort of the NCI, CDC, American Cancer Society, Substance Abuse and Mental Health Services Administration, Agency for Healthcare Research and Quality, and Commission on Cancer [53]. Featured program areas include breast and colorectal cancer screening, diet/nutrition, obesity, and physical activity. A searchable database of research-tested intervention programs (RTIPs) focused on cancer control is available online at: http://rtips.cancer.gov. An example of a featured program designed outside of Appalachia is StrongWomen, an evidence-based strength training program. This program was successfully modified to address physical activity as well as breast cancer awareness, screening and survivorship among women in Appalachian Pennsylvania [54]. The resulting 12-week program, New STEPS (Strength Through Education, Physical fitness and Support) was implemented using existing community resources and networks, which has afforded the ability for programs to extend beyond the study period [54].

The Appalachia Community Cancer Network (ACCN), a joint effort of the University of Kentucky, The Ohio State University, Pennsylvania State University, Virginia Tech University, and West Virginia University, is currently pursuing a research project in collaboration with over 20 churches in the Appalachian region of Kentucky, Ohio, Pennsylvania, Virginia, and West Virginia [55]. Using a group-randomized study design, half of the churches receive *Walk by Faith*, a dietary and physical activity faith-based intervention program, while the other half receive *Ribbons of Faith*, a comparison condition focused on cancer screening. *Walk by Faith* uses eHealth technology to address individual and environmental level changes to increase physical activity and to improve healthy food choices among participants [55]. Changes in physical activity levels, diet, and blood pressure will be assessed to determine the effectiveness of the faith-based intervention [55]. The sustainability of the intervention effects will also be tested and the results will be used to disseminate the intervention to the comparison churches, churches located in other Appalachian and rural areas, and the RTIPs Web site.

It is important to note that the selection, adaption and implementation of evidence-based programming to meet the needs of individuals living in Appalachia are not simple or necessarily straight-forward undertakings [56, 57]. The realities of Appalachian communities (e.g., limited community resources, competing time demands for potential participants and staff, reduced access to technology, low rates of participation) may present significant barriers to the success of evidence-based programs, which are often derived from well-funded, highly controlled research settings outside of the region [56, 57].

Suggestions for the Future

In summary, there are disparities in cancer rates, overweight and obesity prevalence, risk factors for overweight and obesity, and interventions in the Appalachian region of the USA. Thus, efforts to develop and test interventions to improve risk factors are urgently needed. Potential interventions that could be implemented in Appalachia to address disparities in obesity include efforts to partner with the USDA-sponsored cooperative extension service agents; focus on the entire family; utilize culturally relevant activities and develop advertisements with a focus on "health" versus "appearance" [34].

In addition, good surveillance data to monitor trends in cancer rates and risk factors are needed. Appalachia-specific health data are limited in terms of both volume and quality [58, 22]. Most notably, there is a lack of public health surveillance efforts that span the entire 13-state Appalachian region. For example, compared to other subregions, the information available on the South Central and Southern regions of Appalachia (e.g., North Carolina, South Carolina, Georgia, Alabama, and Mississippi) is more limited. Despite the efforts of state cancer registries and multi-state collaborations like the ACCN, there is a need for data that captures the heterogeneity of Appalachia. For example, an Appalachian variable could be added to BRFSS and NHANES to facilitate region-wide studies.

To make a difference in overweight and obesity prevalence in Appalachia, culturally appropriate interventions need to be developed, tested, and if effective, disseminated through trusted channels. Lastly, for effective strategies to be adopted the community needs to be involved from the start and take ownership of the problem as well the solution. Efforts are currently underway using community based participatory research (CBPR) strategies and will be able to inform researchers and the community about possible approaches that could be disseminated to reduce disparities in overweight and obesity and cancer rates in Appalachia.

References

1. US Department of Agriculture, US Department of Health and Human Services (2010) Dietary guidelines for Americans, 2010. US Government Printing Office
2. Centers for Disease Control and Prevention (2013) Behavioral risk factor surveillance system survey data
3. Ogden CL, Carroll MD, Kit BK, Flegal KM (2012) Prevalence of obesity in the United States, 2009–2010. NCHS Data Brief 82:1–8
4. Calle EE, Rodriguez C, Walker-Thurmond K, Thun MJ (2003) Overweight, obesity, and mortality from cancer in a prospectively studied cohort of U.S. adults. N Engl J Med 348 (17):1625–1638
5. Hursting SD, Digiovanni J, Dannenberg AJ, Azrad M, Leroith D, Demark-Wahnefried W, Kakarala M, Brodie A, Berger NA (2012) Obesity, energy balance, and cancer: new opportunities for prevention. Cancer Prev Res (Phila) 5(11):1260–1272. doi:10.1158/1940-6207. CAPR-12-0140, Epub 2012 Oct 3
6. Vucenik I, Stains JP (2012) Obesity and cancer risk: evidence, mechanisms, and recommendations. Ann N Y Acad Sci 1271:37–43. doi:10.1111/j.1749-6632.2012.06750.x
7. National Cancer Institute Obesity and Cancer Risk. http://www.cancer.gov/cancertopics/ factsheet/Risk/obesity. Accessed 10 July 2013
8. Vandelanotte C, Spathonis KM, Eakin EG, Owen N (2007) Website-delivered physical activity interventions a review of the literature. Am J Prev Med 33(1):54–64
9. National Cancer Institute (2011) Overview of health disparities research: health disparities definition, vol 2011
10. Appalachian Regional Commission (1966) Appalachia: a report by the President's Appalachian Regional Commission, 1964. Appalachian Regional Commission
11. Newman M (1972) The political economy of Appalachia; a case study in regional integration, vol Book, Whole. Lexington Books, Lexington, MA
12. Appalachian Regional Commission (2009) Subregions in Appalachia, vol 2013. Appalachian Regional Commission
13. Pollard K, Jacobsen L (2013) The Appalachian region: a data overview from the 2007–2011 American Community Survey
14. Gregg E, Kirtland KC, Cadwell BL, Rios Burrows N, Barker L, Thompson T, Geiss L (2009) Estimated county-level prevalence of diabetes and obesity—U.S., 2007. MMWR Morb Mortal Wkly Rep 58(45):1259–1263
15. Appalachia Community Cancer Network. The cancer burden in Appalachia, 2009
16. Berlin KS, Hamel-Lambert J, DeLamatre C (2013) Obesity and overweight status health disparities among low-income rural Appalachian preschool children. Child Health Care 42 (1):15–26
17. Montgomery-Reagan K, Bianco JA, Heh V, Rettos J, Huston RS (2009) Prevalence and correlates of high body mass index in rural Appalachian children aged 6-11 years. Rural Remote Health 9(4):1234, Epub 2009 Oct 16
18. Griffith BN, Lovett GD, Pyle DN, Miller WC (2011) Self-rated health in rural Appalachia: health perceptions are incongruent with health status and health behaviors. BMC Public Health 11:229. doi:10.1186/1471-2458-11-229
19. Ely GE, Miller K, Dignan M (2011) The disconnect between perceptions of health and measures of health in a rural Appalachian sample: implications for public health social workers. Soc Work Health Care 50(4):292–304. doi:10.1080/00981389.2010.534342
20. Pagidipati NJ, Gaziano TA (2013) Estimating deaths from cardiovascular disease: a review of global methodologies of mortality measurement. Circulation 127(6):749–756. doi:10.1161/ CIRCULATIONAHA.112.128413
21. Blackley D, Behringer B, Zheng S (2012) Cancer mortality rates in Appalachia: descriptive epidemiology and an approach to explaining differences in outcomes. J Community Health 37 (4):804–813. doi:10.1007/s10900-011-9514-z

22. Wingo PA, Tucker TC, Jamison PM, Martin H, McLaughlin C, Bayakly R, Bolick-Aldrich S, Colsher P, Indian R, Knight K, Neloms S, Wilson R, Richards TB (2008) Cancer in Appalachia, 2001-2003. Cancer 112(1):181–192

23. Borak J, Salipante-Zaidel C, Slade MD, Fields CA (2012) Mortality disparities in Appalachia: reassessment of major risk factors. J Occup Environ Med 54(2):146–156. doi:10.1097/JOM. 0b013e318246f395

24. Siegel R, Naishadham D, Jemal A (2013) Cancer statistics, 2013. CA Cancer J Clin 63 (1):11–30. doi:10.3322/caac.21166, Epub 2013 Jan 17

25. Naishadham D, Lansdorp-Vogelaar I, Siegel R, Cokkinides V, Jemal A (2011) State disparities in colorectal cancer mortality patterns in the United States. Cancer Epidemiol Biomarkers Prev 20(7):1296–1302. doi:10.1158/1055-9965.EPI-11-0250

26. Fleming ST, Love MM, Bennett K (2011) Diabetes and cancer screening rates among Appalachian and non-Appalachian residents of Kentucky. J Am Board Fam Med 24 (6):682–692. doi:10.3122/jabfm.2011.06.110094

27. Berry DA, Cronin KA, Plevritis SK, Fryback DG, Clarke L, Zelen M, Mandelblatt JS, Yakovlev AY, Habbema JDF, Feuer EJ (2005) Effect of screening and adjuvant therapy on mortality from breast cancer. N Engl J Med 353(17):1784–1792. doi:10.1056/NEJMoa050518

28. Yao N, Lengerich EJ, Hillemeier MM (2012) Breast cancer mortality in Appalachia: reversing patterns of disparity over time. J Health Care Poor Underserved 23(2):715–725. doi:10.1353/hpu.2012.0043

29. Centers for Disease Control and Prevention (2013) Diabetes interactive atlases. Accessed 10 July 2013

30. Hortz B, Stevens E, Holden B, Petosa RL (2009) Rates of physical activity among Appalachian adolescents in Ohio. J Rural Health 25(1):58–61. doi:10.1111/j.1748-0361.2009.00199.x

31. Zizzi S, Goodrich D, Wu Y, Parker L, Rye S, Pawar V, Mangone C, Tessaro I (2006) Correlates of physical activity in a community sample of older adults in Appalachia. J Aging Phys Act 14(4):423–438

32. Wilcox S, Castro C, King AC, Housemann R, Brownson RC (2000) Determinants of leisure time physical activity in rural compared with urban older and ethnically diverse women in the United States. J Epidemiol Community Health 54(9):667–672

33. Sloane D, Nascimento L, Flynn G, Lewis L, Guinyard JJ, Galloway-Gilliam L, Diamant A, Yancey AK (2006) Assessing resource environments to target prevention interventions in community chronic disease control. J Health Care Poor Underserved 17(2 Suppl):146–158

34. Kruger TM, Swanson M, Davis RE, Wright S, Dollarhide K, Schoenberg NE (2012) Formative research conducted in rural Appalachia to inform a community physical activity intervention. Am J Health Promot 26(3):143–151. doi:10.4278/ajhp.091223-QUAL-399

35. Swanson M, Schoenberg NE, Davis R, Wright S, Dollarhide K (2013) Perceptions of healthful eating and influences on the food choices of Appalachian youth. J Nutr Educ Behav 45 (2):147–153. doi:10.1016/j.jneb.2011.07.006, Epub 2012 Jan 24

36. Michimi A, Wimberly MC (2010) Spatial patterns of obesity and associated risk factors in the conterminous U.S. Am J Prev Med 39(2):e1–e12. doi:10.1016/j.amepre.2010.04.008

37. Economic Research Service, US Department of Agriculture Food Access Research Atlas. http://www.ers.usda.gov/data-products/food-access-research-atlas.aspx. Accessed 10 July 2013

38. Fleischhacker SE, Evenson KR, Rodriguez DA, Ammerman AS (2011) A systematic review of fast food access studies. Obes Rev 12(5):e460–e471. doi:10.1111/j.1467-789X.2010.00715.x

39. Guthrie JF, Lin BH, Frazao E (2002) Role of food prepared away from home in the American diet, 1977-78 versus 1994-96: changes and consequences. J Nutr Educ Behav 34(3):140–150

40. Hovland JA, McLeod SM, Duffrin MW, Johanson G, Berryman DE (2010) School-based screening of the dietary intakes of third graders in rural Appalachian Ohio. J Sch Health 80 (11):536–543. doi:10.1111/j.1746-1561.2010.00539.x

41. US Department of Agriculture Archived MyPyramid Materials. http://www.choosemyplate.gov/print-materials-ordering/mypyramid-archive.html. Accessed 5 Sept 2013

42. Schoenberg NE, Howell BM, Swanson M, Grosh C, Bardach S (2013) Perspectives on healthy eating among Appalachian residents. J Rural Health 29 Suppl 1:s25–s34. doi: 10.1111/jrh. 12009, Epub 2013 Feb 21

43. Wenrich TR, Brown JL, Miller-Day M, Kelley KJ, Lengerich EJ (2010) Family members' influence on family meal vegetable choices. J Nutr Educ Behav 42(4):225–234. doi:10.1016/j. jneb.2009.05.006

44. Wu T, Stoots JM, Florence JE, Floyd MR, Snider JB, Ward RD (2007) Eating habits among adolescents in rural Southern Appalachia. J Adolesc Health 40(6):577–580, Epub 2007 Mar 9

45. Quandt SA, Popyach JB, DeWalt KM (1994) Home gardening and food preservation practices of the elderly in rural Kentucky. Ecol Food Nutr 31(3–4):183–199

46. Cuyun Carter GB, Katz ML, Ferketich AK, Clinton SK, Grainger EM, Paskett ED, Bloomfield CD (2011) Dietary intake, food processing, and cooking methods among Amish and non-Amish adults living in Ohio Appalachia: relevance to nutritional risk factors for cancer. Nutr Cancer 63(8):1208–1217. doi:10.1080/01635581.2011.607547, Epub 2011 Oct 25

47. Dalton WT 3rd, Schetzina KE, Pfortmiller DT, Slawson DL, Frye WS (2011) Health behaviors and health-related quality of life among middle school children in Southern Appalachia: data from the winning with wellness project. J Pediatr Psychol 36(6):677–686. doi:10.1093/jpepsy/ jsq108, Epub 2010 Dec 3

48. Williams KJ, Taylor CA, Wolf KN, Lawson RF, Crespo R (2008) Cultural perceptions of healthy weight in rural Appalachian youth. Rural Remote Health 8(2):932, Epub 2008 May 22

49. Schetzina KE, Dalton WT 3rd, Pfortmiller DT, Robinson HF, Lowe EF, Stern HP (2011) The Winning with Wellness pilot project: rural Appalachian elementary student physical activity and eating behaviors and program implementation 4 years later. Fam Community Health 34 (2):154–162. doi:10.1097/FCH.0b013e31820e0dcb

50. Smith LH, Holloman C (2013) Comparing the effects of teen mentors to adult teachers on child lifestyle behaviors and health outcomes in Appalachia. J Sch Nurs 29:386–396

51. Gardner T, Gavaza P, Meade P, Adkins DM (2012) Delivering free healthcare to rural Central Appalachia population: the case of the Health Wagon. Rural Remote Health 12:2035, Epub 2012 Mar 20

52. Vyas A, Madhavan S, Kelly K, Metzger A, Schreiman J, Remick S (2013) Do Appalachian women attending a mobile mammography program differ from those visiting a stationary mammography facility? J Community Health 38(4):698–706. doi:10.1007/s10900-013-9667-z

53. Sood R, Ho P, Tornow C, Frey W (2007) Cancer control P.L.A.N.E.T. evaluation final report. WESTAT, Rockville, MD

54. Gallant NR, Corbin M, Bencivenga MM, Farnan M, Wiker N, Bressler A, Camacho F, Lengerich EJ (2013) Adaptation of an evidence-based intervention for Appalachian women: new STEPS (Strength Through Education, Physical fitness and Support) for breast health. J Cancer Educ 28(2):275–281

55. Appalachia Community Cancer Network Research Program—full research project. http:// www.accnweb.com/FullResearchProject.aspx. Accessed 5 Sept 2013

56. Bencivenga M, DeRubis S, Leach P, Lotito L, Shoemaker C, Lengerich EJ (2008) Community partnerships, food pantries, and an evidence-based intervention to increase mammography among rural women. J Rural Health 24(1):91–95. doi:10.1111/j.1748-0361.2008.00142.x

57. Vanderpool RC, Gainor SJ, Conn ME, Spencer C, Allen AR, Kennedy S (2011) Adapting and implementing evidence-based cancer education interventions in rural Appalachia: real world experiences and challenges. Rural Remote Health 11(4):1807, Epub 2011 Oct 10

58. Krause DD, May WL, Cossman JS (2012) Overcoming data challenges examining oral health disparities in Appalachia. Online J Public Health Inform 4(3) pii: ojphiv4i34279. doi:10.5210/ ojphi.v4i3.4279, Epub 2012 Dec 19

59. University of Wisconsin Population Health Institute, Robert Wood Johnson Foundation (2013) County health rankings and roadmaps 2011, 2012, vol 2013

Chapter 2
Disparities in Cancer Outcomes: A UK Perspective

Donald J. Nicolson, Una Macleod, and David Weller

> *It is the cry of men who feel themselves the victims of blind economic forces beyond their control. . . The feeling of despair and hopelessness that pervades people who feel with justification that they have no real say in shaping or determining their own destinies (Reid 1972).*

Abstract The social problem described by Jimmy Reid in 1972 [1] is still prevalent in the UK in the twenty-first century. Many people who are socio-economically disadvantaged do not have the capacity to influence their freedom, and as a consequence, they do not have control over the destiny of their own health. In this chapter we examine how socially disadvantaged people in the UK are at greater risk of poorer outcomes when they have cancer. That is, socio-economic factors determine disparities in cancer outcomes, incidence, mortality, and survival rates, in the UK.

Keywords The UK • Cancer incidence • Health inequalities • Black report • Acheson report • Marmot review • National Health Service cancer plan • Carstairs deprivation index • Socio-economic status • Inverse care law

Preface

While the focus of this book is Energy Balance and its relation to cancer disparities, this chapter takes a broader look at health inequalities and cancer with a UK perspective; it draws on UK and international research and policy work spanning the last 30 years and more. Energy balance is a key factor in cancer outcomes; the

D.J. Nicolson • U. Macleod
Centre for Health and Population Sciences, Hull York Medical School (HYMS), University of Hull, Hertford Building, Cottingham Road, Hull HU6 7RX, UK
e-mail: donald.nicolson@hyms.ac.uk; una.macleod@hyms.ac.uk

D. Weller (✉)
Edinburgh Cancer Research UK Centre, The University of Edinburgh, Crewe Road South, Edinburgh EH4 2XR, UK
e-mail: David.Weller@ed.ac.uk

D.J. Bowen et al. (eds.), *Impact of Energy Balance on Cancer Disparities*,
Energy Balance and Cancer 9, DOI 10.1007/978-3-319-06103-0_2,
© Springer International Publishing Switzerland 2014

UK has a rich literature on health inequalities, and we hope that by examining the multiple contributing factors to cancer outcome disparities, the role of energy balance can be better understood. We have considered "health inequalities" to be synonymous with "disparities" (a term that is more commonly used in the USA). Several UK Governments have commissioned significant documents over the last few decades and these form the principle overview and understanding of health inequalities in the UK. The first of these was the Black Report [2], commissioned in the late 1970s. This illustrated extensive health inequalities in the UK, despite the advent of the National Health Service in 1948. Similarly, two decades later the Acheson Report [3] reported a relationship between health disparities and social class, with the higher social classes having greater decline in mortality than the rest of the population. These landmark reports have added to our understanding of how health inequalities arise from social inequalities.

A more recent report, the Marmot Review [4] noted that health inequalities are a profound social justice issue for the UK; highlighting how there is a social gradient in health and health inequalities, and concluding that addressing health inequalities is a matter of fairness. The Marmot Review also noted that, based on deprivation categories (a score constructed around communities access to resources, relationships in society, income, housing, and employment), people from more deprived backgrounds not only have a higher rate of cancer, but men from the most deprived category have nearly double the risk of cancer than men from the least deprived background. This is a powerful example of the relationship between cancer mortality and level of deprivation.

Alongside these reports on health inequalities, there has been a policy drive to improve cancer outcomes. This was initially formalised in the NHS Cancer Plan in 2000 [5] and the Cancer Reform Strategy in 2007 [6]. These set in place a national cancer programme for England with a focus on saving more lives to ensure that people with cancer got the right support, care, and treatments; that inequalities in health and cancer were tackled; to invest in strong research; and to prepare for the genetics revolution.

Incidence, Mortality, and Survival

Before we examine health inequalities in cancer outcomes, it is necessary to understand the epidemiology of cancer outcomes. The outcomes we are interested in are the rates of newly diagnosed cancers (incidence), the numbers of people dying from cancer (mortality), and the survival rates for people living with a cancer. The data reported in this section have been largely produced by Cancer Research UK, a highly reputable source of cancer statistics in the UK, who make cancer data available on their website (www.cancerresearchuk.org) [6]. Unless otherwise stated, figures below have been obtained from this source.

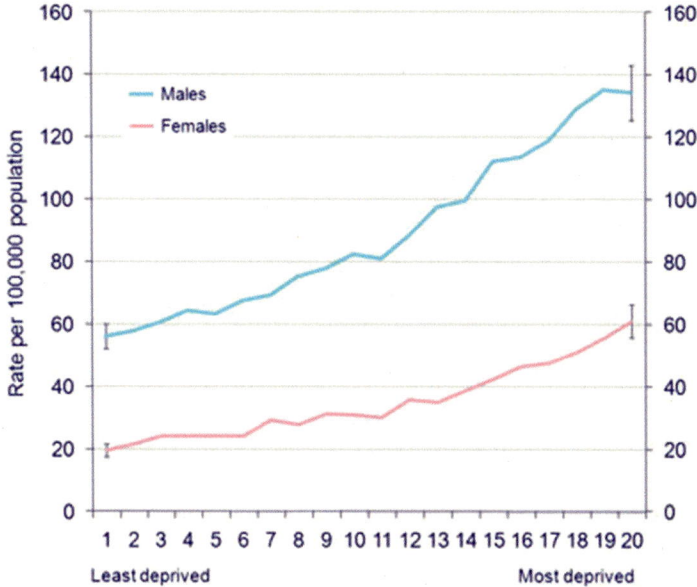

Fig. 2.1 Lung cancer (C33-C34), age-standardised incidence rates by deprivation category, England and Wales, 1993

Incidence of Cancer in the UK

In the UK in 2010, around 325,000 people were newly diagnosed with a cancer. This included a similar number of males and females, around 160,000 each. However when the rates were standardised for age, considerably more men (426 per 100,000) compared to women (374 per 100,000) were newly diagnosed. The incidence of cancer in the UK has steadily risen for men and women since the mid-1970s by 22 %. However, the rate of increase has slowed down from the period 2001–2010, with just a 2 % increase for men and a 6 % increase for women.

Data from 1993 for the incidence of lung cancer showed clear evidence of the impact of deprivation. Two and a half times as many men and three times as many women from the most deprived groups compared to the least deprived groups were diagnosed with lung cancer. Figure 2.1 below shows how the age standardised rates of lung cancer increase across deprivation categories. Although these data are old, more recent work confirms no change [8].

Four types of cancer: breast, lung, bowel, and prostate, accounted for 54 % of all new cases of cancer in 2010. The most commonly diagnosed cancer in men is prostate—one in four cases. The more commonly diagnosed cancer in women is of the breast—just under one in three cases.

Worldwide there were approximately 12.7 million new cases of cancer in 2008. The rate was considerably greater for North America and Europe compared to the developing world (Cancer Research UK).

Scotland has the worse rates of cancer in the UK; reflecting the all-cause mortality gap between Scotland and England which grew from 1981 to 2001 [9]. This is not necessarily determined by social inequalities; the Carstairs deprivation index (a measure of deprivation), declined during the same period [10]. Other factors, such as the "Scottish Effect" (a factor related to living in Scotland, independent of other risk factors) have been proposed to explain poor outcomes in Scotland [11, 12].

Mortality from Cancer in the UK

Around 82,000 men and 75,000 women in the UK died from a cancer between 2007 and 2009, i.e. 427 per 100,000 men, and 371 per 100,000 women [13]. The 157,000 people who died from a cancer in the UK in 2010 accounted for more than one in four (28 %) of all deaths. The most common cause of cancer mortality was due to lung cancer 19,410 cases (24 %) in men; and 15,449 cases (21 %) in women. Death from a cancer becomes more likely with age and is more common for men than women.

In recent years in the UK, more men than women have been newly diagnosed with a cancer. However, overall the rates of newly diagnosed cancers have been falling. More men than women die from a cancer each year in the UK. Deaths from cancer accounted for more than one quarter of all deaths in the UK in 2010. Lung cancer was the most common cause of a cancer death in both men and women. More people are now dying from cancer of the liver than in previous decades.

Mortality rates from cancer have been declining in the UK since the early 1990s. Between 2001 and 2008, there was a 12 % and 9 % decrease in all cancers for men and women respectively. However the rates of cancer mortality from liver cancer have increased in both sexes, which may be due to trends in increased alcohol intake. While deaths from lung cancer have decreased for men by 19 % they have increased for women by 6 % (Cancer Research UK).

Surviving Cancer

Coleman et al. [14] analysed data from population-based cancer registries in six countries for two to four million adults diagnosed with a cancer during 1995–2007 and found survival rates were lower in the UK (and Denmark) than in Australia, Canada, and Sweden.

Rachet et al. [15] have found survival rates for patients with cancer was significantly higher in the most affluent groups compared to the most deprived groups. However, the relationship is complex, due to the interplay between the type of cancer, patient personal factors, and the role of the health service [16].

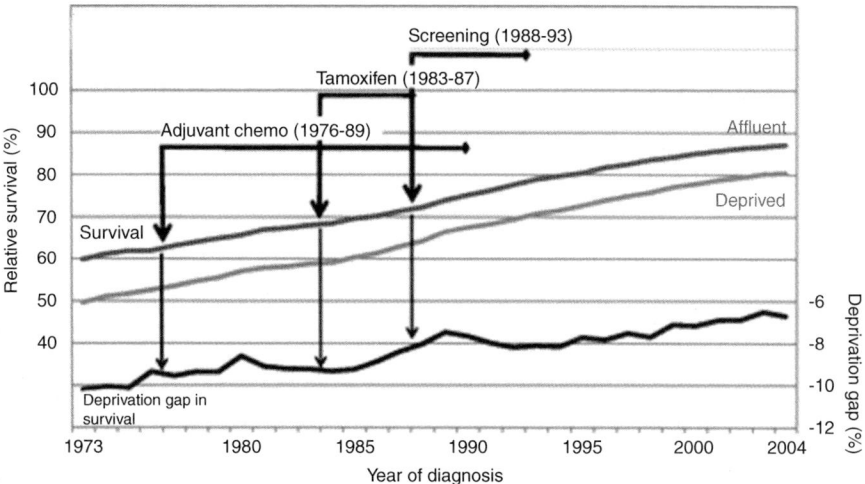

Fig. 2.2 Trends in 5-year relative survival (%) from breast cancer in women in the most affluent and most deprived groups and deprivation gap (%) in survival: 5-year moving average values, England and Wales, 1973–2004. Periods of emergence of evidence about the efficacy of new interventions are denoted on the graph [Reproduced with permission]

Trends in survival differences, by deprivation category, are also complex; Lyratzopoulos and colleagues [17] examined changes in socio-economic inequalities in survival from breast cancer for women, and from rectal cancer for men in England and Wales from 1973 to 2004. They found survival rates increased over this period from 55 % to 85 % for women with breast cancer, while the survival gap between the two deprivation groups narrowed slightly from −10 % to −6 % (Fig. 2.2). For men they found 5-year relative survival rates from rectal cancer improved from 29 % to 53 % between 1973 and 2004; but the survival gap between the two deprivation groups increased from −5 % to −11 % (Fig. 2.3).

These authors conclude that the cause of inequalities in survival rates remains unknown, but may partly reflect differences in clinical management (the "health care factors" hypotheses). If so, socio-economic inequalities should be largely determined by socio-economic differences in the quality of treatment received, with deprived patients more often managed suboptimally.

Coleman et al. [18] clearly highlighted the link between socio-economic disadvantage and poorer cancer outcomes, finding a difference in 1 and 5 year survival rates for all cancers combined when comparing people from deprivation categories between 1986 and 1990. People from the more affluent groups had higher survival rates after diagnosis than people from the most deprived category. The difference remained fairly stable between 1 and 5 year survival; 12.7 % and 11.1 % respectively. Figure 2.4 shows this gap in survival rates.

In a related study, Abdel-Rahman et al. [19] found that compared with data from countries in continental Europe, socio-economic differences in survival in Britain may account for half the avoidable premature mortality from cancers.

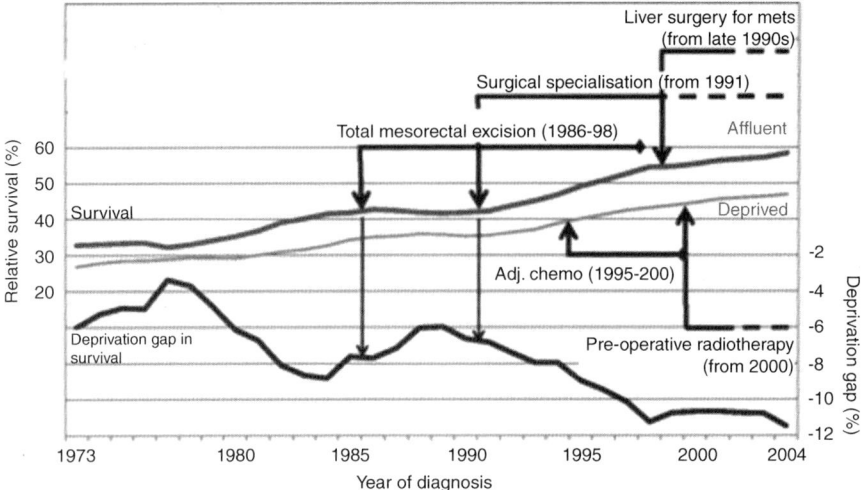

Fig. 2.3 Trends in 5-year relative survival (%) from rectal cancer in men in the most affluent and most deprived groups and deprivation gap (%) in survival: 5-year moving average values, England and Wales, 1973–2004. Periods of emergence of evidence about the efficacy of new interventions are denoted on the graph. Increasing use of flexible sigmoidoscopy occurred throughout the study period, and is not denoted on the graph [Reproduced with permission]

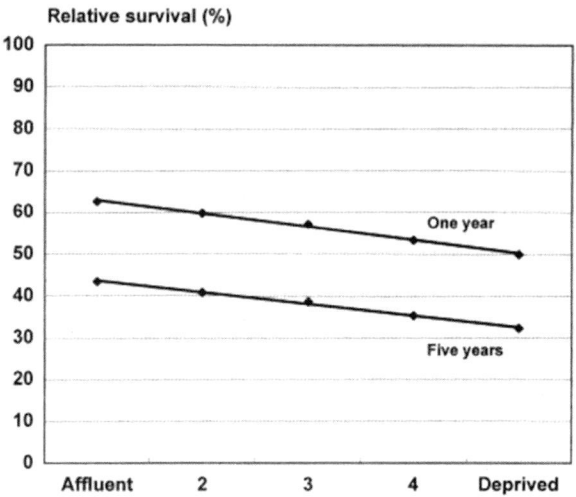

Fig. 2.4 Relative survival rates 1 and 5 years after diagnosis by deprivation category, all cancers combined: England and Wales, adults diagnosed 1986–1990

Explaining Disparities in Health Outcomes

It is estimated that only 5–10 % of cancers are attributable to genetic variation [20]. If this is the case, then most cancers might be preventable if people avoid specific environmental risks, or practise health promoting behaviours. It is thought that about half of all cases of cancer could be prevented by lifestyle changes [21]. This indicates how social and behavioural factors, e.g. gender, ethnic group, income, geographical, education, and social class, are important determinants of cancer.

The Black Report

The Black Report [2] has played a fundamental role in explaining health-care disparities in the UK. It examined four ways to understand health inequalities; they remain a useful framework in understanding disparities in people's cancer outcomes.[1]

1. Artefact: This argument proposes that a relationship between class and health is spurious; that there is no real relationship and that the findings are a product of the way the data were measured. Macintyre [22] suggests this relationship is not straightforward because the level of class influence on illness will depend on how both class and illness are measured. A failing of this hypothesis is that evidence of health inequalities are consistent across populations and periods of time [23], which suggests the finding reflects reality and is not a social construct.

2. Social selection: This model proposes that health determines class [22]; thus health inequalities are thought to produce health-related social inequalities; that is, for example, people with illness tend to suffer downward social mobility from loss of employment and/or income. This is also known as the "reverse causation" or "drift" hypothesis [24]. At best, this model can only partially explain health disparities. For example the link between cancers and education cannot be accounted for by social selection because people have usually completed their education in early adulthood before succumbing to a cancer [24]. There is little evidence to support this theory and it does not have widespread support in the international literature [25].

3. Cultural/behavioural: This model proposes that health damaging behaviours (e.g. smoking, excessive alcohol intake, or poor diet) are more common among the socially disadvantaged. The more extreme version of this argument suggests that individual ignorance, lifestyle choices, and neglect are the cause of illness [26]. Individuals from lower socio-economic status (SES) groups are also

---

[1] Macintyre [22] noted that each explanation has a "hard" (extreme) and "soft" (moderate) version for explaining the relationship between social class and health.

more likely to be exposed involuntarily to environmental pollutants [27] and occupational hazards [28]; factors which put them at high risk of developing a cancer. This argument however ignores the social context of people's lives and can be said to blame the victims of health inequalities for their poor health, although this argument in itself does not discredit this model. Critics have said that little has been done to disentangle the relationship between social disadvantage and health damaging behaviour [25]. We know there is a relation, but we are unclear why this is.

4. Material/structural: This proposal suggests that health is determined by a person's wealth—at its simplest, whether a person is "rich" or "poor". One such explanation proposes that health status is determined by income inequality; in particular that negative exposure and lack of resources combine to produce health inequalities [29]. A softer version acknowledges that psychosocial and other influences mediate this relationship. Coleman et al. [30] spoke of a "deprivation gap", e.g. the deficit in a cancer outcome between the rich and the poor.

There is clear and considerable evidence showing socio-economically disadvantaged people have significant health problems and poor access to health care. For example there is a gradient in the relationship between class and mortality: as a whole, people from lower classes have a lower life expectancy and die earlier than people from more affluent backgrounds. As we will later show, there is much evidence showing cancer incidence, mortality, and survival are related to social class. The material/structural argument helps explain national and international health disparity at a population level, but it remains a challenge to understand how socio-structural factors influence health inequalities [25], and at an individual level.

SES, Cancer and Pathways

The models discussed in "Incidence of Cancer in the UK" offer generic explanations of health inequalities. Kawachi and Kroenke [24] have sought to explain the mechanism linking SES and cancer by means of two possible pathways. In the first pathway, people from higher SES groups are able to access various resources to help prevent them developing cancer, or improving their outcome following cancer onset. They give the example of people who, through better education, are more "health literate" and consequently better able to understand options for cancer treatments. The second pathway suggests that people with higher SES have a differential exposure to psychosocial mediators (compared to people from poorer backgrounds), which benefits their health outcomes (see below for further details).

Examples of Disparities

Disparities are observed across a range of categories:

- Gender disparity: women have a longer life expectancy than men [31].
- Ethnic group disparity: there is a higher rate of cardiovascular disease in the UK amongst people from South Asia [32].
- Income disparity: people with higher levels of income tend to have better health overall than people with lower incomes [33].
- Geographic disparity: the Scottish city of Glasgow has nearly half of the 10 % most health deprived areas in Scotland [34]. These areas have higher rates of morbidity and mortality than more affluent areas in the same city and elsewhere.
- Education disparity: people with better education opportunities tend to have better health and well-being than people who have not had the same level of education [35].
- Social class disparity: people from lower social classes tend to have poorer health, and receive poorer health care than people from higher social classes [33].

It is important to recognise that individuals can face inequality across a number of these categories.

The Impact of Socio-economic Disadvantage on Cancer Outcomes

Having explored rates of people living with, surviving, and dying from cancer; and examined how socio-economic disadvantage impacts on people's health and access to health care in general; we now examine the evidence that socio-economic disparities impact on cancer outcomes. We consider lifestyle factors, public perception of cancer, issues related to cancer screening, awareness and recognition of cancer, health-care factors, and psychosocial factors.

Lifestyle Risk Factors

People who are socio-economically disadvantaged are often at greater risk of exposure to lifestyle risk factors than people from more affluent backgrounds. This may be seen to reflect a cultural/behavioural explanation for cancer inequalities. Lifestyle is intricately woven with socio-economic conditions and so it does not solely reflect someone's "choices".

1. Tobacco smoking: Smoking is an unequivocal risk factor for cancer and other diseases. For example it is considered to be the main determinant of lung cancer, with 90 % of people with lung cancer having smoked [24]. Smoking is the main cause of difference in morbidity and mortality between wealthy and poor individuals [36]. Accordingly, tackling smoking among people from the lowest socio-economic groups might reduce the incidence of smoking-related cancers and other smoking-related diseases. Much has been done in recent times in the UK to encourage and support people to stop smoking. The Scottish Government banned smoking in public places in 2006, with the rest of the UK doing so a year later. The National Health Service in the UK also runs a "Smokefree" service which offers people who want to stop smoking support via telephone, the Internet, and paper-based materials. However poorer people have less success in stopping smoking than more affluent people [37]. Therefore smoking related health inequalities will likely continue.

2. Poor diet: Poor diet has been linked to around one third of cancer deaths [21]. Diets rich in fats and red meat, high in calories and low in vegetables, are commonly related to lower SES [24]. Diets that have greater amounts of fruit and vegetables are more often consumed by people from an affluent background [38]. People from lower socio-economic backgrounds are at further disadvantaged because of the link between the availability and cost of food [39].

3. Physical activity: Minimal physical activity is related to the risk of several cancers [40], as well as obesity. Recreational physical activity tends to be strongly correlated with higher income households [41]. This is related to lower levels of obesity linked cancers [24]. The affordability and accessibility of recreational physical activity may be beyond many people from poorer backgrounds.

4. Weight and obesity: As expressed in other chapters of this book, there are major disparities in levels of obesity, between different social classes [42]. Given the growing body of evidence linking overweight and obesity with unfavourable cancer outcomes [43], poor dietary and energy balance trends in the UK must play a significant role in cancer disparities. It is suggested that if individuals maintained a healthy body weight, up to 12,000 cases of cancer could be prevented (Cancer Research UK; Cancer and Health Inequalities: an introduction to current evidence). People from lower socio-economic backgrounds, because they are more likely to be obese, are disadvantaged and so at greater risk of acquiring a cancer. Being obese increases the risk of several cancers, including cancer of the uterus, kidney, or colon [44]. Obesity levels in the UK have trebled over the last 20 years [21], indicating that this is a recent risk factor.

5. Alcohol consumption: Excessive alcohol intake is related to various diseases, including liver disease, heart disease, stroke, and cancers of the liver and the head and neck [21]. While the evidence of a link between alcohol and illnesses is clear, there is no conclusive evidence that people from a disadvantaged background are more at risk of misusing alcohol. It is thought in the UK that this is because excessive alcohol intake has no class pattern—in contrast to smoking [21].

Public Perceptions of Cancer

People's perceptions of cancer are probably relevant in their decisions to take screening tests for cancer in the absence of symptoms of cancer, or to attend for care when they develop symptoms. Dein [45] noted that beliefs about cancer can determine the perception of risk of developing cancer, and therefore have implications for the perceived urgency for patients to participate in screening, their decisions about treatment, and emotional responses to the disease.

It is not only the perception of someone's risk of cancer that can impact on their outcome, but their opinion of the likelihood that treatment would be successful [46]. For example, Powe & Finnie [47] have spoken of "Cancer Fatalism", where death from cancer is considered inevitable. This can be seen to reflect the observation by Susan Sontag [48] that some people held the belief that "cancer equals death". It is not difficult to see that if someone perceived this, she/he may not appreciate an urgency or benefit from early diagnosis and treatment, because they would not perceive any benefits from this.

Screening for Cancer

There are three national screening campaigns in the UK: for breast, bowel, and cervical cancer. These are available through the National Health Service, which is funded by taxation and so the tests are free to everyone. While there is no economic barrier to their uptake, other factors intervene for each cancer screened.

There are socio-economic differences in who is screened, with poorer people less likely to take up screening. Moser et al. [49] found a correlation between "indicators of wealth" (e.g. an owner occupied house, or a household with a car), and women having had breast screening. Women, who lived in a bought house or lived in a household with one or more cars, were more likely to have had a mammogram than women living in rented accommodation and not having a car. Reduced uptake of cervical screening has also been found among lower socio-economic groups [50]. Moss et al. [51] found people from lower SES less often took up the opportunity for bowel cancer screening compared to people from higher SES, despite it being free at the point of access. They proposed that great effort would be necessary to avoid significant disparities in screening uptake between deprived and wealthy people.

Despite the best efforts of national screening programmes to promote equitable uptake of screening, significant inequalities exist across all the programmes; the reasons for this are complex [51], and resistant to interventions. Consequently, cancer screening has the potential to enhance disparities in cancer outcome.

Awareness and Recognition of Cancer

There is a great deal of interest in the UK in awareness and recognition of cancer by patients and family doctors. In countries with strong primary health-care systems, such as the UK, family practice is typically the first point of contact for the majority of patients. In order for timely diagnosis to take place people need to recognise that their symptoms may be serious and so worthy of contacting a doctor, and then the doctor needs to recognise these symptoms as potential cancer symptoms [53].

If people do not present as early as possible with cancer symptoms, an opportunity may be lost to diagnose and treat the cancer early (and, potentially, improve survival). A link between prolonged diagnostic intervals and deprivation is challenging to prove; although we know that there are differences in stage of diagnosis for many cancers, based on whether someone is from a deprived or better-off background, [54] this may not be because there was any delay in presenting with symptoms. Rather, the nature of the illness may be such that the symptoms duration was short. There is, nevertheless, a policy drive in the UK to seek to ensure that patients recognise symptoms as early as possible and for practitioners to refer appropriately [55].

Systematic reviews of the evidence have been carried out to seek to understand the factors associated with timely recognition of cancer by patients and family doctors [53]. These reviews have concluded that, for many cancers, non-recognition of symptom seriousness is the main patient-related factor resulting in increased time to presentation. There is strong evidence of an association between older age and patient delay for breast cancer, between lower SES and delay for upper gastrointestinal and urological cancers and between lower education level and delay for breast and colorectal cancers [53]. Fear of cancer is a contributor to delayed presentation, while sanctioning of help seeking by others can be a powerful mediator of reduced time to presentation [53].

These findings have resulted in an interest in awareness of cancer, even though it is clear that awareness is insufficient in and of itself. The evidence does, however, suggest that many people appear to have very limited knowledge about cancers. This may be based on how they are asked about cancers. For example, a study examined the awareness of cancer of patients from both an affluent residential and deprived inner-city area in the same city in the North of England, and found that people had very poor open recall, but better prompted recognition [56].

On the whole people tend to have poor awareness about the warning signs of cancer for all symptoms (except lumps and swelling). Robb et al. [57] asked people to freely recall and then to recognise a set of cancer symptoms, and found recognition, which studies of memory have shown to be a more effective means of retrieval of information, was much higher for cancer symptoms, e.g. mole, lump, or swelling, than free recall. This was a general finding across the population: in particular men, younger people, people from an ethnic minority, and people from the lower end of the socio-economic spectrum had poorer awareness.

Further, it appears that people from ethnic minorities, who are often amongst the most socially disadvantaged, have poor awareness of the warning signs for cancers [58]—these authors suggest poor understanding of English may be a contributing factor, as people from ethnic minorities in the UK have higher levels of deprivation.

Evidence on cancer disparities has prompted considerable policy interest and activity regarding early detection of cancer. In England, a key programme is the National Awareness and Early Diagnosis Initiative (NAEDI) [55]; in Scotland there is a similar initiative—the Detect Cancer Early programme. Both these programmes seek to join up expertise from the NHS, the academic sector and the NHS in order to improve cancer survival outcomes (The Scottish Government. Detect cancer early. http://www.scotland.gov.uk/Topics/Health/Services/Cancer/Detect-Cancer-Early; accessed Sept 2013) [59].

Health Service Factors

So far, we have shown how a person's cancer outcomes are disadvantaged by socio-economic factors. However, the patient can also be disadvantaged through poor provision and/or poor quality of health services. Julian Tudor Hart [60] proposed the Inverse Care Law; this states that the accessibility of good medical care is inclined to vary inversely with the need for it by the population. Thus, people with cancer from poorer backgrounds may be disadvantaged by the poor availability of good quality care as much as by their own personal circumstances.

The first important health service factor is the response of the family doctor when a patient presents himself/herself with a new symptom. The evidence for factors associated with delay by family doctors is mixed [53]. In family practice many patients present with symptoms that may be indicative of cancer, but diagnostic tests later exclude cancer. On the other hand, family doctors assigning a diagnosis other than cancer to a set of symptoms can introduce delay in the pathway to referral [53].

Some work has also considered whether patients from poorer regions experience different care once diagnosed with cancer than those from better neighbourhoods and in general this has been found not to be the case [61]. However the presence of other coexisting illnesses occurring more commonly in socio-economically deprived patients may in part explain the poorer outcomes.

An individual with cancer receives care across several stages, from when they first present with their symptoms to a health-care professional, through living with cancer and then either surviving or dying from cancer. Lewis et al. [62] noted how SES impacts on four dimensions of access to palliative care: its availability, affordability, accessibility, and acceptability. Broadly speaking, palliative care is less available to people from the lower social classes and is less affordable for them; they have less access to it, and they are less accepting of it.

While care is free at the point of delivery to all in the UK, differences in care remain. Raine et al. [63] found patients from deprived areas, older people, and women were more likely to be admitted as emergencies for their cancer. People living in deprived areas and males were less likely to receive their preferred surgical procedures for cancers. They also found that older people were more likely to receive their preferred surgical procedure for rectal cancer but less likely to receive breast conserving surgery and lung cancer resection.

Psychosocial Factors

As we have shown, the evidence points to people from poor backgrounds being differentially exposed to environmental stressors compared to people from more affluent backgrounds. This adversely affects their health outcomes in general. White and Macleod [64] have noted three psychological consequences from having cancer: the patient can experience depression; the patient can feel anxiety, fear, and panic; or if the patient has a cancer that spreads to the brain, she/he can suffer neuropsychiatric problems.

A follow-up study of women with breast cancer showed that affluent women were more likely to have received information from their hospital specialist and from a breast care nurse than deprived women, but deprived women had poorer SF-36 scores (Short-Form 36, self-reported survey of health status) than affluent women, and reported greater anxiety about money, other health problems, and family problems [65]. In a recent study of cancer survivors in England, individuals from most socio-economically deprived areas reported lower quality-of-life scores [66].

Conclusion

People with cancer from disadvantaged socio-economic backgrounds have poorer health in general, poor access to health care, and poorer outcomes. The reasons for this are undoubtedly multifactorial; in this chapter we have emphasised that the relationship between inequalities and cancer is complex and probably not unidirectional. People from lower SES groups may be in a poorer position to cope with hardship resulting from living with cancer, while people from a more affluent background will have the resources and knowledge to cope with cancer. In other words, understanding context is key.

Because socio-economic inequalities can determine people's health in general and particularly for cancer, people from poorer backgrounds do not always have access to the same quality of care as more affluent people. Reflecting Wilkinson and Pickett [67] we argue that political efforts need to be made to rebalance social and health inequalities. Heath [68] has likewise argued the need to confront causes of health inequalities. Reducing disparities is difficult; there was hope that the NHS Cancer Plan [5], with a number of measures focused on deprived sectors of the population, would improve cancer survival rates and reduce disparities. While it resulted in a decrease in the deprivation gap for cancer outcomes at 1 year, this was not maintained, and the reason for this is unclear [15]. It may be that changes enacted around this time needed longer to impact on morbidity and mortality from a cancer. Or perhaps social class (and resultant social inequalities) are so well entrenched within society and so less responsive to policy initiatives.

The problem of health inequalities was recognised in 1997 by the UK Secretary of State for Health Frank Dobson, He stated that:

> Inequality in health is the worst inequality of all. There is no more serious inequality than knowing that you'll die sooner because you're badly off. (Dobson and Department of Health 1997) [69].

Health inequalities prevail in the UK and have a significant impact on people with a cancer. To ensure that everyone has the best possible outcome from a cancer, regardless of whether they are affluent or poor, will probably require great effort at a national policy level.

References

1. Reid J (1972) Alienation: rectoral address. University of Glasgow Publications, Glasgow
2. Department of Health and Social Security (1980) Inequalities in health: report of a working group chaired by Sir Douglas Black. DHSS, London
3. Acheson D (1998) Independent inquiry into inequalities in health. Stationery Office, London
4. Marmot M (2010) Strategic review of health inequalities in England post-2010. Marmot review final report, University College London. http://www.instituteofhealthequity.org/
5. Department of Health (2000) The NHS cancer plan: a plan for investment, a plan for reform. Department of Health, London
6. Department of Health (2007) Cancer reform strategy. Department of Health, London
7. Cancer Research UK. http://info.cancerresearchuk.org/cancerstats/causes/lifestyle/bodyweight/. Accessed July 2013
8. Shack L, Jordan C, Thomson CS, Mak V, Moller H (2008) Variation in incidence of breast, lung and cervical cancer and malignant melanoma of skin by socioeconomic group in England. BMC Cancer 8:271
9. McCartney G, Shipley M, Hart C, Davey-Smith G, Kivimäki M, Walsh D, Watt GC, Batty GD (2012) Why do males in Scotland die younger than those in England? Evidence from three prospective cohort studies. PLoS One 7:e38860. doi:10.1371/journal.pone.0038860
10. Carstairs V, Morris R (1989) Deprivation: explaining differences in mortality between Scotland and England and Wales. BMJ 299(6704):886–889
11. Hanlon P, Lawder R, Buchanan D, Redpath A, Walsh D, Wood R et al (2005) Why is mortality higher in Scotland than in England & Wales? Decreasing influence of socioeconomic deprivation between 1981 and 2001 supports the existence of a 'Scottish effect'. J Public Health 27:199–204
12. Norman P, Boyle P, Exeter D, Feng Z, Popham F et al (2011) Rising premature mortality in the UK's persistently deprived areas: only a Scottish phenomenon? Soc Sci Med 73:1575–1584. doi:10.1016/j.socscimed.2011.09.034
13. Office for National Statistics (2012) Cancer incidence and mortality in the UK, 2007–2009, Statistical Bulletin

14. Coleman MP, Forman D, Bryant H, Butler J, Rachet B, Maringe C et al (2011) Cancer survival in Australia, Canada, Denmark, Norway, Sweden and the UK, 1995–2007 (the International Cancer Benchmarking Partnership): an analysis of population-based cancer registry data. Lancet 377(9760):127–138. doi:10.1016/S0140-6736(10)62231-3

15. Rachet B, Ellis L, Maringe C, Chu T, Nur U, Quaresma M et al (2010) Socioeconomic inequalities in cancer survival in England after the NHS cancer plan. Br J Cancer 103:446–453. doi:10.1038/sj.bjc.6605752

16. Woods LM, Rachet B, Coleman MP (2006) Origins of socio-economic inequalities in cancer survival: a review. Ann Oncol 17:5–19

17. Lyratzopoulos G, Barbiere JM, Rachet B, Baum M, Thompson MR, Coleman MP (2011) Changes over time in socioeconomic inequalities in breast and rectal cancer survival in England and Wales during a 32-year period (1973–2004): the potential role of health care. Ann Oncol 22:1661–1666

18. Coleman MP, Babb P, Sloggett A, Quinn M, De Stavola B (2001) Socioeconomic inequalities in cancer survival in England and Wales. Cancer 91(1 Suppl):208–216

19. Abdel-Rahman MA, Stockton DL, Rachet B, Hakulinen T, Coleman MP (2009) What if cancer survival in Britain were the same as in Europe: how many deaths are avoidable? Br J Cancer 101(Suppl 2):S115–S124. doi:10.1038/sj.bjc.6605401

20. Anand P, Kunnumakara AB, Sundaram C, Harikumar KB, Tharakan ST, Lai OS et al (2008) Cancer is a preventable disease that requires major lifestyle changes. Pharm Res 25(9):2097–2116. doi:10.1007/s11095-008-9661-9

21. Gordon-Dseagu V (2006) Cancer and health inequalities: an introduction to current evidence. Cancer Research UK, London

22. Macintyre S (1997) The black report and beyond what are the issues? Soc Sci Med 44:723–745

23. Merletti F, Galassi C, Spadea T (2011) The socioeconomic determinants of cancer. Environ Health 10(Suppl 1):S7

24. Schottenfeld D, Fraumeni JF, Jr. (2006) Socioeconomic disparities in cancer incidence and mortality. In: David Schottenfeld, Fraumeni JF (eds) Cancer epidemiology and prevention. Oxford University Press, Oxford, pp 174–188

25. Asthana S, Halliday J (2006) What works in tackling health inequalities? Pathways, policies and practice through the lifecourse. Policy Press, Bristol

26. Peacock M, Bissell P (2011) The social determinants of health inequalities: implications for research and practice. University of Sheffield and Sheffield Teaching Hospitals NHS Foundation Trust

27. Woodward A, Boffetta P (1997) Environmental exposure, social class, and cancer risk. IARC scientific publications no. 138. International Agency for Research on Cancer, Lyon

28. Buffetta R, Kogevinas M, Westerholm P, Saracci R (1997) Exposure to occupational carcinogens and social class differences in cancer occurrence. IARC scientific publications no. 138. International Agency for Research on Cancer, Lyon

29. Lynch J (2000) Income inequality and health: expanding the debate. Soc Sci Med 51:1001–1005, discussion 1009–10

30. Coleman MP, Rachet B, Woods LM, Mitry E, Riga M, Cooper N et al (2004) Trends and socioeconomic inequalities in cancer survival in England and Wales up to 2001. Br J Cancer 90:1367–1373

31. Barford A, Dorling D, Davey Smith G, Shaw M (2006) Life expectancy: women now on top everywhere. BMJ 332(7545):808

32. Cappuccio FP (1997) Ethnicity and cardiovascular risk: variations in people of African ancestry and South Asian origin. J Hum Hypertens 11(9):571–576

33. Lantz PM, House JS, Lepkowski JM, Williams DR, Mero RP, Chen J (1998) Socioeconomic factors, health behaviors, and mortality: results from a nationally representative prospective study of US adults. JAMA 279:1703–1708

34. The Scottish Government (2004) Scottish index of multiple deprivation 2004: summary technical report. http://www.scotland.gov.uk/Publications/2004/06/19429/38161

35. Ross CE, Wu C (1995) The links between education and health. Am Sociol Rev 60:719–745
36. Jarvis MJ, Wardle J (2006) Social patterning of health behaviours: the case of cigarette smoking. In: Marmot M, Wilkinson RG (eds) Social determinants of health. Oxford University Press, Oxford, pp 224–237
37. Jarvis M (2001) The challenge for reducing inequalities: analysis of the General Household Survey 1998 presentation to a Department of Health seminar. January 2001
38. Centers for Disease Control and Prevention (2000) Behavioral risk factor surveillance system: prevalence data: nationwide—2000 nutrition. CDC, Atlanta
39. James WP, Nelson M, Ralph A, Leather S (1997) Socioeconomic determinants of health: the contribution of nutrition to inequalities in health. BMJ 24:1545–1549
40. Colditz GA, Cannuscio CC, Frazier AL (1997) Physical activity and reduced risk of colon cancer: implications for prevention. Cancer Causes Control 8(4):649–667
41. Department of Health (1998) Health survey for England. Department of Health, London
42. McLaren L (2007) Socioeconomic status and obesity. Epidemiol Rev 29:29–48
43. Calle EE, Kaaks R (2004) Overweight, obesity and cancer: epidemiological evidence and proposed mechanisms. Nat Rev Cancer 4(8):579–591
44. Reeves GK, Pirie K, Beral V, Green J, Spencer E, Bull D (2007) Cancer incidence and mortality in relation to body mass index in the Million Women Study: cohort study. BMJ 335:1134
45. Dein S (2004) Explanatory models of and attitudes towards cancer in different cultures. Lancet Oncol 5:119–124
46. Soler-Vila H, Kasl SV, Jones BA (2005) Cancer-specific beliefs and survival: a population-based study of African-American and White breast cancer patients. Cancer Causes Control 16:105–114
47. Powe BD, Finnie R (2003) Cancer fatalism: the state of the science. Cancer Nurs 26:454–465
48. Sontag S (1978) Illness as metaphor. Farrar, Straus & Giroux, New York
49. Moser K, Patnick J, Beral V (2009) Inequalities in reported use of breast and cervical screening in Great Britain: analysis of cross sectional survey data. BMJ 338:b2025. doi:10.1136/bmj. b2025
50. Baker D, Middleton E (2003) Cervical screening and health inequality in England in the 1990s. J Epidemiol Community Health 57(6):417–423
51. Moss M, Campbell C, Melia J, Coleman D, Smith S, Parker R et al (2012) Performance measures in three rounds of the English bowel cancer screening pilot. Gut 61:101–107. doi:10. 1136/gut.2010.236430
52. Wardle J, McCaffery K, Nadel M, Atkin W (2004) Socioeconomic differences in cancer screening participation: comparing cognitive and psychosocial explanations. Soc Sci Med 59:249–261
53. Macleod U, Mitchell E, Burgess C, Macdonald S, Ramirez AJ (2009) Risk factors for delayed presentation and referral of symptomatic cancer: evidence for common cancers. Br J Cancer 101(Suppl 2):S92–S101
54. Lyratzopoulos G, Abel GA, Brown CH, Rous BA, Vernon SA, Roland M et al (2013) Socio-demographic inequalities in stage of cancer diagnosis: evidence from patients with female breast, lung, colon, rectal, prostate, renal, bladder, melanoma, ovarian and endometrial cancer. Ann Oncol 24:843–850
55. Richards MA (2009) The national awareness and early diagnosis initiative in England: assembling the evidence. Br J Cancer 101:S1–S4. doi:10.1038/sj.bjc.6605382
56. Adlard JW, Hume MJ (2003) Cancer knowledge of the general public in the United Kingdom: survey in a primary care setting and review of the literature. Clin Oncol (R Coll Radiol) 15 (4):174–180
57. Robb K, Stubbings S, Ramirez A, Macleod U, Austoker J, Waller J et al (2009) Public awareness of cancer in Britain: a population-based survey of adults. Br J Cancer 101(Suppl 2):S18–S23. doi:10.1038/sj.bjc.6605386

58. Waller J, Robb K, Stubbings S, Ramirez A, Macleod U, Austoker J et al (2009) Awareness of cancer symptoms and anticipated help seeking among ethnic minority groups in England. Br J Cancer 101(Suppl 2):S24–S30

59. The Scottish Government. Detect cancer early. http://www.scotland.gov.uk/Topics/Health/Services/Cancer/Detect-Cancer-Early. Accessed Sept 2013

60. Hart JT (1971) The inverse care law. Lancet 297(7696):405–412

61. Macleod U, Ross S, Twelves C, George WD, Gillis C, Watt GC (2000) Primary and secondary care management of women with early breast cancer from affluent and deprived areas: a retrospective review of hospital and general practice records. BMJ 320:1442–1445

62. Lewis JM, DiGiacomo M, Currow DC, Davidson PM (2011) Dying in the margins: understanding palliative care and socioeconomic deprivation in the developed world. J Pain Symptom Manage 42:105–118

63. Raine R, Wong W, Scholes S, Ashton C, Obichere A, Ambler G (2010) Social variations in access to hospital care for patients with colorectal, breast, and lung cancer between 1999 and 2006: retrospective analysis of hospital episode statistics. BMJ 340:b5479. doi:10.1136/bmj.b5479

64. White CA, Macleod U (2002) ABC of psychological medicine: cancer. Br Med J 325:377–380

65. Macleod U, Ross S, Fallowfield L, Watt GC (2004) Anxiety and support in breast cancer: is this different for affluent and deprived women? A questionnaire study. Br J Cancer 91:879–883

66. Glaser AW, Fraser LK, Corner J, Feltbower R, Morris EJ, Hartwell G et al (2013) Patient-reported outcomes of cancer survivors in England 1–5 years after diagnosis: a cross-sectional survey. BMJ Open 3(4):e002317. doi:10.1136/bmjopen-2012-002317

67. Wilkinson R, Pickett K (2010) The spirit level: why equality is better for everyone. Penguin, London

68. Heath F (2007) Let's get tough on the causes of health inequality. BMJ 334(7607):1301

69. Dobson F, Department of Health (1997) Government takes action to reduce health inequalities. Press release in response to the Joseph Rowntree publication Death in Britain. DoH press release 97/192 11 August

Chapter 3
Behavioral Differences Leading to Disparities in Energy Balance and Cancer

Donna Spruijt-Metz, Lauren Cook, C.K. Freddy Wen, Robert Garcia, Gillian A. O'Reilly, Ya-Wen Hsu, Jennifer B. Unger, and Selena T. Nguyen-Rodriguez

D. Spruijt-Metz (✉)
USC mHealth Collaboratory, Center for Economic and Social Research, University of Southern California, 12015 Waterfront Drive, Playa Vista, CA 90094-2536, USA
e-mail: dmetz@usc.edu

L. Cook
Department of Preventive Medicine, Keck School of Medicine, University of Southern California, 2250 Alcazar Street, CSC-200, Los Angeles, CA 90089, USA
e-mail: laurenco@usc.edu

C.K.F. Wen
Department of Preventive Medicine, Keck School of Medicine, University of Southern California, 2001N. Soto Street, Los Angeles, CA 90033, USA
e-mail: chengkuw@usc.edu

R. Garcia
Institute for Prevention Research, Department of Preventive Medicine, Keck School of Medicine, University of Southern California, 556 Black Hills Drive, Claremont, CA 91711, USA
e-mail: garc617@usc.edu

G.A. O'Reilly
Department of Preventive Medicine, Keck School of Medicine, University of Southern California, 2001N. Soto Street, SSB 3rd Floor, Los Angeles, CA 90032, USA
e-mail: goreilly@usc.edu

Y.-W. Hsu
Department of Hospital and Health Care Administration, Chia Nan University of Pharmacy & Science, 60, Erh-Jen Road, Sec.1, Build Q, Room 413-3, Jen-Te, Tainan 71710, Taiwan
e-mail: janiceywhsu@gmail.com

J.B. Unger
Institute for Health Promotion and Disease Prevention Research, Department of Preventive Medicine, Keck School of Medicine, University of Southern California, Soto Street Building, Suite 330A, 2001N Soto Street, MC 9239, Los Angeles, CA 90089-9239, USA
e-mail: unger@usc.edu; jenniferbethunger@gmail.com

S.T. Nguyen-Rodriguez
Department of Health Science, California State University, Long Beach, 1250 Bellflower Boulevard, HHS2-115, Long Beach, CA 90840, USA
e-mail: selena.nguyen-rodriguez@csulb.edu

D.J. Bowen et al. (eds.), *Impact of Energy Balance on Cancer Disparities*, Energy Balance and Cancer 9, DOI 10.1007/978-3-319-06103-0_3, © Springer International Publishing Switzerland 2014

Abstract The patterns of racial/ethnic, gender, and socio-demographic disparities in cancer incidence patterns are complex. While susceptibility, exposure, environment, access to and attitudes towards screening and medical treatment influence cancer incidence and mortality, there are strong behavioral influences on racial/ethnic and socioeconomic disparities in incidence and mortality by cancer site. These behaviors are intertwined with culture and acculturation. In this chapter, we discuss disparities in four central areas of behavior that are related to both energy balance and cancer. These include dietary intake (broken down by key nutrients), disparities in physical activity and sedentary behavior, disparities in sleep, and disparities in smoking.

Keywords Racial/ethnic disparities • Socioeconomic disparities • Diet • Physical activity • Sleep • Smoking • Sugar consumption • Red meat consumption • Fish consumption • Dietary fat • Fruit and vegetable consumption

Introduction

Racial/ethnic and socioeconomic disparities in cancer screening, incidence, treatment, and mortality are both glaring and complex. For all cancer sites combined, African-American men have a 14 % higher incidence rate and a 33 % higher death rate than white men, whereas African-American women have a 6 % lower incidence rate but a 16 % higher death rate than white women [1]. However, specific cancers are more prevalent in particular groups. For instance, stomach and liver cancer incidence and death rates are twice as high in Asian Americans/Pacific Islanders as in Whites [2]. Another example is that kidney cancer incidence and death rates are the highest among American Indians/Alaskan Natives, which may reflect the high prevalence of obesity and smoking in this population [3]. In some cases, equal treatment for similar disease and tumor status has been shown to yield similar outcomes between racial/ethnic minorities and Whites [4, 5]. Nonetheless, racial disparities continue to exist in cancer treatment [4, 5]. Several systemic and doctor-related barriers contribute to these differences, as do factors that influence patient freedom of choice, decision-making, and ultimately, patient behaviors.

Broad disparities between subgroups within racial and ethnic groupings reflect possible differences in exposure, susceptibility, and access, but also between cultures and behaviors. For instance, the incidence rate for invasive cervical cancer, much of which is preventable by screening, is four times higher among Vietnamese women than in all Asian American/Pacific Islander populations combined. Another example of subgroup disparities is the regional influence on cancer rates among some American Indian populations, which increases with proximity to reservations [6].

The strong socioeconomic gradients in cancer incidence differ by racial/ethnic group, gender, country/region and type of cancer [7]. For instance, in the USA, lung cancer incidence is associated with markers of lower socioeconomic status in

Whites, Blacks, and Asians but with markers of higher socioeconomic status in Hispanics. One study in California Hispanics found that higher neighborhood socioeconomic status was associated with increased lung cancer incidence in women, but weakly associated in men, and ever-smoking rates were higher with increased acculturation [8].

Thus, although susceptibility, exposure, environment, access to and attitudes towards screening and medical treatment influence cancer incidence and mortality, there are strong behavioral influences on racial/ethnic and socioeconomic disparities in incidence and mortality by cancer site. These behaviors are intertwined with culture, acculturation, and socioeconomic status [9]. The links and mechanisms between energy balance-related behaviors and cancer have been discussed previously [10]. In this chapter, we discuss disparities in four central areas of behavior that are related to both energy balance and cancer. These include dietary intake (broken down by key nutrients), disparities in physical activity and sedentary behavior, disparities in sleep, and disparities in smoking.

Racial/Ethnic and Socioeconomic Disparities in Dietary Intake

Sugar Intake

Much research aimed at elucidating dietary contributions to energy balance has focused on the role of dietary sugar, particularly added sugar. There is substantial evidence that excess sugar intake contributes to positive energy balance [11, 12]. However, the contribution of sugar intake to positive energy balance may depend upon whether calories from sugar replace other calories or add to them [13]. Additionally, the form in which added sugar is consumed may contribute differentially to energy balance. There is evidence that energy consumed from beverages, such as soda and fruit juice, is regulated by different mechanisms than energy consumed from food, such that energy consumed from beverages may lead to a greater positive energy balance [13, 14]. This is supported by experimental studies that have shown a causal association between consumption of sugar-sweetened beverages and weight gain [15–17]. One potential explanation for the contribution of sugar-sweetened beverages to positive energy balance is that consumption of these beverages leads to extra energy intake before adequate feedback is provided by physiological satiety signals [18]. Given that most of the added sugar consumed by Americans is in the form of sugar-sweetened beverages, the contribution of sugar-sweetened beverages to positive energy balance is a particularly important public health concern [13, 19].

Dietary sugar has also received attention for its potential contribution to increased risk for certain types of cancer. High intake of sugar-laden foods may be associated with increased risk for pancreatic cancer [20, 21]. Added sugar intake has also been associated with increased risk for colorectal cancer [22–24]. The increased risk for these types of cancers may be explained by the impact that dietary sugars have on insulin sensitivity, body fat distribution, and the concentration of growth factors that can contribute to the growth of cancers [25].

There are notable differences in sugar consumption across ethnicities and socioeconomic strata, but studies provide mixed evidence about which racial/ethnic groups consume the highest amounts of added sugar. In one recent study among a nationally representative sample of preschool-aged children, Hispanic children consumed less added sugar than all other ethnic groups, while non-Hispanic black and non-Hispanic white children consumed the most added sugar [26]. However, evidence indicates that the diets of low socioeconomic status Hispanic children exceed guidelines for added sugar intake [27]. Findings are similarly mixed for adults. A recent study using dietary data from a nationally representative sample of adults indicated that Hispanics consumed less added sugar than all other ethnicities except for Asian-Americans [28]. In this sample, non-Hispanic Blacks had the highest added sugar intake, followed by American Indian/Alaskan Natives and non-Hispanic Whites [28]. However, there is also evidence that Hispanics in particular consume high amounts of sugar-sweetened beverages [29]. There is more consistent evidence of an association between socioeconomic status and dietary sugar intake. A number of studies have indicated that individuals with low income and low education consume more sugar than individuals from higher socioeconomic backgrounds [28, 30, 31].

Acculturation and food cost may provide explanations for the differences in sugar consumptions between racial/ethnic and socioeconomic groups. While there are mixed findings from studies regarding the relative sugar consumption of Hispanics in the USA, studies on acculturated Hispanics provide more consistent findings. Hispanic individuals who are more acculturated to US culture have been consistently shown to have higher sugar intake than other racial/ethnic groups [32–34]. This indicates that acculturation status may explain the differences in findings of studies on sugar intake among Hispanics. The higher sugar intake exemplified by groups from lower socioeconomic strata may be explained by the difference in cost of high energy, low nutrient dense foods versus low energy, high nutrient dense foods [28]. Foods with high energy density but low nutrient density, such as those high in added sugars, are typically less expensive than foods with low energy density but high nutrient density, such as fruits and vegetables [28, 35, 36]. Added sugar intake has been shown to be directly related to the amount of income one has to spend on food [12], so individuals from low socioeconomic status groups are particularly vulnerable to consuming diets that are high in added sugar.

Beneficial Dietary Fats and Proteins

Although findings are inconsistent [37, 38], evidence suggest that consumption of omega-3 polyunsaturated fatty acids (PUFA, found in fatty fish like salmon, sardines, and herring) may be protective of various cancers by altering the carcinogenic process [39], especially those that are hormone-related [38]. A major limitation of this area of research is that total fish intake is often examined as a risk factor, and lean fish such as cod or halibut may not offer the same health benefits [38] (methodological limitations will also likely bias findings towards null). Some specific ethnic groups have a rich history of consuming diets high in fish and marine mammals, such as the Inuit (natives of Alaska, Canada, and Greenland) [40], yet these individuals have a higher risk for all cancers, except breast and prostate cancer, than non-Inuits [41]. The protective effects of a native diet are likely mitigated once native cultures adopt a more Western diet with smaller amounts of these potentially protective foods [42].

Similarly, an assessment of dietary intake by varying ethnic groups in Hawaii and Los Angeles found that relative to Whites, Japanese-Americans and Native Hawaiians consumed the greatest quantities of fish, whereas African-Americans and especially Latinos had lower consumption [43]. Given the relationship between socioeconomic status and ethnicity, it is likely that the lack of fish consumption is due to cost, as this was identified as a barrier to consumption in Belgian adults [44]. Perceived inconvenience was also identified as a barrier in another European study [45]. Yet a study of fishing and subsequent consumption in South Carolina found that African-Americans consumed more fish than Whites, as did those who did not finish high school (it should be noted that this study was conducted in an area with a fishing advisory due to high fish mercury content) [46]. Given this evidence, it is likely that fish availability and accessibility (indicated by proximity to coasts and rivers and/or historical prominence in the diet of certain regional groups) are highly predictive of intake.

It should also be briefly noted that soy products (specific isoflavones contained in these foods) have also emerged as protective for breast cancer among Asian women [47], and possibly for prostate cancer, although the evidence is not so strong [48]. This protective relationship has not been observed with those following a Western diet, possibly due to the predominant use of soy as an additive rather than whole food [47].

Deleterious Dietary Fats and Proteins

Similar to findings on fish and PUFA intake, studies examining the impact of red meat consumption on various cancers are mixed, with stronger support for the impact on certain cancers (such as esophageal) than others (including prostate, gastric, pancreatic, and colorectal cancers) [49–56]. There are two components of red meat that make this a risk factor: higher levels of saturated fat [57, 58], and frequent use with cooking methods (such as char-grilling) than may impart carcinogens into the food [59].

Data from the Los Angeles/Hawaii multiethnic cohort indicate that Latinos consume about 30 % more red meat than African-Americans and Whites

[43]. NHANES (National Health and Nutrition Examination Survey) data also show that Latinos consume the greatest amounts of red meat, although this is not statistically higher than other ethnic groups [60]. Studies of meat intake and acculturation indicate that Latinos with the lowest US assimilation had the lowest avoidance of foods high in saturated fats, relative to more assimilated Latinos and Whites [61]; and that less assimilated Latinos were more likely to eat meat, relative to more assimilated Latinos [62]. Qualitative data suggest that Latina mothers find it more easy to procure red meat in the US relative to their home countries [63], so perhaps this higher degree of intake in this lower assimilated group is due to the interaction of accessibility and cultural value placed on meat intake (as a symbol of prosperity, as in many parts of the developing world) [64]. A separate qualitative study found that Latina mothers perceived meat to be a healthy food type [65], so it is possible that higher assimilated Latinos decrease intake of red meat given a gain in knowledge of potential negative healthy effects. Supporting this hypothesis, NHANES data indicate that both the perceived benefit of diet quality and use of food labels are negatively associated with red meat intake [66].

Whole Grains, Fruits, and Vegetables

Fruit and vegetable (FV) intake is strongly associated with cancer prevention, especially those related to the gastrointestinal system, lungs, and pancreas [67, 68]. Raw FV are the most beneficial, likely because they have the greatest concentration of antioxidants, including vitamins, minerals, and polyphenols [67]. Another protective component of FV is fiber, which has been found to be beneficial for several different types of cancers [69–73]. Whole grain products, which are also high in fiber, have also been shown to reduce cancer risk [74, 75].

A comparison of FV intake between different ethnic groups found that relative to Whites, African-Americans consumed approximately 1 serving/day less of FV, and Mexican Americans consumed about 0.3 servings/day less than Whites per day, with the majority of this disparity attributable to differences in intake of vegetables [76]. After adjusting for neighborhood SES, the difference between African-Americans and Whites remained [76]. NHANES data also indicate that African-Americans are less likely to consume the recommended number of FV servings/day, compared to Whites [77]. However, other neighborhood factors (adjusting for socioeconomic variables) may still contribute to this disparity. A study in Brooklyn, NY found that supermarkets were located in one third of the predominantly White US census tracts, while predominantly African-American census tracks had no supermarkets, and had less fresh produce available [78]. Similarly, in a study of African-American and Latina women eligible for the Women, Infants and Children (WIC) supplemental nutrition program, it was found that Latina women and children consumed a significantly greater amount of whole grains, compared to African-Americans [79]. This difference could be attributable to the lower levels of US assimilation among the Latina women (and likely manifested in a higher intake of corn tortillas, a dietary staple), or possibly because they were significantly more likely to be married or living with a partner (perhaps attributable to cultural or religious values, which could enhance social support; it is unclear if this was controlled for in analyses) [79].

Racial/Ethnic and Socioeconomic Disparities in Physical Activity and Energy Balance

Evidence indicates that regular physical activity is protective against several types of cancer, including breast, colon, endometrium, prostate, and pancreatic cancer [25, 80–83]. In fact, the American Cancer Society points to physical activity as one of the most important modifiable determinants of caner [25]. Physical activity reduces the risk of cancer through several direct and indirect mechanisms, including body weight and energy balance regulation, immune system functioning, and regulation of sex hormones, insulin, and prostaglandins [25, 84, 85]. Physical activity plays an important role in energy balance because the movements of skeletal muscles during physical activity result in energy expenditure [86]. The intensity, duration, and frequency of physical activity all affect the extent to which physical activity contributes to energy balance [87]. For example, vigorous intensity activities (>6.0 metabolic equivalents or METs, a metric for estimating energy expenditure during physical activity) lead to greater energy expenditure than moderate-intensity activities (3.0–6.0 METs) carried out over the same duration and frequency [87]. National guidelines suggest that adults accrue at least 150 min a week of moderate-intensity, or 75 min a week of vigorous-intensity aerobic physical activity [88].

Although physical activity engagement provides great promise for preventing cancer incidence and recurrence, there are disparities in physical activity among many groups, especially ethnic minority communities. Disparities in physical activity engagement can be attributed to environmental and individual factors. Physical features in the built environment, including but not limited to streetlights, infrastructure that facilitates active transportation, and park availability, predict physical activity levels. The skewed availability of these features in areas with high ethnic minority density and/or low socioeconomic status (SES) may put minority populations at a disadvantage for opportunities to engage in physical activity [89]. There is also a wide array of individual level factors that influence physical activity, including perceived neighborhood safety [90, 91], psychosocial barriers [92]. While these factors influence behavior at an individual level, the interaction between individual factors and the environment also contribute to disparities in individual physical activity engagement. In the following section, we present examples illustrating the intertwined relationship between neighborhood environment and individual behavior.

Moderate and Vigorous Physical Activity (MVPA)

Physical activity engagement differs across age and gender. A cross-sectional study using NHANES data highlighted the differences in time spent in MVPA across gender and age by finding that females are less active than males and, overall, MVPA engagement declines with increasing age [93]. The lower physical activity levels observed among females are believed to be influenced by cultural and psychosocial factors, for example, body image issues, especially among ethnic minority groups [94]. Challenges that limit physical activity participation in culturally and linguistically diverse groups include: cultural and religious beliefs, issues within social relationships, socioeconomic status, environmental barriers, and culture-related perceptions of physical activity outcomes. Several strategies to overcome these challenges have been suggested, including culturally sensitive programs, education sessions addressing healthy behavior, and improving access to environments that promote physical activity both at work and in the community.

Occupation and Physical Activity

Within the context of one's occupation, socioeconomic status and environmental factors can interact to influence physical activity. Certain physical activities associated with low-wage, labor-intensive occupations qualify as MVPA. According to Troiano et al., despite the decline in MVPA with increasing age demonstrated in the overall population, Hispanic and African-American males exhibit higher activity levels than non-Hispanic white males [93]. This may be because labor-intensive jobs are more prevalent among African-American and Hispanic males than non-Hispanic white males [92]. While low-wage, labor intensive jobs may have a positive impact on physical activity for workers, their family members, especially children and adolescents, may experience negative effects on physical activity because such families have a higher likelihood of living in low socioeconomic neighborhoods [95].

Access to Physical Activity Facilities

A body of literature has documented the disparities in access to physical activity facilities influence MVPA. From an ecological perspective, however, factors that influence disparities in MVPA engagement may not be applicable only to ethnic minorities. Studies from various parts of the world indicate that affluent neighborhoods have better access to physical activity facilities than low SES neighborhoods [95–98]. Whether neighborhood demographics is a better predictor of physical activity engagement than other factors that can be applicable to all ethnic backgrounds, such as socioeconomic status, is not well understood.

Leisure Time or Recreational Physical Activity

Leisure time physical activity (LTPA) engagement is also influenced by multiple factors. LTPA encompasses activities that people engage in during discretionary time. Examples of such activities include: gardening, walking, and recreational sports such as cycling. Like MVPA, LTPA engagement is also influenced largely by socioeconomic position both in an individual and an ecological perspective. Similar to how occupation influences disparities in activity engagement, time constraints due to work schedule and lack of energy after work also hinder adults with labor-intensive occupations from engaging in LTPA [99]. From an ecological perspective LTPA engagement is influenced by the SES of one's neighborhood. Features that are associated with lower SES neighborhoods, such as high crime rate, limited streetlights, and increased traffic density negatively impact PA engagements [90, 91].

Sedentary Behavior

Increasing attention has been paid to reducing sedentary behavior, as sedentary behavior is not simply a "lack of physical activity". However, there is no clear evidence that there are disparities in sedentary behavior among individuals from different ethnic and SES groups [100]. This may be because there is increasingly universal access to activities that promote sedentary behavior, such as television watching and video game playing. Environmental features that negatively impact physical activity engagement are also found to be associated with increased likelihood of time spent in sedentary behaviors [90].

Racial/Ethnic and Socioeconomic Disparities in Sleep, Obesity, and Cancer

Sleep is important for restoring physical and mental health. A growing body of literature indicates that inadequate sleep increases the risk of a range of chronic diseases including obesity [101, 102], diabetes [103], and hypertension [104, 105]. The associations between sleep duration and obesity may differ by age group. Based on a meta-analysis in children and adolescents (≤ 18 years) [101], it was summarized that short sleep duration was inversely related to risks of childhood overweight/obesity in pediatric population (OR: 1.58, 95 % CI: 1.26, 1.98) [101]. A recent prospective cohort study supports these findings, showing that short sleep among 0–4 year olds led to a subsequent 80 % increased odds of overweight or obesity [106]. Among adults, however, sleep duration was shown to have U-shaped associations with obesity [107, 108], suggesting that both short and long sleep duration were associated with concurrent and future obesity and weight gain.

Inadequate sleep is a risk factor for not only obesity, diabetes, and CVD [109, 110], but also to some types of cancer. Individuals with less than 6 h per night had an almost 50 % increase in risk of colorectal adenomas (OR = 1.47; 95 % CI = 1.05–2.06) as compared with individuals sleeping at least 7 h per night [111]. According to a meta-analysis on the relationship between sleep duration and cancer risk [112], there was a positive association between long sleep duration and colorectal cancer, and an inverse relationship with incidence of hormone related cancers like those in the breast.

Ethnic and SES Differences in Sleep

Considering the multiple deleterious effects of poor sleep, exploration of demographic patterns may shed light on methods to increase healthful sleep. Race/ethnic and socioeconomic factors may play a substantial role in sleep patterns and related disease. For example, in a sample of multiethnic US adults, insufficient sleep was related to increased odds of diabetes in all races, except non-Hispanic Blacks [113]. This again highlights the complexity of the impact of sleep on health; further compounding these intricacies is the difficulty in disentangling the influence of race/ethnicity versus socioeconomic status (SES).

In an epidemiological review, Bixler [114] highlights a body of research indicating low SES as a culprit for short sleep. In a 34-year longitudinal study of residents from Alameda County, CA, low SES led to fewer than 7 h of sleep, and short sleep was more common for African-Americans and Hispanics, as well as those with less education and lower SES (adjusting for other health factors related to poor sleep) [115]. A cross-sectional study on a national sample of US adults found that non-Hispanic Blacks were at increased odds for both short and long sleep, attenuated after controlling for SES, but remained significant [116]. Mexicans had increased odds for long sleep, although that association became non-significant after adjusting for SES. The authors speculated that these high risk sleep patterns may be attributable to substantial stressors experienced by urban minorities, and they conclude these differences in sleep may contribute to overall health disparities experienced by minorities [116]. Patel and colleagues [117] reported that poverty and race also contributed to poor sleep quality, and that employment and education mediated this relationship, highlighting that poorer individuals were most vulnerable.

Similar to the aforementioned study, a national study of over one-hundred thousand Americans found that education, employment and SES were inversely associated with sleep complaints [118]. Interestingly, this research showed Black and Hispanic women had fewer sleep complaints (e.g., trouble falling asleep, staying asleep, or sleeping too much) than their White counterparts, but these differences were not observed in men. Interaction analyses revealed more detailed patterns: the employment-sleep complaint association was inverse for African-American men (e.g., homemakers reported fewer sleep complaints compared to the combined male group); the positive income-sleep complaint relationship held for Hispanic men reporting less than $50,000 annual income; multiracial men had higher complaints if they were in the low SES group; non-college graduate Asian and Other women reported significantly increased sleep complaints and multiracial women showed a similar pattern; in contrast, Latina women were less likely to report complaints if they did not finish high school [118]. Goodin, McGuire and Smith [119], found ethnicity to be a moderator, reporting that lower perceived social status (perception of SES) was related to reduced sleep quality in Asians and African-Americans, but not Caucasians. Again, these findings illustrate the intricacies among ethnicity and SES factors in their influence on sleep.

These patterns are also seen in youth populations. A diary study of a nationally representative sample of children and adolescents found that Asian children (aged 5–11 years) and African-American adolescents (12–19 years old) reported shorter sleep durations during the week, which African-American and Hispanic adolescents also engaged in fewer hours of sleep on the weekend [120]. Crosby, LeBourgeois, and Harsh [121] found differences in sleep distribution by race in children as young as 3 years old. Caretaker reports indicated that 2–8 year old Black children napped more often, had shorter sleep durations, less sleep during the week than on weekends compared to their non-Hispanic white counterparts.

Understanding why these disparities occur may help to identify methods to improve sleep for ethnic minorities and people with low SES. Hicken and colleagues [122] found that Black adults experienced higher levels of sleep difficulties than Whites, and that this was fully mediated by racism-related vigilance, a marker of racially salient chronic stress. Although a similar pattern was found for Hispanics, it was not statistically significant. The authors suggest that racial discrimination plays a significant role in ethnic health disparities [122]. Tomfohr et al. [123] found that African-Americans experience more time in lighter sleep stages (sleep architecture) than their Caucasian counterparts, and that increased perceived discrimination was a partial mediator of these differences.

Family interactions and stressors may be responsible for sleep disturbance among youth. A study of urban Hispanic American infants and children (aged 6–48 months) found that frequent all-night cosleeping was more prevalent among minority families (21 %) than white American urban children (6 %); this practice was also associated with single parents and living in multiple households [124], which may be markers of lower SES. A longitudinal study found that marital conflict reported later sleep disruption in children, and that this association was stronger among African-American children and those from families of lower SES [125].

Substantial research shows the significant negative impact of poor sleep on health, including positive energy balance, obesity, and cancer. However, research identifying determinants of poor sleep is limited. Reducing sleep disparities could help the field of health promotion progress toward equalizing health outcomes for all.

Racial/Ethnic and Socioeconomic Disparities in Smoking

Adult Current Smokers

The prevalence of smoking in the USA has declined since the first Surgeon General's Report documented the health hazards of smoking in 1964. When the report: *Smoking and Health*, was released approximately 43 % of the US adult population were current smokers [126]. Fifty years later the prevalence of adult current smokers has decreased to 19 % [126, 127]. These advances are remarkable but we have hit a plateau and the rates have not decreased significantly in recent years. In 2011, over 43 million adults reported as current cigarette smokers, of whom 77.3 % smoked daily and 22.2 % smoked intermittently [126, 127]. Furthermore, deaths from smoking and tobacco use remain the number one preventable killer of US adults [126, 127]. Cigarette smoking has also led to annual financial losses such as the $96 billion lost to direct medical expenses and $97 billion lost in productivity [126–129]. Important for this review, smoking is related to negative energy balance [130] making smoking an attractive tool for weight loss, while quitting smoking is related to weight gain [131]. Maternal smoking has also been shown to predict offspring obesity [132].

Large disparities in smoking and tobacco use remain for the racial and ethnic groups; these are further exacerbated when broken down by socioeconomic status (SES) and region [126, 129]. Prevalence of current smokers in 2011 was highest amongst the American Indians and Alaska Natives (31.5 %), followed by African-Americans (24.2 %), Hispanics (17.0 %), and the group with the lowest rate were Asians (14.9 %) [126, 127]. The prevalence of smoking in non-Hispanic Whites for the same year was 22.6 % [127]. While the differences in prevalence rates might be lower for most of the groups and non-significantly higher for African-Americans, racial and ethnic minorities carry most of the burden of tobacco related diseases [133], including lung cancer, other cancers, and cardiovascular disease. African-American males have shown a higher incidence of lung cancer (122.8 per 100,000) compared to non-Hispanic Whites (81.5 per 100,000) [134]. The other racial groups show much lower incidence; Asian Americans (61.2), American Indian/Alaska Native (49.8), and Hispanics (47.2) [134, 135].

Paradoxically, although racial and ethnic minorities are more likely to die from tobacco-related diseases, they are also more likely to report as intermittent or light smokers [133]. African-Americans, for example, show higher rates of lung cancer but only 8.0 % report as heavy smokers, 25 or more cigarettes per day, compared to non-Hispanic Whites (28.3 %) [136]. In a recent study Trinidad and colleagues found that among racial and ethnic minorities who had any history of smoking behavior there were significantly higher rates of current intermittent smoking than non-Hispanic Whites (8.5 %): African-Americans (15.9 %), Asian Americans (16.1 %), and Hispanics (20.8 %) [133]. The 1998 Surgeon General's report on racial and ethnic minority smoking provides a breakdown of the number of cigarettes consumed per day for four groups: African-Americans, American Indian/Alaska Natives, Asians, and Hispanics. The report shows that the prevalence of consuming 25 or more cigarettes has declined since 1976 and the proportion of smokers who consume 15 or fewer cigarettes per day has been increasing [137]. While this data show signs of progress, tobacco-related disparities persist [126, 127, 133, 138–140].

Adult Smoking Cessation

Recent data from the National Health Interview Surveys show that 68.8 % of adult smokers would like to quit [141]. The same data show that in the previous year: 52.4 % attempted to quit by ceasing to smoke by 1 or more days, 6.2 % had recently quit, 48.3 % had a physician advise them to quit, and 31.7 % had used either medications or a counseling service to assist in their attempt to quit [141]. Smokers between the age of 25 and 64 all showed increased rates of quit attempts from 2001 to 2010 [141]. Quit attempts are most common among younger smokers and college graduates [141]. When cessation statistics are broken down by race and ethnicity we get a clearer picture of where we should focus our efforts.

In 2010, 75.6 % of African-Americans reported interest in quitting smoking which was higher than non-Hispanic Whites (69.1 %) and Hispanics (61.0 %) [141, 142]. Furthermore, attempts at smoking cessation were higher for African-Americans (59.1 %) than those of non-Hispanic Whites (50.7 %), Hispanics (56.5 %), and other non-Hispanic races (53.8 %) [141, 142]. This and other evidence shows that African-Americans and Hispanics are more likely to make cessation attempts, but investigation into cessation success rates paints a different picture. The quit ratio (percentage of lifetime smokers who have stopped) in 2000 was lower for African-Americans (37.5 %) and Hispanics (42.9 %) as compared to their non-Hispanic White counterparts (50.4 %) [142]. These differences can be attributed to many factors such as SES, education, access to health care, quality of health care, type of health insurance, smoking behaviors, access to cessation resources, and perceptions of evidence-based cessation methods [134, 141, 142].

One of the leading hypotheses as to why some racial/ethnic minority groups do not succeed in cessation is the higher prevalence of menthol cigarettes [141,

143]. Menthol cigarettes anesthetize the throat and allow the smoker to inhale more nicotine per puff, leading to increased nicotine dependence [143]. The low use of evidence-based cessation programs is another prominent factor in racial/ethnic disparities in successful smoking cessation [141]. The rates of use for cessation counseling were lower for Hispanics (15.9 %) and African-Americans (21.6 %) than for non-Hispanic Whites (36.1 %) [141]. Racial/ethnic minorities are also shown to be less likely to be advised by a physician to stop smoking or about the health consequences [133, 144]. Nicotine replacement therapy has been shown as a promising method to increase success in smoking cessation attempts but also has low uptake by racial/ethnic minorities due to low prescription rates and utilization [133, 142]. There is plenty of evidence that shows the success of these and other interventions to increase the success of quit attempts [144]. There is also, however, a lack of evidence-based programs and outreach intended for racial/ethnic minorities decreasing chances of success [144]. Future research needs to be dedicated to creating targeted and culturally appropriate cessation interventions for these vulnerable populations.

Youth Tobacco Use

The significance of youth tobacco use has garnered attention from investigators and the Surgeon General. The use of tobacco among youth in the USA is of importance since 80 % of adult smokers report initiation before the age of 18 and 99 % before age 26 [128, 145]. It is estimated that over 3,800 youth under the age of 18 begin smoking each day in the USA [128]. In 2012 the prevalence for current tobacco use in middle school students was 6.7 % and 23.3 % for those in high school [145]. Youth also engage in cigar use with 2.8 % of middle school students reporting use and 12.6 % of high school students [145]. The use of cigars by youth can be explained by the popularity of small cigars, cigarillos, by this group and their lower price compared to cigarettes [145]. Youth tobacco use has shown a similar pattern as adults in that the progress has stalled in recent years [128, 145]. In the master settlement with the USA and several state governments the tobacco companies were obligated to create prevention programs for youth but their efforts have not shown any documented evidence of success [128]. Since the master settlement, however, tobacco companies have increased their efforts to reduce prices as they are aware that youth are more price conscious than adults [128, 145]. It also been shown that tobacco companies focus such strategies in areas high with racial/ethnic minorities [128].

Overall tobacco use for middle school students was higher in 2012 among Hispanics (10.5 %) than non-Hispanic Whites (5.1 %), African-Americans (7.7 %), and all other non-Hispanic groups (3.1 %) [145]. Among high school students, African-Americans (22.6 %) and Hispanics (22.5 %) had similar current tobacco prevalence while non-Hispanic Whites (24.6 %) had the highest prevalence [145]. Cigarette use in high school students was highest among non-Hispanic

Whites (15.4 %) and cigar use was highest among African-Americans (16.7 %) [128, 145]. Concurrent tobacco product use is prevalent across all racial groups in high school [128, 145]. Furthermore, over half of current Hispanic female tobacco users report of using more than one type of tobacco product on a regular basis [128]. In terms of susceptibility it has been shown that youth of Mexican descent are more vulnerable to initiate smoking than other youths [146]. These statistics show the need to better fund prevention programs for youth in order to prevent the premature deaths of one in three current young smokers [128].

Youth Tobacco Cessation

Smoking cessation is a rare occurrence in youth and young adults (16–24 years); only 4 % per year quit smoking [147]. The rates for attempts to quit, however, are higher for youth (58 %) than for adults (52.4 %) [147]. Youth and young adults between 16 and 24 years of age were more likely to attempt to quit without assistance [147]. Only 20 % of current youth smokers sought advice from a nurse or physician prior to their quit attempts, with females (24.9 %) being more likely to seek the help than males (15.6 %) [147]. The only two groups of current high school smokers who showed differences in attempts to quit were African-Americans (68.1 %) and Hispanics (54.1 %) compared to non-Hispanic Whites (62.8 %) [148]. In general there is limited data on youth cessation with even less for racial/ethnic minorities; therefore it is important that future research address this gap. Public health officials also need to create and adequately fund prevention and intervention programs that are both relevant and appropriate for all groups of youth.

New and Emerging Tobacco Products

Since the groundbreaking Surgeon General's Report on Smoking in 1964 there have been many advances but now we must adapt our efforts to include new and emerging tobacco products. Products such as electronic cigarettes and hookah are gaining in popularity especially among youth [145]. Tobacco companies have noticed this gain in popularity for e-cigarettes and have responded by increasing smokeless tobacco product marketing by 277 % compared to 48 % increase for cigarettes [126]. Over 6 % of adults in the USA, including 21 % of current smokers, have tried e-cigarettes [138]. From 2011 to 2012 there were significant increases in e-cigarette use among middle (0.6–1.1 %) and high school students (1.5–2.8 %) [149]. Hookah use also increased among high school student from 2011 (4.1 %) to 2012 (5.4 %) [145]. E-cigarettes are more popular among current high school non-Hispanic White (3.4 %) tobacco users, followed by Hispanics (2.7 %), other non-Hispanic groups (2.2 %), and African-Americans (1.1 %) [145].

While the statistics for e-cigarettes and other emerging products are low in comparison to traditional cigarettes, public health officials and researchers need to investigate the health consequences of such products. Little is currently know about these new products and currently most are not regulated by the Food and Drug Administration. The lack of regulation allows tobacco companies to make claims about the use of these products for harm reduction or smoking cessation without the need for scientific review. Claims such as health benefits could be detrimental especially for our vulnerable populations such as racial/ethnic minorities by making it easier to fall unto nicotine dependence and harder to achieve cessation [150].

Some Brief Conclusions

This overview of four central behavioral domains that influence both energy balance and cancer shows clear racial/ethnic disparities in all four areas, although patterns are not always straightforward. It is clear that, in the USA, racial/ethnic "minorities" are at higher risk for poor diet, exercise, sleep and smoking behaviors, at higher risk for obesity, and at higher risk for most cancers, than their white counterparts. Mechanisms for these disparities likely differ according to a complicated network of influences from cell to society, from genes to cultural views on the specific behaviors. However, some commonalities can be noted.

Socioeconomic Status and Health Behavior

In each of the sections above, the inverse relationships between low socioeconomic status and unhealthy behaviors related to energy balance and cancer have been demonstrated empirically. The underlying mechanisms for these relationships remain unclear. In the USA, racial/ethnic minorities are disproportionately represented in lower socioeconomic strata [151], but even after correction for race/ethnicity, socioeconomic differences in health-related behaviors persist [152]. There is no doubt that gaps in socioeconomic status impact access to insurance, adequate health care, healthy food, and safe places to exercise, among many other important needs related to attaining and maintaining a healthy energy balance. However, unlike disparities in many other components of health, disparities in health behaviors appear to involve something *more* than the ability to use income to purchase good health [152].

Stress: One Possible Missing Link

Stress has been related to poor sleep [116], unhealthy eating patterns [153], lower physical activity and increased sedentary time [154], and the initiation and maintenance of smoking [155]. Research has shown that low socioeconomic status and minority status are strongly related to increased stress. For example, in a population of 3,105 adults (34 % white), Sternthal et al. [156] found significant racial differences in exposure to eight stress domains, e.g., acute life events, employment, financial, life discrimination, job discrimination, relationship, early life, and community stressors.

Although disentangling the roles of race/ethnicity and socioeconomic status in health behaviors remains complex [157], there exist acute disparities in energy-balance and cancer-related behaviors. There are also racial/ethnic differences in stress, as well as a socioeconomic gradient in stress and stressful experiences. The documented interrelationships between these core behaviors and stress highlight the role that social stressors uniquely experienced by minority populations may play in existing health disparities. Considering public health's mandate to achieve health equity for our communities, efforts must be put into reducing the stressors associated with poverty and racism. These efforts may help to ameliorate the racial/ethnic and socioeconomic disparities in energy balance and cancer in the USA.

References

1. Siegel R, Naishadham D, Jemal A (2013) Cancer statistics, 2013. CA Cancer J Clin 63(1):11–30
2. Shavers VL, Brown ML (2002) Racial and ethnic disparities in the receipt of cancer treatment. J Natl Cancer Inst 94(5):334–357
3. Espey DK, Wu X-C, Swan J, Wiggins C, Jim MA, Ward E, Wingo PA, Howe HL, Ries LAG, Miller BA, Jemal A, Ahmed F, Cobb N, Kaur JS, Edwards BK (2007) Annual report to the nation on the status of cancer, 1975–2004, featuring cancer in American Indians and Alaska Natives. Cancer 110(10):2119–2152. doi:10.1002/cncr.23044
4. Aziz H, Hussain F, Edelman S, Cirrone J, Aral I, Fruchter R, Homel P, Rotman M (1996) Age and race as prognostic factors in endometrial carcinoma. Am J Clin Oncol 19(6):595–600
5. Boyer-Chammard A, Taylor TH, Anton-Culver H (1999) Survival differences in breast cancer among racial/ethnic groups: a population-based study. Cancer Detect Prev 23 (6):463–473
6. Ward E, Jemal A, Cokkinides V, Singh GK, Cardinez C, Ghafoor A, Thun M (2004) Cancer disparities by race/ethnicity and socioeconomic status. CA Cancer J Clin 54(2):78–93. doi:10.3322/canjclin.54.2.78
7. Chu KC, Miller BA, Springfield SA (2007) Measures of racial/ethnic health disparities in cancer mortality rates and the influence of socioeconomic status. J Natl Med Assoc 99 (10):1092–1104
8. Wong ML, Clarke CA, Yang J, Hwang J, Hiatt RA, Wang S (2013) Incidence of non-small-cell lung cancer among California Hispanics according to neighborhood socioeconomic status. J Thorac Oncol 8(3):287–294

 9. Siegel R, Ward E, Brawley O, Jemal A (2011) Cancer statistics, 2011. CA Cancer J Clin 61 (4):212–236
10. Spruijt-Metz D, Nguyen-Rodriguez ST, Davis JN (2010) Behavior, energy balance, and cancer: an overview. In: Berger N (ed) Cancer and energy balance, epidemiology and overview, vol 1, Energy balance and cancer. Springer, New York, pp 233–266
11. Kranz S, Hartman T, Siega-Riz AM, Herring AH (2006) A diet quality index for American preschoolers based on current dietary intake recommendations and an indicator of energy balance. J Am Diet Assoc 106(10):1594–1604. doi:10.1016/j.jada.2006.07.005
12. Drewnowski A, Specter SE (2004) Poverty and obesity: the role of energy density and energy costs. Am J Clin Nutr 79(1):6–16
13. Hill JO (2006) Understanding and addressing the epidemic of obesity: an energy balance perspective. Endocr Rev 27(7):750–761. doi:10.1210/er.2006-0032
14. DiMeglio DP, Mattes RD (2000) Liquid versus solid carbohydrate: effects on food intake and body weight. Int J Obes Relat Metab Disord 24(6):794–800
15. Raben A, Vasilaras TH, Moller AC, Astrup A (2002) Sucrose compared with artificial sweeteners: different effects on ad libitum food intake and body weight after 10 wk of supplementation in overweight subjects. Am J Clin Nutr 76(4):721–729
16. Ludwig DS, Peterson KE, Gortmaker SL (2001) Relation between consumption of sugar-sweetened drinks and childhood obesity: a prospective, observational analysis. Lancet 357 (9255):505–508
17. Malik VS, Schulze MB, Hu FB (2006) Intake of sugar-sweetened beverages and weight gain: a systematic review. Am J Clin Nutr 84(2):274–288
18. Saris WH (2003) Sugars, energy metabolism, and body weight control. Am J Clin Nutr 78 (4):850S–857S
19. Bray GA, Nielsen SJ, Popkin BM (2004) Consumption of high-fructose corn syrup in beverages may play a role in the epidemic of obesity. Am J Clin Nutr 79(4):537–543
20. Michaud DS, Liu S, Giovannucci E, Willett WC, Colditz GA, Fuchs CS (2002) Dietary sugar, glycemic load, and pancreatic cancer risk in a prospective study. J Natl Cancer Inst 94 (17):1293–1300
21. Larsson SC, Bergkvist L, Wolk A (2006) Consumption of sugar and sugar-sweetened foods and the risk of pancreatic cancer in a prospective study. Am J Clin Nutr 84(5):1171–1176
22. La Vecchia C, Franceschi S, Dolara P, Bidoli E, Barbone F (1993) Refined-sugar intake and the risk of colorectal cancer in humans. Int J Cancer 55(3):386–389
23. Michaud DS, Fuchs CS, Liu S, Willett WC, Colditz GA, Giovannucci E (2005) Dietary glycemic load, carbohydrate, sugar, and colorectal cancer risk in men and women. Cancer Epidemiol Biomarkers Prev 14(1):138–147
24. Terry PD, Jain M, Miller AB, Howe GR, Rohan TE (2003) Glycemic load, carbohydrate intake, and risk of colorectal cancer in women: a prospective cohort study. J Natl Cancer Inst 95(12):914–916
25. Kushi LH, Doyle C, McCullough M, Rock CL, Demark-Wahnefried W, Bandera EV, Gapstur S, Patel AV, Andrews K, Gansler T (2012) American Cancer Society guidelines on nutrition and physical activity for cancer prevention. CA Cancer J Clin 62(1):30–67
26. Kranz S, Siega-Riz AM (2002) Sociodemographic determinants of added sugar intake in preschoolers 2 to 5 years old. J Pediatr 140(6):667–672. doi:10.1067/mpd.2002.124307
27. Wilson TA, Adolph AL, Butte NF (2009) Nutrient adequacy and diet quality in non-overweight and overweight Hispanic children of low socioeconomic status: the Viva la Familia Study. J Am Diet Assoc 109(6):1012–1021. doi:10.1016/j.jada.2009.03.007
28. Thompson FE, McNeel TS, Dowling EC, Midthune D, Morrissette M, Zeruto CA (2009) Interrelationships of added sugars intake, socioeconomic status, and race/ethnicity in adults in the United States: National Health Interview Survey, 2005. J Am Diet Assoc 109(8):1376–1383. doi:10.1016/j.jada.2009.05.002
29. Perez-Escamilla R, Putnik P (2007) The role of acculturation in nutrition, lifestyle, and incidence of type 2 diabetes among Latinos. J Nutr 137(4):860–870

30. Rehm CD, Matte TD, Van Wye G, Young C, Frieden TR (2008) Demographic and behavioral factors associated with daily sugar-sweetened soda consumption in New York City adults. J Urban Health 85(3):375–385. doi:10.1007/s11524-008-9269-8

31. Beydoun MA, Wang Y (2008) How do socio-economic status, perceived economic barriers and nutritional benefits affect quality of dietary intake among US adults? Eur J Clin Nutr 62 (3):303–313. doi:10.1038/sj.ejcn.1602700

32. Cullen KW, Ash DM, Warneke C, de Moor C (2002) Intake of soft drinks, fruit-flavored beverages, and fruits and vegetables by children in grades 4 through 6. Am J Public Health 92 (9):1475–1478

33. Himmelgreen DA, Bretnall A, Perez-Escamilla R, Peng Y, Bermudez A (2005) Birthplace, length of time in the US, and language are associated with diet among inner-city Puerto Rican women keywords. Ecol Food Nutr 44(2):105–122

34. Bermudez OL, Falcon LM, Tucker KL (2000) Intake and food sources of macronutrients among older Hispanic adults: association with ethnicity acculturation, and length of residence in the United States. J Am Diet Assoc 100(6):665–673

35. Drewnowski A (2007) The real contribution of added sugars and fats to obesity. Epidemiol Rev 29:160–171. doi:10.1093/epirev/mxm011

36. Popkin BM, Nielsen SJ (2003) The sweetening of the world's diet. Obes Res 11(11):1325–1332. doi:10.1038/oby.2003.179

37. MacLean CH, Newberry SJ, Mojica WA, Khanna P, Issa AM, Suttorp MJ, Lim YW, Traina SB, Hilton L, Garland R, Morton SC (2006) Effects of omega-3 fatty acids on cancer risk: a systematic review. JAMA 295(4):403–415. doi:10.1001/jama.295.4.403

38. Terry PD, Rohan TE, Wolk A (2003) Intakes of fish and marine fatty acids and the risks of cancers of the breast and prostate and of other hormone-related cancers: a review of the epidemiologic evidence. Am J Clin Nutr 77(3):532–543

39. Larsson SC, Kumlin M, Ingelman-Sundberg M, Wolk A (2004) Dietary long-chain n-3 fatty acids for the prevention of cancer: a review of potential mechanisms. Am J Clin Nutr 79 (6):935–945

40. Bjerregaard P (2013) The association of n-3 fatty acids with serum High Density Cholesterol (HDL) is modulated by sex but not by Inuit ancestry. Atherosclerosis 226(1):281–285. doi:10.1016/j.atherosclerosis.2012.10.071

41. Kelly J, Lanier A, Santos M, Healey S, Louchini R, Friborg J (2008) Cancer among the circumpolar Inuit, 1989–2003. II. Patterns and trends. Int J Circumpolar Health 67(5):408–420

42. Dewailly É, Blanchet C, Gingras S, Lemieux S, Holub BJ (2003) Fish consumption and blood lipids in three ethnic groups of Québec (Canada). Lipids 38(4):359–365

43. Kolonel LN, Henderson BE, Hankin JH, Nomura AM, Wilkens LR, Pike MC, Stram DO, Monroe KR, Earle ME, Nagamine FS (2000) A multiethnic cohort in Hawaii and Los Angeles: baseline characteristics. Am J Epidemiol 151(4):346–357

44. Verbeke W, Vackier I (2005) Individual determinants of fish consumption: application of the theory of planned behaviour. Appetite 44(1):67–82. doi:10.1016/j.appet.2004.08.006

45. Olsen SO, Scholderer J, Brunsø K, Verbeke W (2007) Exploring the relationship between convenience and fish consumption: a cross-cultural study. Appetite 49(1):84–91. doi:10.1016/j.appet.2006.12.002

46. Burger J, Stephens WL, Boring CS, Kuklinski M, Gibbons JW, Gochfeld M (1999) Factors in exposure assessment: ethnic and socioeconomic differences in fishing and consumption of fish caught along the Savannah River. Risk Anal 19(3):427–438

47. Wu AH, Lee E, Vigen C (2013) Soy isoflavones and breast cancer. Am Soc Clin Oncol Ed Book 33:102–106. doi:10.1200/EdBook_AM.2013.33.102

48. van Die MD, Bone KM, Williams SG, Pirotta MV (2013) Soy and soy isoflavones in prostate cancer: a systematic review and meta-analysis of randomised controlled trials. BJU Int. doi:10.1111/bju.12435

49. Di Maso M, Talamini R, Bosetti C, Montella M, Zucchetto A, Libra M, Negri E, Levi F, La Vecchia C, Franceschi S, Serraino D, Polesel J (2013) Red meat and cancer risk in a network of case-control studies focusing on cooking practices. Ann Oncol 24(12):3107–3112. doi:10.1093/annonc/mdt392

50. Alexander DD, Cushing CA (2011) Red meat and colorectal cancer: a critical summary of prospective epidemiologic studies. Obes Rev 12(5):e472–e493. doi:10.1111/j.1467-789X.2010.00785.x

51. Choi Y, Song S, Song Y, Lee JE (2013) Consumption of red and processed meat and esophageal cancer risk: meta-analysis. World J Gastroenterol 19(7):1020–1029. doi:10.3748/wjg.v19.i7.1020

52. Alexander DD, Mink PJ, Cushing CA, Sceurman B (2010) A review and meta-analysis of prospective studies of red and processed meat intake and prostate cancer. Nutr J 9:50. doi:10.1186/1475-2891-9-50

53. Chan DS, Lau R, Aune D, Vieira R, Greenwood DC, Kampman E, Norat T (2011) Red and processed meat and colorectal cancer incidence: meta-analysis of prospective studies. PLoS One 6(6):e20456. doi:10.1371/journal.pone.0020456

54. Larsson SC, Wolk A (2012) Red and processed meat consumption and risk of pancreatic cancer: meta-analysis of prospective studies. Br J Cancer 106(3):603–607. doi:10.1038/bjc.2011.585

55. Zhu H, Yang X, Zhang C, Zhu C, Tao G, Zhao L, Tang S, Shu Z, Cai J, Dai S, Qin Q, Xu L, Cheng H, Sun X (2013) Red and processed meat intake is associated with higher gastric cancer risk: a meta-analysis of epidemiological observational studies. PLoS One 8(8): e70955. doi:10.1371/journal.pone.0070955

56. Smolińska K, Paluszkiewicz P (2010) Risk of colorectal cancer in relation to frequency and total amount of red meat consumption. Systematic review and meta-analysis. Arch Med Sci 6 (4):605–610. doi:10.5114/aoms.2010.14475

57. Bingham SA, Luben R, Welch A, Wareham N, Khaw KT, Day N (2003) Are imprecise methods obscuring a relation between fat and breast cancer? Lancet 362(9379):212–214. doi:10.1016/S0140-6736(03)13913-X

58. Giovannucci E, Rimm EB, Colditz GA, Stampfer MJ, Ascherio A, Chute CG, Chute CC, Willett WC (1993) A prospective study of dietary fat and risk of prostate cancer. J Natl Cancer Inst 85(19):1571–1579

59. Sinha R, Rothman N (1999) Role of well-done, grilled red meat, heterocyclic amines (HCAs) in the etiology of human cancer. Cancer Lett 143(2):189–194

60. Daniel CR, Cross AJ, Koebnick C, Sinha R (2011) Trends in meat consumption in the USA. Public Health Nutr 14(4):575–583. doi:10.1017/S1368980010002077

61. Elder JP, Castro FG, de Moor C, Mayer J, Candelaria JI, Campbell N, Talavera G, Ware LM (1991) Differences in cancer-risk-related behaviors in Latino and Anglo adults. Prev Med 20 (6):751–763

62. Otero-Sabogal R, Sabogal F, Pérez-Stable EJ, Hiatt RA (1995) Dietary practices, alcohol consumption, and smoking behavior: ethnic, sex, and acculturation differences. J Natl Cancer Inst Monogr 18:73–82

63. Sussner KM, Lindsay AC, Greaney ML, Peterson KE (2008) The influence of immigrant status and acculturation on the development of overweight in Latino families: a qualitative study. J Immigr Minor Health 10(6):497–505. doi:10.1007/s10903-008-9137-3

64. Drewnowski A, Popkin BM (1997) The nutrition transition: new trends in the global diet. Nutr Rev 55(2):31–43

65. Gomel JN, Zamora A (2007) English- and Spanish-speaking Latina mothers' beliefs about food, health, and mothering. J Immigr Minor Health 9(4):359–367. doi:10.1007/s10903-007-9040-3

66. Wang Y, Beydoun MA, Caballero B, Gary TL, Lawrence R (2010) Trends and correlates in meat consumption patterns in the US adult population. Public Health Nutr 13(9):1333–1345. doi:10.1017/S1368980010000224

67. Steinmetz KA, Potter JD (1996) Vegetables, fruit, and cancer prevention: a review. J Am Diet Assoc 96(10):1027–1039. doi:10.1016/S0002-8223(96)00273-8

68. Block G, Patterson B, Subar A (1992) Fruit, vegetables, and cancer prevention: a review of the epidemiological evidence. Nutr Cancer 18(1):1–29. doi:10.1080/01635589209514201

69. Coleman HG, Murray LJ, Hicks B, Bhat SK, Kubo A, Corley DA, Cardwell CR, Cantwell MM (2013) Dietary fiber and the risk of precancerous lesions and cancer of the esophagus: a systematic review and meta-analysis. Nutr Rev 71(7):474–482. doi:10.1111/nure.12032

70. Aune D, Chan DS, Greenwood DC, Vieira AR, Rosenblatt DA, Vieira R, Norat T (2012) Dietary fiber and breast cancer risk: a systematic review and meta-analysis of prospective studies. Ann Oncol 23(6):1394–1402. doi:10.1093/annonc/mdr589

71. Zhang Z, Xu G, Ma M, Yang J, Liu X (2013) Dietary fiber intake reduces risk for gastric cancer: a meta-analysis. Gastroenterology 145(1):113.e113–120.e113. doi:10.1053/j.gastro.2013.04.001

72. Nomura AM, Hankin JH, Henderson BE, Wilkens LR, Murphy SP, Pike MC, Le Marchand L, Stram DO, Monroe KR, Kolonel LN (2007) Dietary fiber and colorectal cancer risk: the multiethnic cohort study. Cancer Causes Control 18(7):753–764. doi:10.1007/s10552-007-9018-4

73. Trock B, Lanza E, Greenwald P (1990) Dietary fiber, vegetables, and colon cancer: critical review and meta-analyses of the epidemiologic evidence. J Natl Cancer Inst 82(8):650–661

74. Jacobs DR, Marquart L, Slavin J, Kushi LH (1998) Whole-grain intake and cancer: an expanded review and meta-analysis. Nutr Cancer 30(2):85–96. doi:10.1080/01635589809514647

75. Slavin JL (2000) Mechanisms for the impact of whole grain foods on cancer risk. J Am Coll Nutr 19(3 Suppl):300S–307S

76. Dubowitz T, Heron M, Bird CE, Lurie N, Finch BK, Basurto-Dávila R, Hale L, Escarce JJ (2008) Neighborhood socioeconomic status and fruit and vegetable intake among whites, blacks, and Mexican Americans in the United States. Am J Clin Nutr 87(6):1883–1891

77. Patterson BH, Block G, Rosenberger WF, Pee D, Kahle LL (1990) Fruit and vegetables in the American diet: data from the NHANES II survey. Am J Public Health 80(12):1443–1449

78. Morland K, Filomena S (2007) Disparities in the availability of fruits and vegetables between racially segregated urban neighbourhoods. Public Health Nutr 10(12):1481–1489. doi:10.1017/S1368980007000079

79. Kong A, Odoms-Young AM, Schiffer LA, Berbaum ML, Porter SJ, Blumstein L, Fitzgibbon ML (2013) Racial/ethnic differences in dietary intake among WIC families prior to food package revisions. J Nutr Educ Behav 45(1):39–46. doi:10.1016/j.jneb.2012.04.014

80. Wiseman M (2008) The second World Cancer Research Fund/American Institute for Cancer Research expert report. Food, nutrition, physical activity, and the prevention of cancer: a global perspective. Proc Nutr Soc 67(3):253–256

81. Patel AV, Rodriguez C, Bernstein L, Chao A, Thun MJ, Calle EE (2005) Obesity, recreational physical activity, and risk of pancreatic cancer in a large US Cohort. Cancer Epidemiol Biomarkers Prev 14(2):459–466

82. Patel AV, Calle EE, Bernstein L, Wu AH, Thun MJ (2003) Recreational physical activity and risk of postmenopausal breast cancer in a large cohort of US women. Cancer Causes Control 14(6):519–529

83. Giovannucci EL, Liu Y, Leitzmann MF, Stampfer MJ, Willett WC (2005) A prospective study of physical activity and incident and fatal prostate cancer. Arch Intern Med 165(9):1005

84. McTiernan A, Tworoger SS, Ulrich CM, Yasui Y, Irwin ML, Rajan KB, Sorensen B, Rudolph RE, Bowen D, Stanczyk FZ (2004) Effect of exercise on serum estrogens in postmenopausal women a 12-month randomized clinical trial. Cancer Res 64(8):2923–2928

85. McTiernan A, Tworoger SS, Rajan KB, Yasui Y, Sorenson B, Ulrich CM, Chubak J, Stanczyk FZ, Bowen D, Irwin ML (2004) Effect of exercise on serum androgens in postmenopausal women: a 12-month randomized clinical trial. Cancer Epidemiol Biomarkers Prev 13(7):1099–1105

86. Caspersen CJ, Powell KE, Christenson GM (1985) Physical activity, exercise, and physical fitness: definitions and distinctions for health-related research. Public Health Rep 100 (2):126–131

87. Haskell WL, Lee IM, Pate RR, Powell KE, Blair SN, Franklin BA, Macera CA, Heath GW, Thompson PD, Bauman A (2007) Physical activity and public health: updated recommendation for adults from the American College of Sports Medicine and the American Heart Association. Med Sci Sports Exerc 39(8):1423–1434, doi:10.1249/mss.0b013e3180616b2700005768-200708000-00027 [pii]

88. Physical Activity Guidelines Advisory Committee (2008) Physical activity guidelines advisory committee report. Department of Health and Human Services, Washington, DC

89. Popkin BM, Duffey K, Gordon-Larsen P (2005) Environmental influences on food choice, physical activity and energy balance. Physiol Behav 86(5):603–613

90. Gordon-Larsen P, McMurray RG, Popkin BM (2000) Determinants of adolescent physical activity and inactivity patterns. Pediatrics 105(6):E83

91. Foster S, Giles-Corti B (2008) The built environment, neighborhood crime and constrained physical activity: an exploration of inconsistent findings. Prev Med 47(3):241–251. doi:10.1016/j.ypmed.2008.03.017

92. Marquez DX, Neighbors CJ, Bustamante EE (2010) Leisure time and occupational physical activity among racial or ethnic minorities. Med Sci Sports Exerc 42(6):1086–1093. doi:10.1249/MSS.0b013e3181c5ec05

93. Troiano RP, Berrigan D, Dodd KW, Masse LC, Tilert T, McDowell M (2008) Physical activity in the United States measured by accelerometer. Med Sci Sports Exerc 40(1):181–188. doi:10.1249/mss.0b013e31815a51b3

94. Caperchione CM, Kolt GS, Mummery WK (2009) Physical activity in culturally and linguistically diverse migrant groups to Western society: a review of barriers, enablers and experiences. Sports Med 39(3):167–177. doi:10.2165/00007256-200939030-00001

95. Maher CA, Olds TS (2011) Minutes, MET minutes, and METs: unpacking socio-economic gradients in physical activity in adolescents. J Epidemiol Community Health 65(2):160–165. doi:10.1136/jech.2009.099796

96. Carlson JA, Mignano AM, Norman GJ, McKenzie TL, Kerr J, Arredondo EM, Madanat H, Cain KL, Elder JP, Saelens BE, Sallis JF (2014) Socioeconomic disparities in elementary school practices and children's physical activity during school. Am J Health Promot 28 (3 Suppl):S47–S53. doi:10.4278/ajhp.130430-QUAN-206

97. De Meester F, Van Dyck D, De Bourdeaudhuij I, Deforche B, Sallis JF, Cardon G (2012) Active living neighborhoods: is neighborhood walkability a key element for Belgian adolescents? BMC Public Health 12:7. doi:10.1186/1471-2458-12-7

98. Lamb KE, Ogilvie D, Ferguson NS, Murray J, Wang Y, Ellaway A (2012) Sociospatial distribution of access to facilities for moderate and vigorous intensity physical activity in Scotland by different modes of transport. Int J Behav Nutr Phys Act 9:55. doi:10.1186/1479-5868-9-55

99. Marquez DX, McAuley E, Overman N (2004) Psychosocial correlates and outcomes of physical activity among Latinos: a review. Hisp J Behav Sci 26(2):195–229

100. Whitt-Glover MC, Taylor WC, Floyd MF, Yore MM, Yancey AK, Matthews CE (2009) Disparities in physical activity and sedentary behaviors among US children and adolescents: prevalence, correlates, and intervention implications. J Public Health Policy 30(Suppl 1): S309–S334. doi:10.1057/jphp.2008.46

101. Chen X, Beydoun MA, Wang Y (2008) Is sleep duration associated with childhood obesity? A systematic review and meta-analysis. Obesity (Silver Spring) 16:265–274

102. Lumeng J, Somashekar D, Appugliese D, Kaciroti N, Corwyn R, Bradley R (2007) Shorter sleep duration is associated with increased risk for being overweight at ages 9 to 12 years. Pediatrics 12(5):1020–1029

103. Vgontzas AN, Liao D, Pejovic S, Calhoun S, Karataraki M, Bixler EO (2009) Insomnia with objective short sleep duration is associated with type 2 diabetes: a population-based study. Diabetes Care 32(11):1980–1985. doi:10.2337/dc09-0284

104. Gangwisch JE, Heymsfield SB, Boden-Albala B, Buijs RM, Kreier F, Pickering TG, Rundle AG, Zammit GK, Malaspina D (2006) Short sleep duration as a risk factor for hypertension: analyses of the first National Health and Nutrition Examination Survey. Hypertension 47 (5):833–839. doi:10.1161/01.HYP.0000217362.34748.e0

105. Vgontzas AN, Liao D, Bixler EO, Chrousos GP, Vela-Bueno A (2009) Insomnia with objective short sleep duration is associated with a high risk for hypertension. Sleep 32 (4):491–497

106. Bell JF, Zimmerman FJ (2010) Shortened nighttime sleep duration in early life and subsequent childhood obesity. Arch Pediatr Adolesc Med 164(9):840–845

107. Ohkuma T, Fujii H, Iwase M, Kikuchi Y, Ogata S, Idewaki Y, Ide H, Doi Y, Hirakawa Y, Nakamura U, Kitazono T (2013) Impact of sleep duration on obesity and the glycemic level in patients with type 2 diabetes: the Fukuoka Diabetes Registry. Diabetes Care 36(3):611–617. doi:10.2337/dc12-0904

108. Taheri S (2007) The interactions between sleep, metabolism, and obesity. Int J Sleep Wakefulness 1(1):20–29

109. Cappuccio FP, D'Elia L, Strazzullo P, Miller MA (2010) Sleep duration and all-cause mortality: a systematic review and meta-analysis of prospective studies. Sleep 33(5):585–592

110. Chien KL, Chen PC, Hsu HC, Su TC, Sung FC, Chen MF, Lee YT (2010) Habitual sleep duration and insomnia and the risk of cardiovascular events and all-cause death: report from a community-based cohort. Sleep 33(2):177–184

111. Thompson CL, Larkin EK, Patel S, Berger NA, Redline S, Li L (2011) Short duration of sleep increases risk of colorectal adenoma. Cancer 117(4):841–847. doi:10.1002/cncr.25507

112. Zhao H, Yin JY, Yang WS, Qin Q, Li TT, Shi Y, Deng Q, Wei S, Liu L, Wang X, Nie SF (2013) Sleep duration and cancer risk: a systematic review and meta-analysis of prospective studies. Asian Pac J Cancer Prev 14(12):7509–7515

113. Vishnu A, Shankar A, Kalidindi S (2011) Examination of the association between insufficient sleep and cardiovascular disease and diabetes by race/ethnicity. Int J Endocrinol 2011:789358. doi:10.1155/2011/789358

114. Bixler E (2009) Sleep and society: an epidemiological perspective. Sleep Med 10;S3–S6

115. Stamatakis KA, Kaplan GA, Roberts RE (2007) Short sleep duration across income, education, and race/ethnic groups: population prevalence and growing disparities during 34 years of follow-up. Ann Epidemiol 17(12):948–955

116. Hale L, Do DP (2007) Racial differences in self-reports of sleep duration in a population-based study. Sleep 30(9):1096–1103

117. Patel NP, Grandner MA, Xie D, Branas CC, Gooneratne N (2010) "Sleep disparity" in the population: poor sleep quality is strongly associated with poverty and ethnicity. BMC Public Health 10:475

118. Grandner MA, Patel NP, Gehrman PR, Xie D, Sha D, Weaver T, Gooneratne N (2010) Who gets the best sleep? Ethnic and socioeconomic factors related to sleep complaints. Sleep Med 11(5):470–478

119. Goodin BR, McGuire L, Smith MT (2010) Ethnicity moderates the influence of perceived social status on subjective sleep quality. Behav Sleep Med 8(4):194–206

120. Adam EK, Snell EK, Pendry P (2007) Sleep timing and quantity in ecological and family context: a nationally representative time-diary study. J Fam Psychol 21(1):4

121. Crosby B, LeBourgeois MK, Harsh J (2005) Racial differences in reported napping and nocturnal sleep in 2-to 8-year-old children. Pediatrics 115(Supplement 1):225–232

122. Hicken MT, Lee H, Ailshire J, Burgard SA, Williams DR (2013) "Every shut eye, ain't sleep": the role of racism-related vigilance in racial/ethnic disparities in sleep difficulty. Race Soc Probl 5:100–112

123. Tomfohr L, Pung MA, Edwards KM, Dimsdale JE (2012) Racial differences in sleep architecture: the role of ethnic discrimination. Biol Psychol 89(1):34–38
124. Schachter FF, Fuchs ML, Bijur PE, Stone RK (1989) Cosleeping and sleep problems in Hispanic-American urban young children. Pediatrics 84(3):522–530
125. Kelly RJ, El-Sheikh M (2011) Marital conflict and children's sleep: reciprocal relations and socioeconomic effects. J Fam Psychol 25(3):412–422
126. US Department of Health and Human Services (2014) The health consequences of smoking-50 years of progress: a report of the Surgeon General. US Department of Health and Human Services, Centers for Disease Control and Prevention, National Center for Chronic Disease Prevention and Health Promotion, Office of Smoking and Health, Atlanta, GA
127. Centers for Disease Control and Prevention (2012) Current cigarette smoking among adults—United States, 2011. Morb Mortal Wkly Rep 61(44):889–894
128. U.S. Department of Health and Human Services (2012) Preventing tobacco use among youth and young adults: a report of the Surgeon General. U.S. Department of Health and Human Services, Centers for Disease Control and Prevention, National Center for Chronic Disease Prevention and Health Promotion, Office on Smoking and Health, Atlanta, GA
129. Centers for Disease Control and Prevention (2010) Racial disparities in smoking-attributable mortality and years of potential life lost—Missouri, 2003–2007. Morb Mortal Wkly Rep 59 (46):1518–1522
130. de Morentin PBM, Whittle AJ, Fernø J, Nogueiras R, Diéguez C, Vidal-Puig A, López M (2012) Nicotine induces negative energy balance through hypothalamic AMP-activated protein kinase. Diabetes 61(4):807–817
131. Aubin H-J, Farley A, Lycett D, Lahmek P, Aveyard P (2012) Weight gain in smokers after quitting cigarettes: meta-analysis. BMJ 345:e4439
132. Behl M, Rao D, Aagaard K, Davidson TL, Levin ED, Slotkin TA, Srinivasan S, Wallinga D, White MF, Walker VR (2013) Evaluation of the association between maternal smoking, childhood obesity, and metabolic disorders: a national toxicology program workshop review. Environ Health Perspect 121(2):170
133. Trinidad DR, Pérez-Stable EJ, White MM, Emery SL, Messer K (2011) A nationwide analysis of US racial/ethnic disparities in smoking behaviors, smoking cessation, and cessation-related factors. Am J Public Health 101(4):699–706. doi:10.2105/AJPH.2010. 191668
134. Fagan P, Moolchan ET, Lawrence D, Fernander A, Ponder PK (2007) Identifying health disparities across the tobacco continuum. Addiction 102(s2):5–29. doi:10.1111/j.1360-0443. 2007.01952.x
135. Moolchan ET, Fagan P, Fernander AF, Velicer WF, Hayward MD, King G, Clayton RR (2007) Addressing tobacco-related health disparities. Addiction 102(s2):30–42. doi:10.1111/ j.1360-0443.2007.01953.x
136. Haiman CA, Stram DO, Wilkens LR, Pike MC, Kolonel LN, Henderson BE, Le Marchand L (2006) Ethnic and racial differences in the smoking-related risk of lung cancer. N Engl J Med 354(4):333–342
137. US Department of Health and Human Services (1998) 1998 Surgeon General's report-tobacco use among U.S. racial/ethnic minority groups. US Department of Health and Human Services. US Department of Health and Human Services, Centers for Disease Control and Prevention, National Center for Chronic Disease Prevention and Health Promotion, Office on Smoking and Health, Atlanta, GA
138. Health CsOoSa (2012) Trends in current cigarette smoking among high school students and adults, United States, 1965–2011. http://www.cdc.gov/tobacco/data_statistics/tables/trends/ cig_smoking/index.htm. Accessed 1 Feb 2014
139. Fagan P, King G, Lawrence D, Petrucci SA, Robinson RG, Banks D, Marable S, Grana R (2011) Eliminating tobacco-related health disparities: directions for future research. Am J Public Health 94(2):211–217

140. Centers for Disease Control and Prevention (2010) Racial/ethnic disparities and geographic differences in lung cancer incidence—38 States and the District of Columbia, 1998–2006. Morb Mortal Wkly Rep 59(44):1434–1438
141. Centers for Disease Control and Prevention (2011) Quitting smoking among adults—United States, 2001–2010. Morb Mortal Wkly Rep 60(44):1513–1519
142. Fu SS, Kodl MM, Joseph AM, Hatsukami DK, Johnson EO, Breslau N, Wu B, Bierut L (2008) Racial/ethnic disparities in the use of nicotine replacement therapy and quit ratios in lifetime smokers ages 25 to 44 years. Cancer Epidemiol Biomarkers Prev 17:1640–1647. doi:10.1158/1055-9965.EPI-07-2726
143. Okuyemi KS, Program in Health Disparities Research UoMMS, USA, Department of Family Medicine and Community Health UoM, USA, Faseru B, Department of Preventive Medicine and Public Health UoKMC, USA, Sanderson Cox L, Department of Preventive Medicine and Public Health UoKMC, USA, Bronars CA, Department of Preventive Medicine and Public Health UoKMC, USA, Ahluwalia JS, Program in Health Disparities Research UoMMS, USA, Office of Clinical Research UoMAHC, USA, Department of Medicine UoMMS, USA. Relationship between menthol cigarettes and smoking cessation among African American light smokers. Addiction 102(12):1979–1986. doi:10.1111/j.1360-0443.2007.02010.x
144. Cokkinides V (2008) Racial and ethnic disparities in smoking-cessation interventions: analysis of the 2005 National Health Interview Survey. Am J Prev Med 34(5):404–412. doi:10.1016/j.amepre.2008.02.003
145. Centers for Disease Control and Prevention (2013) Tobacco product use among middle and high school students-United States, 2011 and 2012. Morb Mortal Wkly Rep 62(45):892–897
146. Centers for Disease Control and Prevention (2006) Racial/ethnic differences among youths in cigarette smoking and susceptibility to start smoking—United States, 2002–2004. Morb Mortal Wkly Rep 55(47):1275–1277
147. Centers for Disease Control and Prevention (2006) Use of cessation methods among smokers aged 16–24 years—United States, 2003. Morb Mortal Wkly Rep 55(50):1351–1354
148. Centers for Disease Control and Prevention (2009) High school students who tried to quit smoking cigarettes—United States, 2007. Morb Mortal Wkly Rep 58(16):428–431
149. Centers for Disease Control and Prevention (2005) Tobacco use, access, and exposure to tobacco in media among middle and high school students—United States, 2004. Morb Mortal Wkly Rep 54(12):297–301
150. Hsu R, Myers AE, Ribisl KM, Marteau TM (2013) An observational study of retail availability and in-store marketing of e-cigarettes in London: potential to undermine recent tobacco control gains? BMJ Open 3(12):e004085. doi:10.1136/bmjopen-2013-004085
151. Association AP (2007) Report of the APA task force on socioeconomic status. Association AP, Washington, DC
152. Pampel FC, Krueger PM, Denney JT (2010) Socioeconomic disparities in health behaviors. Annu Rev Sociol 36(1):349–370. doi:10.1146/annurev.soc.012809.102529
153. Dallman MF (2010) Stress-induced obesity and the emotional nervous system. Trends Endocrinol Metab 21(3):159–165
154. Holmes M, Ekkekakis P, Eisenmann J (2010) The physical activity, stress and metabolic syndrome triangle: a guide to unfamiliar territory for the obesity researcher. Obes Rev 11(7):492–507
155. Richards JM, Stipelman BA, Bornovalova MA, Daughters SB, Sinha R, Lejuez C (2011) Biological mechanisms underlying the relationship between stress and smoking: state of the science and directions for future work. Biol Psychol 88(1):1–12
156. Sternthal MJ, Slopen N, Williams DR (2011) Racial disparities in health. Du Bois Rev. Social Science Research on Race 8(01):95–113
157. LaVeist TA (2005) Disentangling race and socioeconomic status: a key to understanding health inequalities. J Urban Health 82(2 Suppl 3):iii26–iii34

Chapter 4
Impact of Obesity, Race, and Ethnicity on Cancer Survivorship

Kathryn H. Schmitz, Tanya Agurs-Collins, Marian L. Neuhouser, Lisa Pollack, and Sarah Gehlert

Abstract It is estimated that between 1971 and 2002, the population of cancer survivors grew from approximately three million to ten million. Currently, it is estimated that there are over 13.7 million cancer survivors in the USA and this number is expected to increase to 18 million by 2022. The seminal Institute of Medicine's report on cancer survivorship that outlines the need to develop strategies to address the unique issues faced by this growing clinical population was published 8 years ago. However, long-term cancer survivors are still a relatively new clinical population in the field of oncology, borne of successes in improved cancer screening and treatment approaches. There continues to be a need to define and clarify the factors that contribute significantly to outcomes in cancer survivors

K.H. Schmitz (✉)
Division of Clinical Epidemiology, Center for Clinical Epidemiology and Biostatistics,
Perelman School of Medicine, University of Pennsylvania, 423 Guardian Drive, 8th Floor
Blockley Hall, Philadelphia, PA 19104-6021, USA
e-mail: Schmitz@upenn.edu

T. Agurs-Collins
Health Behaviors Research Branch, Division of Cancer Control and Population Sciences,
National Cancer Institute/NIH/DHHS, Rockville, MD, USA
e-mail: collinsta@mail.nih.gov

M.L. Neuhouser
Division of Public Health Sciences, Fred Hutchinson Cancer Research Center, MS M4-B402,
1100 Fairview Avenue North, PO Box 19024, Seattle, WA 98109, USA
e-mail: mneuhous@fhcrc.org

L. Pollack
George Warren Brown School of Social Work, Washington University, St. Louis, Campus Box
1196, One Brookings Drive, St. Louis, MO 63130, USA
e-mail: lpollack@wustl.edu

S. Gehlert
George Warren Brown School of Social Work, Washington University, St. Louis, 1010 Saint
Charles, #804, St. Louis, MO 63101, USA
e-mail: sgehlert@wustl.edu

D.J. Bowen et al. (eds.), *Impact of Energy Balance on Cancer Disparities*,
Energy Balance and Cancer 9, DOI 10.1007/978-3-319-06103-0_4,
© Springer International Publishing Switzerland 2014

in order to develop effective and efficient intervention strategies. Within this chapter we address the independent and interactive contributions of two issues thought to substantively influence the length and quality of cancer survivorship: obesity and race/ethnicity.

Keywords Cancer survivorship • Breast cancer survival • Endometrial cancer survival • Colorectal cancer survival • Prostate cancer survival • Adverse treatment effects • Poverty • Affordable care act • Prevention and public health fund • Quality of life • Obesity • Patient Protection and Affordable Health Care Act • Cancer-related fatigue

Introduction

From 1971 to 2002, the number of cancer survivors in the USA has grown from three million to ten million [1], and the number of survivors is expected to reach 18 million by 2022 [2]. This growth has occurred over the same decades during which the prevalence of obesity has increased as well. The prevalence of over-weight (BMI = 25.0–29.9 kg/m^2) and obesity (BMI $\geq$ 30.0 kg/m^2) in the USA rose from 13.5 % in the 1960s to 35.9 % in 2010 [2, 3]. Data from the 2007 Health Information National Trends Survey (HINTS) indicate that cancer survivors are no more or less likely to be obese than those who have not experienced cancer [4]. However, the combined experience of cancer and obesity may influence the length and quality of life after completing treatments for common cancers, including breast, colon, prostate, and endometrial cancer [5]. In section "Obesity and Cancer" of this chapter, we review this evidence.

The racial and ethnic diversity in the USA has also shifted over the same decades during which the cancer survivorship population has grown. In 1970, when there were approximately three million cancer survivors living in the USA, 11.1 % of the country's population identified their race or ethnicity as African American, 4.4 % as Hispanic, 0.8 % as Asian, and 87.7 % as non-Hispanic White [7]. By 2010, when there were over 12 million cancer survivors, 13.6 % of the US population identified their race or ethnicity as African American, 16.3 % as Hispanic, 4.9 % as Asian, and 72.4 % as non-Hispanic White. Incidence of cancer does vary by race and ethnicity [8]. Social, economic, behavioral, health access, and other differences might be expected to contribute to variability regarding the burden of cancer across race or ethnicity. In section "Race/Ethnicity and Cancer" of this chapter, we review the evidence that survival after diagnosis and adverse effects of cancer treatment vary by race and ethnicity for four common cancer diagnoses (breast, colon, prostate, and endometrial).

Finally, it can also be observed that prevalence of obesity varies by race and ethnicity, with a higher prevalence of obesity among ethnic and racial minorities in the USA [4]. The racial and ethnic groups for whom cancer survival is worse are the same groups in which obesity is more prevalent, including African Americans and

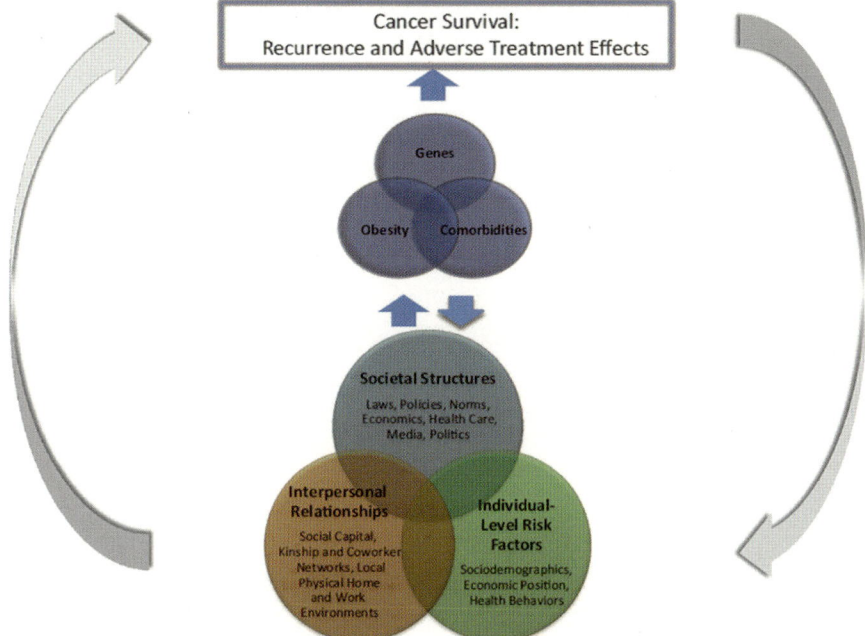

Fig. 4.1 Framework for the combined influence of race and obesity on cancer survivorship. The *bottom* Venn diagram represents distal determinants of disparate cancer survival outcomes. Disparities in these underlying social and physical determinants are embodied and expressed through biological responses and genetic pathways, which lead to disparities in risk for obesity and comorbidities. In the *top* Venn diagram, differential genes, obesity, and comorbidities then lead more proximally to disparities in survivorship outcomes All of these relationships operate in a feedback loop of biological–social–physical environment interactions, making it difficult to disentangle which of the underlying or mediating factors are the greatest contributors to disparities in survivorship

Native American/Pacific Islanders [3, 9]. It could be hypothesized that the disparities by race and ethnicity for cancer survival and treatment outcomes are explained, in part, by disparities in obesity and related comorbidities. A framework for discussing these complexities has been proposed (Fig. 4.1) [6]. Disparities in underlying social and physical determinants are embodied and expressed through biological responses and genetic pathways, which may lead to disparities in risk for cancer, obesity, and comorbidities. Differential risk in obesity and comorbidities may then lead to disparities in cancer survivorship outcomes. All of these relationships operate in a feedback loop of biological–environmental interactions. In section "Future Directions and Summary" of this chapter we review the limited evidence available on this topic and draw from evidence on this interaction of obesity with race/ethnicity for predicting other health outcomes (such as heart disease) to speculate further about the significance of these factors to outcomes among cancer survivors. Section "Future Directions and Summary" also includes commentary on the potential influence of the roll out of the 2010 Patient Protection and Affordable Health Care Act on these issues. The chapter concludes with a summary and comments about possible future directions for better understanding the interactive effects of obesity and race/ethnicity on cancer survivorship in a manner that will lead to expediency with regard to maximizing the health of cancer survivors while minimizing disparities in these outcomes by obesity and race or ethnicity.

Obesity and Cancer

Obesity and Survival/Mortality

Obesity is increasingly recognized as a risk factor for poor survival for many cancers, but particularly so for breast, prostate, endometrial, and colorectal cancers [10–13]. In many published studies the risk of cancer-specific mortality is as much as two times higher for obese patients compared to nonobese patients making treatment of the obese cancer patient a particularly important clinical issue [11, 14, 15]. Obesity is associated with all-cause mortality as well. For example, among endometrial cancer patients, obesity and diabetes at diagnosis are associated with all-cause but not cancer-specific mortality, suggesting the need to address these comorbidities to reduce the burden of mortality in this cancer survivorship population [16]. Another emerging and recognized problem is that many patients may enter cancer treatment at a normal weight, but may gain a significant amount of weight during therapy. This treatment-related weight gain phenomena has been linked to female breast cancer patients receiving systemic adjuvant chemotherapy [12]. However, the causes of weight gain after cancer treatment are not fully understood. Hypothesized causes include a combination of changed diet, reduced physical activity, or altered metabolism that may accompany after exposure to the chemotherapy [12, 17]. It is not yet known whether these patients who gain weight while undergoing treatment have the same mortality risk as those who are obese at the time of diagnosis [18, 19]. Nonetheless, this weight gain during treatment phenomena should be carefully monitored by clinicians with referrals to appropriate weight management plans [12].

Numerous factors have been identified as contributing to the increased mortality rate in obese cancer patients. Many of these patients will have multiple comorbidities. Managing diabetes, cardiovascular issues (including prior stroke or stroke risk), metabolic syndrome, and multitudes of other obesity-related diseases is a clinical challenge from the perspective of polypharmacy and drug–drug interactions. In addition, many obese patients will be viewed as surgical risks due to concurrent poor cardiovascular or respiratory health and as a result, treatment plans must be altered from approaches known to be most effective, as necessary. Further, for those who are able to undergo surgery, clean surgical margins can be very difficult to attain in the obese patient due to the large fat pad around the tumor and surrounding tissue. Obese patients also tend to have disordered metabolism even in the absence of a diagnosis of diabetes or prediabetes. Low levels of adiponectin and high levels of insulin (or C-peptide), IGF-1, glucose, adipose-derived inflammatory factors, cytokines, and other metabolic markers are very common in obese individuals and each of these has been associated with increased risk of cancer-specific mortality [18–21]. Many of the molecules that tend to be higher in obese people (even those without cancer) have direct links to the carcinogenic process by upregulating critical pathways such as mTOR and PI3K and by influencing the local inflammatory environment [22]. Modest evidence exists to suggest that genetic variation in the synthesis or metabolism of certain fatty acids or lipid-related compounds may influence the relationship between obesity and survival [23, 24]. More work is needed to better understand these complex relationships.

Recent and very intriguing data suggests that the tumor phenotype or the tumor microenvironment may be very different for patients who are obese compared to nonobese patients [22, 25, 26]. A recent study examined breast tissue from 30 normal weight and overweight/obese women undergoing surgery for breast cancer [26]. Among the findings, they reported higher aromatase expression in the breast tissue of obese women with subsequent greater local synthesis of estrogen in the breast. They also reported more crown-like structures in the breast tissue from obese women, which is also an indicator of greater levels of local inflammation. Other studies have shown that adipocytes that comprise the tumor microenvironment actively recruit macrophages similar to the peripheral circulation. The macrophages then promote neovascularization and angiogenesis, which sets up the patient for risk of metastasis [22]. Data are suggestive, but by no means definitive, that obese colorectal cancer patients may have tumors with a distinct pathologic phenotype that could be driven by the underlying obesity-driven metabolic disturbances [25, 27].

A critical link to the intersection between obesity and racial/ethnic cancer-related health disparities is that since African Americans and Hispanics have the highest rates of obesity in the USA [4], their body habitus already puts them at disproportionate risk of cancer-related mortality compared to Caucasians. Still, not all studies support an interaction between race and obesity with respect to cancer survival [11], so clearly additional work is needed in order to formulate the most appropriate treatment protocols.

Obesity and Persistent Adverse Treatment Effects

Cancer patients who are obese at diagnosis may require specialized treatment plans and management strategies. It has been observed that obese endometrial cancer patients experience greater loss of blood and longer surgical times than nonobese patients [28].

A persistent question facing clinicians is whether obese cancer patients should receive different chemotherapy doses based on body weight. There has been a long history of controversy on this topic where on the one hand it is thought that obese patients may be underdosed if drug doses are not based on actual body weight while on the other hand the risk of toxicities may be elevated with higher doses. A recent systematic review and meta-analysis had shed some light on this critical clinical issue [29]. Hourdequin et al. reviewed 12 studies that included cohort studies and clinical trials for a variety of cancers. Details of their methods including selection criteria for inclusion in the meta-analysis and patient inclusion criteria are reported in their publication [29]. The prespecified primary outcome was grade 3–4 hematologic toxic effect (on a scale of 1–4) while secondary outcomes included standard blood counts (leukocytes, platelets, hemoglobin), neutropenia, and overall survival. The overall pooled results showed that when chemotherapy doses were based on actual body weight, grade 3–4 hematologic toxic effects were 27 % lower in obese patients compared to nonobese patients [29]. The authors concluded that their evidence suggests that full-dosing based on actual body weight is safe and without clear evidence for risk of excess toxicity. Still, because other studies suggest incomplete pathologic response to chemotherapy among obese patients [30], other reasons besides chemotherapy dosing may be important to examine in future studies.

A relatively new but growing area of clinical and scientific concern is persistent adverse treatment effects from cancer. Below we review the empirical evidence on the association of obesity and common persistent adverse effects of cancer treatment, including lymphedema, quality of life, functional health, cancer related fatigue, chemotherapy induced peripheral neuropathy, and cardiotoxicity.

Lymphedema

Cancer treatments, including removal of lymph nodes and radiation therapy, can damage the lymph system, resulting in a chronic, sometimes progressive condition called lymphedema. Lymphedema is commonly characterized by swelling of the affected body part. However, the swelling associated with lymphedema is distinct from other types of edema due to the enrichment of protein in the accumulation of lymph fluid. This accumulation of protein rich fluid, accompanied by the role of the lymph system in inflammatory and immune responses, sometimes results in lymphedema becoming a systemic issue. Ongoing symptom monitoring is recommended to avoid lymphedema onset and progression [31]. Obesity has been consistently associated with both onset and worsening of lymphedema after breast cancer. Prospective studies report odds ratios of 2.93–3.6 for risk of lymphedema among obese versus normal-weight women [32–35]. There is also evidence of a dose–response relationship between excess weight and lymphedema risk, with an OR of 1.08 for each additional BMI unit above the normal weight category (95 % CIs 1.05–1.12 [32] and 1.0004–1.165 [33]). Further, there is evidence that body fat is associated with lymphedema across a broad range of BMIs. For example, in a study conducted in Hong Kong, higher BMI was noted in breast cancer patients with lymphedema compared to matched controls, even though BMI was low in both groups (22.9 ± 3.6 kg/m^2 for cases vs. 21.8 ± 3.1 kg/m^2 for controls) [36]. One pilot study demonstrated weight loss to reduce lymphedema among overweight breast cancer survivors [37] and a larger study on this topic is ongoing within the Penn Transdisciplinary Research on Energetics and Cancer (TREC) Survivor Center (U54-CA155850 to Schmitz).

Evidence linking obesity with lymphedema is more scant beyond breast cancer. Several studies report no association of BMI and incident lower extremity lymphedema among cervical cancer survivors [38–40], while one other prospective cohort study among cervical cancer survivors found that low BMI (<18.5 kg/m^2) was associated with increased frequency of lymphedema [41]. Finally, in a cross-sectional survey study of 243 Australian women, lymphedema risk was 2.7-fold higher among overweight compared to normal weight endometrial cancer survivors (95 % CI: 1.0–7.5) [38].

Quality of Life and Functional Health

Obesity has been associated with lower physical and functional well-being and poorer quality of life among endometrial cancer [42, 43], breast cancer [44], prostate cancer [45–47], and colorectal cancer survivors [48]. Two other studies with heterogeneous samples of cancer survivors (e.g., breast, colorectal, prostate, bladder, uterine, and melanoma) have also demonstrated reduced quality of life among obese versus nonobese participants [49, 50]. Obesity is also associated with higher prevalence and severity of site-specific symptoms, such as incontinence in prostate cancer survivors [46, 51]. Both cancer [52–54] and obesity [55–57] have been found to be independently associated with functional health. However, the differential impact of cancer and its treatments on functional status among obese versus nonobese survivors remains to be elucidated.

Cancer Related Fatigue (CRF)

Obesity has been positively associated with CRF for a number of cancer sites, including breast [58–61] and endometrial cancers [62]. Factors predicting clinically significant CRF include a BMI > 25 kg/m^2, weight gain, physical inactivity, and low physical functioning [60, 63], and severity of fatigue symptoms is associated with higher BMI [63].

Peripheral Neuropathy and Cardiotoxicity

Very little information is available regarding the potential relationship between obesity and peripheral neuropathy. Three studies have observed no association between obesity and chemotherapy induced peripheral neuropathy after breast cancer [64, 65] and multiple myeloma [66]. However, determining the independent effect of obesity on neuropathy may be challenging, given obesity is also a strong risk factor for diabetes, which is also associated with neuropathy [67]. Because obesity is already a strong risk factor for cardiovascular disease and late effects of chemotherapy may not appear for many years after treatment completion, it is difficult to determine the specific role of obesity in treatment-related cardiac toxicities. Other sources of increased cardiovascular risk, such as weight gain after chemotherapy among breast cancer survivors [68], further complicate the clinical picture. The relationships between excess pretreatment weight, weight gain after treatment, and treatment-related cardiovascular outcomes have not been extensively studied.

As survival from many of the common cancers (i.e., breast, prostate, colorectal) continues to increase, it is critical to evaluate whether any late effects differ for obese versus normal weight patients. Treatment protocols and effective strategies for total patient care will need to be developed so as to treat the late effects from cancer treatment concurrent with treatment to achieve and maintain a healthy weight.

Obesity and the Economics of Survivorship

Health-care costs in the USA are elevated in comparison to other western nations and cancer treatment is no exception to the health-care spending crisis [69]. In 2010, costs for cancer diagnosis, treatment and survival were estimated to be $124.6 billion [70]. These expenditures vary by cancer site where the greatest direct medical costs are for breast, colorectal, lung, and prostate cancers for solid tumors and lymphoma for hematologic cancers [70, 71]. Other expenditures that some may consider hidden include lost productivity and the value of life lost [72, 73]. Bradley and colleagues reported that as many as a third of cancer patients are unable to return to work after the diagnosis and of those who do return, duties must often be limited or task reassignments must be made [73]. Yabroff et al. estimated that by 2020, the value of life lost due to cancer will reach a staggering $1,472.5 billion [72]. Other hidden costs include the loss of productivity of care givers who often must use family medical leave to care for the cancer patient [72, 73]. Over the next several decades, these figures are expected to increase commensurate with the aging population. One study in Washington state reported that cancer patients were 2.65 times more likely to file for bankruptcy compared to non-cancer patients [74] underscoring the fact that a cancer diagnosis often becomes an economic crisis for patients and their families [70, 75]. Many of these critical issues were highlighted (including systematic reviews of site-specific cancers) in an August 2013 Monograph of the Journal of the National Cancer Institute (JNCI Monographs Volume 2013, Issue 46).

While cancer is costly to all patients, an important question is whether subsets of patients, including those who are obese prior to diagnosis share a disproportionate burden of cancer health-care expenditures. For overweight and obese cancer patients, prognosis is worse compared to normal weight patients. However, it is unclear whether these disparities are due to lower health-care access or utilization, lack of follow-up for suspicious findings from screening tests, lower adherence to adjuvant therapies, lack of referrals to medical oncologists, presentation with more advanced disease, or biological differences in tumors or responses to therapies [76–79]. In addition, patients who are overweight or obese may have comorbid conditions (e.g., diabetes, hypertension, dyslipidemia) that need concurrent treatment [15]. Treatments of these concurrent conditions add to the overall economic burden of cancer. The economic disparities that accompany the racial and body habitus disparities in regard to cancer treatment and survival constitutes a problem that has multiple causes and will require a highly coordinated and multilevel approach to the solution [71].

It will be particularly important to monitor the effects of the 2010 Patient Protection and Affordable Health Care Act on the economic challenges for those who are overweight or obese at the time of a cancer diagnosis. An important component of the Affordable Health Care Act is the expansion of the Medicaid Program. One might intuitively think that this expansion will lead to greater access to cancer screenings and treatment that might otherwise be delayed without access to health-care services. Finally, as summarized in an Institute of Medicine Workshop Summary [69] that was specifically focused on delivering affordable cancer care in the twenty-first century, scientists, clinicians, and policy makers must all work together to construct treatment strategies and insurance reimbursement strategies for affordable and effective cancer care, perhaps particularly for those who have experienced health-care disparities in the past.

Race/Ethnicity and Cancer

Race/Ethnicity and Survival/Mortality

It is widely recognized that cancer survival and mortality disparities persist by race and ethnicity. For some cancers, African Americans and Hispanics have a worse disease-free survival than other racial/ethnic groups [80–82]. These disparities are associated with advanced stage of disease, tumor characteristics, comorbidities, suboptimal treatment, obesity, lack of or type of medical insurance, access to high-quality medical care, and low socioeconomic status (SES) [80, 82–86]. For those diagnosed with cancer between 2002 and 2008, the 5-year survival rates from all cancers were 69 %, 65 %, and 65 % for non-Hispanic Whites, Hispanics, and African Americans, respectively [87]. Disparities among racial and ethnic groups have existed for decades, even as overall cancer mortality rates for most cancers have been declining.

Breast Cancer

Breast cancer is the second leading cause of cancer-related death in women. Cancer mortality rates have been declining for the past two decades. However, comparable benefits of these declines have not been equally shared among all racial and ethnic groups. For example, SEER data on 5-year breast cancer survivor rates, using data on women diagnosed with breast cancer between 2002 and 2008, illustrates the disparity among non-Hispanic Whites (92 %), Hispanics (86 %) and African Americans (78 %) [87]. A recent study examined BC survival using SEER-Medicare population data between the years 1991–2005 for 16 SEER sites and found a 12.9 % absolute difference in survival between African Americans and non-Hispanic Whites ($p < 0.001$), which was attenuated, but still significant after matching on presentation characteristics and types of treatment [88]. In this study, cancer-related causes accounted for approximately 2/3 of the difference in 5-year all-cause mortality between African Americans and non-Hispanic Whites. Studies also suggest that the racial disparity in survival and mortality does not discriminate between younger and older breast cancer survivors. African American adolescent and young adult women also have worse survival rates compared to non-Hispanic Whites and Hispanics [89, 90]. It was reported that in a comparison of African American and non-Hispanic White women 60–64 years of age, African American women with luminal A/p53− tumors were reported to have higher all-cause mortality (HR 2.22; 95 %; CI: 1.30–3.79) and breast cancer-specific mortality (HR 1.89; 95 % CI: 0.93, 3.86) [91]. Similar results were reported for older women with Triple Negative Breast Cancer [92]. Data from Southwest Oncology Group (SWOG) phase III trials from 1974 to 2001 found that African American patients with early-stage, premenopausal and postmenopausal breast cancer had significantly worse overall survival than non-Hispanic White patients (HR = 1.41, 95 % CI: 1.10–1.82 and HR = 1.49, 95 % CI: 1.28–1.73), respectfully, after adjusting for stage, demographic and socioeconomic factors, tumor characteristics, and treatment [93].

Hispanics compared to non-Hispanic Whites have higher breast cancer mortality rates [94, 95]. There also is heterogeneity in survival within Hispanic subgroups. When examining Hispanic subgroups, one study revealed that Hispanic-blacks have significantly higher BC-specific mortality compared to Hispanic Whites (HR = 1.4; 95 %: CI: 1.1–1.7) [96], and among Hispanic Whites, Puerto Rican women had the highest risk (HR 1.7, 95 % CI: 1.3–2.1) [95]. Banegas et al. [96] reported that regardless of Hispanic origin, African American women experienced worse breast cancer survival compared to Hispanic and non-Hispanic white women. A meta-analysis of 20 studies completed through 2005 identified African American ethnicity as an independent predictor of higher mortality among breast cancer survivors, including overall survival (HR 1.27; 95 % CI: 1.18–1.38), and breast cancer-specific survival (HR 1.19, 95 % CI, 1.10–1.29) [97].

Endometrial Cancer

Mortality rates for endometrial cancers have declined in the past decades, although there are still racial disparities in both mortality and survival. African American women have lower survival rates for endometrial cancer compared with Hispanics and non-Hispanic White women [98, 99]. A retrospective analysis of data from four Gynecologic Oncology Groups (GOG) randomized treatment trials found that African American women with endometrial cancer had worse overall survival compared with non-Hispanic White women, after adjusting for tumor characteristics and treatment (1.26, 95 % CI: 1.06–1.51) [100]. Others have reported significantly worse survival for African Americans compared to non-Hispanic Whites (HR 1.94) and no significant differences in survival between Hispanics and non-Hispanic Whites [101]. Survival from endometrial cancer was examined in a study in Puerto Rico that revealed African Americans and Puerto Ricans had the lowest rates of survival when compared to non-Hispanic Whites (56.8 % for African Americans, 63.1 % for Puerto Ricans, 78.4 % for non-Hispanic Whites and 79.5 % of US Hispanics [102].

Colorectal Cancer

Colorectal cancer (CRC) remains the third most common cause of cancer-related death in both men and women. The racial disparity in CRC mortality has widened, with 53 % higher mortality in African American men and 46 % higher mortality in African American women compared with non-Hispanic White men and women [80]. Several studies have documented higher CRC mortality for African Americans compared to non-Hispanic Whites [103–105]. A study of resected stage II and stage III colon cancer revealed that African American patients experienced worse overall survival (HR = 1.22, 95 % CI: 1.11–1.34) and recurrence-free survival (HR = 1.14, 95%CI: 1.04–1.24) compared to non-Hispanic Whites [105]. A review of 16 SEER data registries linked to the Medicare database, adjusted for several risk factors including SES, tumor characteristics, treatment, and comorbidities, found that African Americans had a significantly higher risk of CRC death (HR = 1.24; 95 % CI: 1.14, 1.35) compared with non-Hispanic Whites [106]. In this study, Hispanic women had a lower risk of death than non-Hispanic Whites, in the adjusted model. Temporal trends in survival revealed that young Hispanics (20–49 years) improved their 1-year CRC survival from 86 % in 1993–1997 to 91 % in 2003–2007 [107]. Soto-Salgado and colleagues [108] examined CRC mortality in Puerto Rico and revealed that African American women and non-Hispanic White women age >50 years had an increased risk of death from CRC compared with Puerto Rican women. In addition, Puerto Rican women had a similar risk of death compared to US Hispanics, but a lower mortality rate compared to African American women; whereas, the CRC risk was higher for non-Hispanic White men ≥ 60 and US Hispanic men ≥ 80 than Puerto Rican men [108]. The overall 5-year survival from proximal colon cancer was reported as 39.7 % for African Americans, 43.1 % for non-Hispanic Whites, and 46.7 % for Hispanics [109]. An interaction between race and age may influence CRC survival. One study reported that younger African American women (<50 years) with advanced stage proximal tumors had worse survival compared to age-matched non-Hispanic Whites, whereas older African American men had a worse survival compared to older non-Hispanic White men [110]. It should be noted that there were no significant differences in cancer specific and overall survival CRC between African American and non-Hispanic White women enrolled in the Women Health Initiative study [111]. The investigators suggested that equal access to medical care and uniform tumor characteristics may have contributed to this outcome.

Prostate Cancer

Prostate cancer is the second leading cause of cancer-related death in men. Approximately one in six men will be diagnosed with prostate cancer during their lifetime, with increased risk during each decade of life. Several studies have documented an increased mortality among African American and Hispanic men with prostate cancer compared to non-Hispanic Whites [84, 106, 112, 113]. One study reported that Puerto Rican men had a higher mortality than did US Hispanic and non-Hispanic Whites, but lower rates compared to African American men [108]. In recent years, deaths from prostate cancer have narrowed between African Americans and non-Hispanic Whites, but the disparity still exists [80]. One study estimated that the mortality gap for prostate cancer-specific mortality was 1,320 more cases per 100,000 for African American compared with non-Hispanic Whites men [84]. Another study reported that African American men had a higher risk of mortality (HR = 1.23, 95 % CI: 1.04–1.47) relative to non-Hispanic White men, despite treatment on the same protocols [114]. Albain et al. [93] reported similar findings, showing higher mortality among African American men compared to non-Hispanic Whites (HR = 1.19, 95 % CL: 1.05–1.35), after adjustment for all other available factors [93, 114]. Also, the 10-year overall survival estimates were lower for African Americans. However, two meta-analyses had different findings. A meta-analysis of 48 studies found that African American men had lower prostate cancer-specific survival (RR = 1.13, 95 % CI: 1.00–1.27) and recurrence (RR = 1.25, 95 % CI: 1.11–1.41), but no difference in overall survival, after adjustment for clinical predictors and SES [115]. This analysis included several studies from the same registry with overlapping time periods. Sridhar and colleagues [116] conducted a meta-analysis of published studies from 1968 to 2007 and corrected for this methodological concern by including only one publication from the same cancer registry. They found a significant increased risk of mortality among African American men compared with non-Hispanic Whites, but after adjusting for age, clinical, and demographic variables, this association was not significant.

Race/Ethnicity and Persistent Adverse Treatment Effects

Currently, there are 13.7 million cancer survivors in the USA, and this number is expected to risk by 31 % by 2022 [2]. Cancer survivors may have long-term psychological and physical impairments and incidence and severity of these impairments differs by race and ethnicity.

Lymphedema

Racial/ethnic differences are associated with arm lymphedema [32, 117]. The Pathways Study, a prospective cohort of breast cancer survivors found that the risk of transient and persistent lymphedema was higher among African Americans compared to non-Hispanic Whites (HR = 1.93, 95 % CI: 1.00–3.71), adjusting for potentially confounding factors [118]. In this study, advanced stage of cancer, chemotherapy, and radiation therapy were independently associated with the increased risk of arm lymphedema. Others reported greater swelling in non-white women compared with non-Hispanic Whites [119, 120]. Arm lymphedema was studied in 494 African American and non-Hispanic White women with in situ to stage III-A primary breast cancer [32]. African Americans compared to non-Hispanic Whites had a higher prevalence of arm lymphedema (28 % vs. 21 %), but race was not significant in an analysis adjusted for obesity and hypertension [32]. This study also revealed that comorbidities (i.e., hypertension), obesity, surgery, and receipt of chemotherapy were associated with developing lymphedema [32]. Moreover, there appears to be racial differences in the formal diagnoses of lymphedema. In a population-based study of 450 breast cancer survivors, African American women were significantly more likely to have undiagnosed lymphedema than breast cancer survivors of other racial/ethnicity groups (OR = 2.7, 95 % CI: 0.81–0.98) [121]. Additional research investigating the influence of race and ethnicity on lymphedema in other cancer populations is needed.

Quality of Life and Functional Health

Minority cancer survivors experience lower health-related quality of life (HRQOL) than non-Hispanic Whites. One study examined mental health-QOL and physical health-QOL outcomes among 248 African American and 244 non-Hispanic White cancer survivors with a history of breast, prostate, and colorectal cancers [122]. African Americans had significantly poorer mental health-QOL compared to non-Hispanic Whites, after adjusting for SES, clinical, and psychosocial factors. In this study, the authors reported that race moderated the effect of perceived social support, with African Americans reporting higher mental health-QOL if they had high social support.

African American breast cancer survivors report poorer physical functioning and general health, and poorer physical and social well-being compared to non-Hispanic White survivors [123, 124]. African Americans had lower HRQOL outcomes due to higher levels of stress and worry related to recurrence and financial concerns [125–127]. One study found that in a military health-care system, African American women with breast cancer exhibited more physical impairments ≥12 months post-surgery [117]. Fu and colleagues [128] reported that Hispanic women with stage 0-III breast cancer were more likely to report depression, chemotherapy-related symptoms, and pain-related symptoms compared to non-Hispanic Whites, each of which impact QOL. In a multiethnic sample of breast cancer survivors, Hispanic women reported the lowest HRQOL, such as higher physical and emotional burden and socio-ecologic strain, compared to other racial groups [129].

Studies also suggest that the type of treatment and stage of disease can impact QOL in prostate cancer survivors [130]. Palmer and colleagues [131] examined treatment decision making and post-treatment QOL scores among African American men recently treated for prostate cancer. African American prostate cancer survivors reported lower QOL scores for urinary incontinence, sexual function, and bother, but higher scores for bowel and hormonal functions [131]. Another study among prostate cancer survivors found that non-Hispanic Whites reported significantly greater QOL than African American and Hispanic men [132]. The relationship between ethnicity and QOL was partially mediated by SES, medical comorbidity, and health behaviors (e.g., sleep functioning and physical activity) [132]. A systematic review of the literature suggests that Hispanics report poorer mental, physical, and social QOL relative to non-Hispanics [133].

A study that examined differences in QOL in 182 non-Hispanic Whites and 98 Hispanic breast cancer survivors found that Hispanics reported significantly lower levels of total perceived social support and QOL compared to non-Hispanic Whites [134]. In a study that examined physical health and obesity in African Americans, Hispanic-Americans, Asian-Americans, and non-Hispanic White cancer survivors, racial and ethnic differences were identified, with African American and Hispanic-American survivors reporting significantly lower physical health scores compared to Asian-American and non-Hispanic White survivors [135]. In addition, African American survivors had the highest rates of obesity, which was associated with lower physical function scores compared to nonobese survivors [135].

Cancer-Related Fatigue

African American and Hispanic breast cancer survivors report high rates of cancer-related fatigue (CRF) and depression [119, 136]. When comparing African American breast cancer survivors with African American female controls, one study found that breast cancer survivors experience more CRF, worse hot flashes, and worse sleep quality [137]. Pain-related symptoms also were reported to be much higher among Hispanic women and elderly women [119, 128]. Factors predicting clinically significant CRF include a $BMI > 25$ kg/m^2, weight gain, physical inactivity, and low physical functioning [60, 63]. In addition, the severity of fatigue symptoms is associated with increasing BMI [63]. There is a paucity of research on racial differences in CRF. One study examined CRF in African American and non-Hispanic White women, but did not find significant differences between the two groups [126]. Taken together, these findings highlight the important role of race/ethnicity and obesity on poorer HRQOL in cancer survivors, which strongly suggests the need to elucidate the mechanisms leading to poorer HRQOL outcomes in racially and ethnically diverse obese populations, as well as research to examine persistent CRF disparities among racial/ethnic groups.

Peripheral Neuropathy and Cardiotoxicity

Few studies are found in the literature that examined racial differences in peripheral neuropathy and cardiotoxicity among cancer patients. Gewandter et al. [138] studied self-reported chemotherapy-induced peripheral neuropathy (CIPN) in 421 cancer survivors participating in a phase III randomized clinical trial. The authors reported that factors associated with functional impairment included non-white race and greater motor neuropathy scores. Hasan and colleagues conducted a retrospective study of African American patient records to examine cardiotoxicity from doxorubicin-based therapy from 1997–2001 [139]. The investigators found a dose-dependent increase risk of cardiotoxicity in this patient population [139]. Another study reported that low to moderate doses of anthracycline-based chemotherapeutic agents were associated with subclinical abnormalities of cardiovascular function, irrespective of race in patients with breast cancer or a hematologic malignancy [140]. However, one study reported slightly higher trastuzumab-associated cardiac safety events among African Americans (10.9 vs. 7.9) compared to non-Hispanic Whites [86]. Research is needed to examine racial/ethnic differences for peripheral neuropathy and cardiotoxicity related to cancer treatments.

Race/Ethnicity and Economic Impact of Survivorship

Improvements in cancer therapy have led to increases in the cost of treatment, often causing a financial burden for patients and their families. To address this issue, the American Society of Clinical Oncology (ASCO) Cost of Care Task Force developed a Guidance Statement on the Cost of Cancer Care to highlight salient issues to clinicians, provide recommendations, and to identify relevant policy issues [141]. The financial burden associated with treatment is compounded by inadequate insurance coverage, job loss, or the inability to work, as well as higher insurance premiums.

A reported consequence of treatment in breast cancer survivors includes changes in motivation to work, productivity and quality of work, and missed days of work [142]. Racial/ethnicity disparities add another layer of complexity to the cost of treatment. African American and Hispanic survivors who are more often uninsured or receiving Medicaid are likely to present with advanced-stage cancer at diagnosis, live in areas with lower high school graduation rates, and have lower median incomes [143]. As a result of their diagnoses, survivors report increases in insurance premiums at 3 and 6 months from baseline [142].

Private insurance and managed care payer status is associated with improved 5-year overall survival compared to patients with Medicaid, Medicare, or were uninsured/self-pay [144]. For example, African American men with prostate cancer, who tend to receive care from hospitals with higher proportions of African Americans and higher Medicaid admissions, have lower rates of definitive treatment [145], which can impact survival.

Job loss also has been associated with receiving chemotherapy, comorbidities, and a lack of employment support (e.g., paid sick leave and flexible schedules). To illustrate, African American and Hispanic women are more likely to stop working or lose their jobs when compared with non-Hispanic Whites after a cancer diagnosis and during treatment [146, 147]. One study reported that Hispanic women had the highest prevalence of job loss (24.1 %) compared to African Americans (10.1 %), and non-Hispanic Whites (6.9 %) [147]. Another study showed that African American women compared with non-Hispanic White women are more than twice as likely to lose a job due to their diagnoses (6.6 % vs. 2.7 %) [148] and have difficulty paying bills [147].

Race/ethnicity is associated with a higher proportion of income spent on out-of-pocket health-care costs. African Americans often reside in households with a yearly median household income < $35,000, which is negatively associated with lower likelihood of receiving proper treatment and worse survival [144]. Minority breast cancer survivors with yearly incomes ≤ $20,000 and between $20,001 and $40,000 have higher out-of-pocket costs compared with non-Hispanic Whites, 31.4 % versus 12.6 % and 19.5 % versus 8.7 %, respectively [148]. In addition, African Americans are more likely to receive chemotherapy than non-Hispanic White cancer survivors and the cost of chemotherapy treatment significantly impacts the proportion of out-of-pocket costs compared to women not receiving chemotherapy [149]. When examining costs associated with prostate cancer care, African Americans have both higher and incremental costs compared to non-Hispanic Whites [150].

Interaction of Obesity and Race/Ethnicity for Outcomes After a Cancer Diagnosis

Little is known about racial and ethnic differences and obesity's impact on cancer survivorship. Complex theories, models, and frameworks have been developed, and studies have been conducted to try and link the combined influence of race and obesity on cancer survivorship. While incremental progress has been made, the current status of the literature has addressed only fragments of the framework in Fig. 4.1. For example, studies have analyzed underlying social and physical determinants such as the influence of nativity and neighborhoods [151] on cancer survivorship without addressing the impact of obesity. Other studies have considered the impact of obesity [152] or comorbidities [153, 154] on cancer survivorship, but have looked in depth at more distal determinants. What is known is mostly specific to breast cancer, and even there, some of the research findings are inconsistent.

Three large multiethnic cohort studies illustrate this variation. Conroy et al., in a multiethnic prospective cohort study of African American, Native Hawaiian, Japanese American, Latino, and Caucasian women, examined the relationship between self-reported, pre-diagnostic BMI (body mass index (BMI) (weight (kg)/height $(m)^2$) and breast cancer survival [152]. Additionally, they examined whether the association between BMI and risk for breast cancer-specific mortality varied by ethnicity. The study found that while obese women had a modest increased risk of breast cancer-specific mortality, ethnic-specific trends were inconclusive [2]. For example, they found that obese women (BMI ≥ 30 kg/m^2) relative to women of high-normal weight (BMI 25.0–29.9 kg/m^2) had a higher risk of breast cancer-specific mortality (HR $= 1.45$; 95 % CI: 1.05, 2.00), across all ethnic groups except Native Hawaiian. Though ethnic-specific trends were inconclusive, obese Caucasian and Japanese American women had a slightly elevated risk of breast cancer-specific mortality compared with other ethnic groups [152].

Two more recent studies have moved closer to accurately representing the complex relationship between race/ethnicity, obesity, and cancer survivorship by including demographics and lifestyle factors on a larger number of ethnically diverse groups. The California Breast Cancer Survivorship Consortium (CBCSC) combined self-reported interview data regarding demographics and lifestyle factors (e.g., family history of breast cancer, parity, smoking, alcohol consumption) from six California-based breast-cancer epidemiologic studies with amassed cancer registry data on clinical characteristics and mortality [155]. Between 1993 and 2007, a multiethnic cohort of 12,210 women (6,501 non-Latina Whites, 2,060 African Americans, 2,032 Latinas, 1,505 Asian Americans, and 112 other race/ethnicity) were diagnosed with breast cancer. African Americans had higher rates of breast cancer-specific mortality compared with non-Latina Whites (HR $= 1.13$; 95 % CI: 0.97, 1.33). But, the breast cancer-specific mortality rates in Latinas (HR $= 0.84$; 95 % CI: 0.70, 1.00) and Asian Americans (HR $= 0.60$; 95 % CI: 0.37, 0.97) were lower than in non-Latina Whites. In a separate study, and in contrast to findings from the aforementioned CBCSC paper, Kwan et al. further investigated these disparities in survival outcomes by race/ethnicity, accounting for obesity status, and found ethnic variation [156]. Kwan et al. evaluated the association between body size measurements (BMI and waist–hip ratio (WHR)) and breast cancer-specific mortality after breast cancer diagnosis by race/ethnicity using data from questionnaires and the California Cancer Registry between 1993 and 2007, with follow-up through 2009. In the data, 11,351 breast cancer patients were identified. Compared with normal weight (BMI = 18.5–24.9), morbid obesity (BMI ≥ 40) was associated with having an increased risk of breast cancer mortality in non-Latina Whites (HR = 1.43, 95 % CI: 0.84, 2.43) and in Latinas (HR = 2.26, 95 % CI: 1.23, 4.15), though these differences were not statistically significant. No BMI–mortality associations were present in African Americans or Asian Americans. High WHR was associated with breast cancer mortality in Asian Americans (HR = 2.21, 95 % CI: 1.21, 4.03; p for trend = 0.01), but no associations were seen in African Americans, Latinas, or non-Latina Whites. This study demonstrates that obesity and body fat distribution impact cancer survivorship, and this association varies by race/ethnicity. While this study allowed for comparisons of some important determinants in the feedback loop of disparities in cancer survivorship outcomes within and across multiple racial/ethnic groups, additional studies are needed to understand why degree of obesity or body fat distribution at breast cancer diagnosis differentially affects cancer survivorship by race/ethnicity [156].

Due to the limited and incomplete evidence available on the biological-environmental interactions of race/ethnicity, obesity, and cancer survivorship, we draw from evidence of these interactions for predicting other, more well-known health outcomes (i.e., heart disease) to learn and speculate about the significance of these factors to outcomes among cancer survivors. Similar to cancer survivorship, patterns of cardiovascular disease (CVD) and mortality vary by race/ethnicity even after controlling for established risk factors for poor prognosis [157, 158]. In a more recent prospective multiethnic cohort study of adult men and women living in Hawaii and California, Henderson et al. examined the mortality rates from acute myocardial infarction and other heart disease in five racial groups—African American, Native Hawaiians, Japanese Americans, Latinos, and Whites—to investigate whether the observed differences in CVD mortality could be explained by differences in the prevalence of established CVD risk factors [159]. Relative risks for mortality were calculated accounting for established CVD risk factors (BMI, hypertension, diabetes, smoking, alcohol consumption, physical activity, education level, diet, and factors specific to women including type and age at menopause and hormone replacement therapy use), and the authors found that these risk factors explained a large portion of the racial and ethnic variation in risk for acute myocardial infarction and other heart disease mortality. The authors suggested that the unexplained excess risk of mortality in African American women and Native Hawaiians, and the lower than expected risk of mortality in Japanese-Americans and Latinos, compared with Whites, are due to unequal distributions of unmeasured factors: environmental, social or cultural, and genetic risk [159].

The studies reviewed here explore the connections between different ethnic groups and covariates of interest, such as reproductive, lifestyle, sociodemographic, and other cancer-specific (or cardiovascular-specific) risk factors. Additionally, the authors speculate about ways we could improve studies to better understand the impact of obesity on the racial/ethnic variation on risk of mortality. Conroy et al. suggested that larger sample sizes of ethnically diverse groups are needed to definitively evaluate ethnic-specific trends in mortality [152]. Kwan et al. suggested that multiple measurements of obesity are needed to better understand the racial/ethnic differences in body composition and mortality [156]. And Henderson et al. alluded to the significance of unmeasured environmental factors, social or cultural factors, and genetic risk factors [159], which is also recommended in Fig. 4.1 [6]. It might also be true that our theoretical models have advanced far beyond the methodologies being employed [160]. If this is the case, further advances in methodology might be needed to analyze data with complex patterns of variability, especially nested sources of variability [160] as suggested by Fig. 4.1 [6], in order to identify the interactions of combined influences of race/ethnicity, obesity, and cancer survivorship.

The studies reviewed in both cancer and cardiovascular disease strongly suggest that social environmental and cultural factors that have heretofore not been included in analyses might help to explain why degree of obesity or body fat distribution differentially affects cancer survivorship by race/ethnicity. A number of social and cultural factors have been posited to explain differences by race/ethnicity in the USA. We know for example that racial and ethnic minorities are more likely than non-Hispanic Whites to live in poverty [161]. The poverty rate in non-Hispanic Whites was 9.9 % in 2010, compared to 27.4 % and 26.6 % for African Americans and Hispanics, respectively [161].

Poverty and geographic location predispose racial and ethnic minorities to consume foods that are less healthful and more obesogenic. A number of studies have found that lower SES and predominantly African American neighborhoods have fewer supermarkets, more fast-food restaurants, and lower access to fresh fruits and vegetables [162, 163]. Likewise, the nature of the built environment in lower socioeconomic and predominantly African American neighborhoods impedes physical activity [164].

Racial and ethnic minorities may also experience a disproportionate burden of cancer health-care expenditures by virtue of their higher rates of obesity-related comorbid conditions and higher rates of poverty. These comorbidities may pose further limits to physical activity, as well as limits on treatment options for cancer [165]. In addition, the added costs of treating comorbid conditions such as heart disease likely influence decisions and timing to seek treatment among racial and ethnic minorities, resulting in decreased survivorship.

The Affordable Care Act (ACA), through its provisions for expanded health-care coverage for persons living below the federal poverty line through the expansion of Medicaid coverage, increased attention to the social determinants of health through its newly established Prevention and Public Health Fund, and increased linkage to community services holds promise for decreasing cancer disparities by race and ethnicity. Increasing access to treatment should decrease the disproportionate burden of cancer care among racial and ethnic minorities and ensure that social and culture components of health are better attended to by moving the focus of treatment from acute care to community settings. One billion new dollars will become available between Fiscal Year 2012 and Fiscal Year 2017 through the Prevention and Public Health Fund, which can be used to implement community programs to address obesity. Some of these funds are targeted specifically to Medicaid beneficiaries to implement, evaluate, and disseminate preventive health activities through community transformation grants.

Future Directions and Summary

Intervention dollars for reducing the burden of cancer are limited. To make best use of these dollars, it is important that intervention efforts be used as efficiently as possible. Unpacking the complexities of the relationships of obesity, race/ethnicity, and cancer will enable interventions to reduce the burden of cancer to be most effective. We have presented empirical evidence supporting a role for both obesity and race/ethnicity for outcomes after a cancer diagnosis. Further, evidence from heart disease has been used to speculate that upstream influences such as poverty, unavailability of healthy food and low access to physical activity may underlie the combined influence of race/ethnicity and obesity on cancer health disparities. Additional observational research may assist with further clarifying intervention targets, particularly if large multiethnic cohorts can be assembled to clarify remaining questions regarding the independent and interactive roles of obesity and race/ethnicity on the burden of cancer survivorship. Interventions to reduce racial and ethnic disparities among cancer survivors could be targeted upstream, toward addressing poverty, access to health care, and access to healthy foods and physical activity. Concurrently, downstream interventions that are both effective and feasible for improving the length and quality of cancer survival across BMI and racial and ethnic categories also need to be evaluated, perhaps in the context of the rollout of the Affordable Care Act and the Prevention and Public Health Fund. It can be noted that the majority of cancer survivorship interventions focus on Caucasian survivors, so establishing effectiveness and feasibility among racially diverse populations with a broad range of BMI profiles would helpful, given the increasing diversity of the US population.

In summary, the USA has become more racially and ethnically diverse during the same decades in which we have also experienced a sharp rise in obesity prevalence. Obesity prevalence is strongly correlated with race and ethnicity. It is possible that the disparities in cancer survivorship noted by race and ethnicity are due, in part, to differences in obesity. Obesity may be a marker for upstream influences on cancer survivorship such as economic and health access differences by race and ethnicity. It is speculated that obesity contributes to disparities for cancer survivorship outcomes through upstream influences such as poverty and access to health care and lifestyle interventions. Both upstream (system level) and downstream (individual level) interventions for obesity can be better targeted to be effective and feasible to improve outcomes in cancer survivors across a broad spectrum of BMI levels and racial and ethnic groups.

References

1. Hewitt M, Greenfield S, Stovall E (2006) From cancer patient to cancer survivor: lost in transition. National Academies Press, Washington, DC
2. de Moor JS, Mariotto AB, Parry C et al (2013) Cancer survivors in the United States: prevalence across the survivorship trajectory and implications for care. Cancer Epidemiol Biomarkers Prev 22(4):561–570
3. National Center for Health Statistics (2008) Prevalence of overweight, obesity and extreme obesity among adults: United States, trends 1976-80 through 2005-2006. http://www.cdc.gov/nchs/data/hestat/overweight/overweight_adult.pdf
4. Flegal KM, Carroll MD, Kit BK et al (2012) Prevalence of obesity and trends in the distribution of body mass index among US adults, 1999-2010. JAMA 307(5):491–497
5. Moten A, Jeffers K, Larbi D, Smith-White R, Taylor T, Wilson LL, Frederick W, Laiyemo AO (2012) Prevalence of obesity among subjects with a personal history of cancer: a call for action among cancer survivors and care providers. In: AACR 103rd annual meeting 2012. Chicago, IL, 2012: abstract. Cancer Res 72:supplement 1
6. Schmitz KH, Neuhouser ML, Agurs-Collins T et al (2013) Impact of obesity on cancer survivorship and the potential relevance of race and ethnicity. J Natl Cancer Inst 105 (18):1344–1354
7. Gibson C, Jung, K. Historical census statistics on population totals by race, 1790 to 1990, and by Hispanic origin, 1970 to 1990, For The United States, Regions, Divisions, and States. http://www.census.gov/population/www/documentation/twps0056/twps0056.html
8. Humes K, Jones N, Ramirez R (2011) Overview of race and Hispanic origin: 2010. U.S. Census Bureau, Washington, DC
9. Eheman C, Henley SJ, Ballard-Barbash R et al (2012) Annual Report to the Nation on the status of cancer, 1975-2008, featuring cancers associated with excess weight and lack of sufficient physical activity. Cancer 118(9):2338–2366
10. Ewertz M, Jensen MB, Gunnarsdottir KA et al (2011) Effect of obesity on prognosis after early-stage breast cancer. J Clin Oncol 29(1):25–31
11. Lu Y, Ma H, Malone KE et al (2011) Obesity and survival among black women and white women 35 to 64 years of age at diagnosis with invasive breast cancer. J Clin Oncol 29 (25):3358–3365
12. McTiernan A, Irwin M, Vongruenigen V (2010) Weight, physical activity, diet, and prognosis in breast and gynecologic cancers. J Clin Oncol 28(26):4074–4080
13. Chlebowski RT (2012) Obesity and breast cancer outcome: adding to the evidence. J Clin Oncol 30(2):126–128. doi:10.1200/JCO.2011.39.7877, Epub 2011 Dec 12
14. Goodwin PJ, Ennis M, Pritchard KI et al (2012) Insulin- and obesity-related variables in early-stage breast cancer: correlations and time course of prognostic associations. J Clin Oncol 30(2):164–171
15. Griggs JJ, Sabel MS (2008) Obesity and cancer treatment: weighing the evidence. J Clin Oncol 26(25):4060–4062. doi:10.1200/JCO.2008.17.4250
16. Chia VM, Newcomb PA, Trentham-Dietz A et al (2007) Obesity, diabetes, and other factors in relation to survival after endometrial cancer diagnosis. Int J Gynecol Cancer 17(2):441–446
17. Wolin KY, Carson K, Colditz GA (2010) Obesity and cancer. Oncologist 15(6):556–565

18. Meyerhardt JA, Ma J, Courneya KS (2010) Energetics in colorectal and prostate cancer. J Clin Oncol 28(26):4066–4073
19. Meyerhardt JA, Niedzwiecki D, Hollis D et al (2008) Impact of body mass index and weight change after treatment on cancer recurrence and survival in patients with stage III colon cancer: findings from Cancer and Leukemia Group B 89803. J Clin Oncol 26(25):4109–4115
20. Duggan C, Irwin ML, Xiao L et al (2011) Associations of insulin resistance and adiponectin with mortality in women with breast cancer. J Clin Oncol 29(1):32–39
21. Wolpin BM, Meyerhardt JA, Chan AT et al (2009) Insulin, the insulin-like growth factor axis, and mortality in patients with nonmetastatic colorectal cancer. J Clin Oncol 27(2):176–185
22. Arendt LM, McCready J, Keller PJ et al (2013) Obesity promotes breast cancer by CCL2-mediated macrophage recruitment and angiogenesis. Cancer Res 73(19):6080–6093
23. Nguyen PL, Ma J, Chavarro JE et al (2010) Fatty acid synthase polymorphisms, tumor expression, body mass index, prostate cancer risk, and survival. J Clin Oncol 28(25):3958–3964
24. Ogino S, Nosho K, Meyerhardt JA et al (2008) Cohort study of fatty acid synthase expression and patient survival in colon cancer. J Clin Oncol 26(35):5713–5720
25. Sinicrope FA, Foster NR, Yoon HH et al (2012) Association of obesity with DNA mismatch repair status and clinical outcome in patients with stage II or III colon carcinoma participating in NCCTG and NSABP adjuvant chemotherapy trials. J Clin Oncol 30(4):406–412
26. Morris PG, Hudis CA, Giri D et al (2011) Inflammation and increased aromatase expression occur in the breast tissue of obese women with breast cancer. Cancer Prev Res 4(7):1021–1029
27. Lochhead P, Kuchiba A, Imamura Y et al (2013) Microsatellite instability and BRAF mutation testing in colorectal cancer prognostication. J Natl Cancer Inst 105(15):1151–1156
28. Santoso JT, Barton G, Riedley-Malone S et al (2012) Obesity and perioperative outcomes in endometrial cancer surgery. Arch Gynecol Obstet 285(4):1139–1144
29. Hourdequin KC, Schpero WL, McKenna DR et al (2013) Toxic effect of chemotherapy dosing using actual body weight in obese versus normal-weight patients: a systematic review and meta-analysis. Ann Oncol 24(12):2952–2962
30. Litton JK, Gonzalez-Angulo AM, Warneke CL et al (2008) Relationship between obesity and pathologic response to neoadjuvant chemotherapy among women with operable breast cancer. J Clin Oncol 26(25):4072–4077
31. National Lymphedema Network Medical Advisory Committee (2008) Topic: lymphedema risk reduction practices. Position Statement of the National Lymphedema Network. http://www.lymphnet.org/pdfDocs/nlnriskreduction.pdf
32. Meeske KA, Sullivan-Halley J, Smith AW et al (2009) Risk factors for arm lymphedema following breast cancer diagnosis in Black women and White women. Breast Cancer Res Treat 113(2):383–391
33. Helyer LK, Varnic M, Le LW et al (2010) Obesity is a risk factor for developing postoperative lymphedema in breast cancer patients. Breast J 16(1):48–54
34. Ahmed RL, Schmitz KH, Prizment AE et al (2011) Risk factors for lymphedema in breast cancer survivors, the Iowa Women's Health Study. Breast Cancer Res Treat 130(3):981–991
35. Ridner SH, Dietrich MS, Stewart BR et al (2011) Body mass index and breast cancer treatment-related lymphedema. Support Care Cancer 19(6):853–857
36. Mak SS, Yeo W, Lee YM et al (2008) Predictors of lymphedema in patients with breast cancer undergoing axillary lymph node dissection in Hong Kong. Nurs Res 57(6):416–425
37. Shaw C, Mortimer P, Judd PA (2007) A randomized controlled trial of weight reduction as a treatment for breast cancer-related lymphedema. Cancer 110(8):1868–1874
38. Beesley V, Janda M, Eakin E et al (2007) Lymphedema after gynecological cancer treatment: prevalence, correlates, and supportive care needs. Cancer 109(12):2607–2614
39. Kim JH, Choi JH, Ki EY et al (2012) Incidence and risk factors of lower-extremity lymphedema after radical surgery with or without adjuvant radiotherapy in patients with FIGO stage I to stage IIA cervical cancer. Int J Gynecol Cancer 22(4):686–691

40. Ohba Y, Todo Y, Kobayashi N et al (2011) Risk factors for lower-limb lymphedema after surgery for cervical cancer. Int J Clin Oncol 16(3):238–243
41. Kizer NT, Thaker PH, Gao F et al (2011) The effects of body mass index on complications and survival outcomes in patients with cervical carcinoma undergoing curative chemoradiation therapy. Cancer 117(5):948–956
42. Fader AN, Frasure HE, Gil KM et al (2011) Quality of life in endometrial cancer survivors: what does obesity have to do with it? Obstet Gynecol Int 2011:308609
43. Courneya KS, Karvinen KH, Campbell KL et al (2005) Associations among exercise, body weight, and quality of life in a population-based sample of endometrial cancer survivors. Gynecol Oncol 97(2):422–430
44. Milne HM, Gordon S, Guilfoyle A et al (2007) Association between physical activity and quality of life among Western Australian breast cancer survivors. Psychooncology 16 (12):1059–1068
45. Sanda MG, Dunn RL, Michalski J et al (2008) Quality of life and satisfaction with outcome among prostate-cancer survivors. N Engl J Med 358(12):1250–1261
46. Dieperink KB, Hansen S, Wagner L et al (2012) Living alone, obesity and smoking: important factors for quality of life after radiotherapy and androgen deprivation therapy for prostate cancer. Acta Oncol 51(6):722–729
47. Anast JW, Sadetsky N, Pasta DJ et al (2005) The impact of obesity on health related quality of life before and after radical prostatectomy (data from CaPSURE). J Urol 173(4):1132–1138
48. Jansen L, Koch L, Brenner H et al (2010) Quality of life among long-term (>/=5 years) colorectal cancer survivors—systematic review. Eur J Cancer 46(16):2879–2888
49. Mosher CE, Sloane R, Morey MC et al (2009) Associations between lifestyle factors and quality of life among older long-term breast, prostate, and colorectal cancer survivors. Cancer 115(17):4001–4009
50. Blanchard CM, Stein K, Courneya KS (2010) Body mass index, physical activity, and health-related quality of life in cancer survivors. Med Sci Sports Exerc 42(4):665–671
51. Wolin KY, Luly J, Sutcliffe S et al (2010) Risk of urinary incontinence following prostatectomy: the role of physical activity and obesity. J Urol 183(2):629–633
52. Ness KK, Wall MM, Oakes JM et al (2006) Physical performance limitations and participation restrictions among cancer survivors: a population-based study. Ann Epidemiol 16 (3):197–205
53. Sweeney C, Schmitz KH, Lazovich D et al (2006) Functional limitations in elderly female cancer survivors [see comment]. J Natl Cancer Inst 98(8):521–529
54. Michael Y, Kawachi I, Berkman L et al (2000) The persistent impact of breast carcinoma on functional health status: prospective evidence from the Nurses' Health Study. Cancer 89 (11):2176–2186
55. Fontaine KR, Cheskin LJ, Barofsky I (1996) Health-related quality of life in obese persons seeking treatment. J Fam Pract 43(3):265–270
56. Stewart AL, Brook RH (1983) Effects of being overweight. Am J Public Health 73(2):171–178
57. Trakas K, Oh PI, Singh S et al (2001) The health status of obese individuals in Canada. Int J Obes Relat Metab Disord 25(5):662–668
58. Andrykowski MA, Donovan KA, Laronga C et al (2010) Prevalence, predictors, and characteristics of off-treatment fatigue in breast cancer survivors. Cancer 116(24):5740–5748
59. Reinertsen KV, Cvancarova M, Loge JH et al (2010) Predictors and course of chronic fatigue in long-term breast cancer survivors. J Cancer Surviv 4(4):405–414
60. Gerber LH, Stout N, McGarvey C et al (2011) Factors predicting clinically significant fatigue in women following treatment for primary breast cancer. Support Care Cancer 19(10):1581–1591
61. Winters-Stone KM, Bennett JA, Nail L et al (2008) Strength, physical activity, and age predict fatigue in older breast cancer survivors. Oncol Nurs Forum 35(5):815–821

62. Basen-Engquist K, Scruggs S, Jhingran A et al (2009) Physical activity and obesity in endometrial cancer survivors: associations with pain, fatigue, and physical functioning. Am J Obstet Gynecol 200(3):288.e1–288.e8

63. Meeske K, Smith AW, Alfano CM et al (2007) Fatigue in breast cancer survivors two to five years post diagnosis: a HEAL Study report. Qual Life Res 16(6):947–960

64. Schneider BP, Zhao F, Wang M et al (2012) Neuropathy is not associated with clinical outcomes in patients receiving adjuvant taxane-containing therapy for operable breast cancer. J Clin Oncol 30(25):3051–3057

65. Speck RM, Sammel MD, Farrar JT, Hennessy S, Mao JJ, Stineman MG, DeMichele A (2012) Racial disparities in the incidence of dose-limiting chemotherapy induced peripheral neuropathy. In: San Antonio breast cancer symposium. San Antonio, TX

66. Dimopoulos MA, Mateos MV, Richardson PG et al (2011) Risk factors for, and reversibility of, peripheral neuropathy associated with bortezomib-melphalan-prednisone in newly diagnosed patients with multiple myeloma: subanalysis of the phase 3 VISTA study. Eur J Haematol 86(1):23–31

67. de Azambuja E, McCaskill-Stevens W, Francis P et al (2010) The effect of body mass index on overall and disease-free survival in node-positive breast cancer patients treated with docetaxel and doxorubicin-containing adjuvant chemotherapy: the experience of the BIG 02-98 trial. Breast Cancer Res Treat 119(1):145–153

68. Demark-Wahnefried W, Campbell KL, Hayes SC (2012) Weight management and its role in breast cancer rehabilitation. Cancer 118(8 Suppl):2277–2287

69. Ramsey SD, Ganz PA, Shankaran V et al (2013) Addressing the american health-care cost crisis: role of the oncology community. J Natl Cancer Inst 105(23):1777–1781

70. Greenberg D, Earle C, Fang CH et al (2010) When is cancer care cost-effective? A systematic overview of cost-utility analyses in oncology. J Natl Cancer Inst 102(2):82–88

71. Seal BS, Sullivan SD, Ramsey S et al (2013) Medical costs associated with use of systemic therapy in adults with colorectal cancer. J Manag Care Pharm 19(6):461–467

72. Yabroff KR, Bradley CJ, Mariotto AB et al (2008) Estimates and projections of value of life lost from cancer deaths in the United States. J Natl Cancer Inst 100(24):1755–1762

73. Bradley CJ, Yabroff KR, Dahman B et al (2008) Productivity costs of cancer mortality in the United States: 2000-2020. J Natl Cancer Inst 100(24):1763–1770

74. Ramsey S, Blough D, Kirchhoff A et al (2013) Washington State cancer patients found to be at greater risk for bankruptcy than people without a cancer diagnosis. Health Aff 32(6):1143–1152

75. Mariotto AB, Yabroff KR, Shao Y et al (2011) Projections of the cost of cancer care in the United States: 2010-2020. J Natl Cancer Inst 103(2):117–128

76. Goldie SJ, Daniels N (2011) Model-based analyses to compare health and economic outcomes of cancer control: inclusion of disparities. J Natl Cancer Inst 103(18):1373–1386

77. Morris AM, Billingsley KG, Hayanga AJ et al (2008) Residual treatment disparities after oncology referral for rectal cancer. J Natl Cancer Inst 100(10):738–744

78. Simpson DR, Martinez ME, Gupta S et al (2013) Racial disparity in consultation, treatment, and the impact on survival in metastatic colorectal cancer. J Natl Cancer Inst 105(23):1814–1820

79. Goulart BH, Reyes CM, Fedorenko CR et al (2013) Referral and treatment patterns among patients with stages III and IV non-small-cell lung cancer. J Oncol Pract 9(1):42–50

80. DeSantis C, Naishadham D, Jemal A (2013) Cancer statistics for African Americans, 2013. CA Cancer J Clin 63(3):151–166

81. Hershman DL, Unger JM, Barlow WE et al (2009) Treatment quality and outcomes of African American versus white breast cancer patients: retrospective analysis of Southwest Oncology studies S8814/S8897. J Clin Oncol 27(13):2157–2162

82. Siegel R, Naishadham D, Jemal A (2012) Cancer statistics for Hispanics/Latinos, 2012. CA Cancer J Clin 62(5):283–298

83. Krieger N, Chen JT, Kosheleva A et al (2012) Shrinking, widening, reversing, and stagnating trends in US socioeconomic inequities in cancer mortality for the total, black, and white populations: 1960-2006. Cancer Causes Control 23(2):297–319

84. Taksler GB, Keating NL, Cutler DM (2012) Explaining racial differences in prostate cancer mortality. Cancer 118(17):4280–4289

85. Shavers VL, Brown ML (2002) Racial and ethnic disparities in the receipt of cancer treatment. J Natl Cancer Inst 94(5):334–357

86. Rugo HS, Brufsky AM, Ulcickas Yood M et al (2013) Racial disparities in treatment patterns and clinical outcomes in patients with HER2-positive metastatic breast cancer. Breast Cancer Res Treat 141(3):461–470

87. American Cancer Society (2013) Cancer facts & figures 2013. American Cancer Society, Atlanta, GA

88. Silber JH, Rosenbaum PR, Clark AS et al (2013) Characteristics associated with differences in survival among black and white women with breast cancer. JAMA 310(4):389–397

89. Keegan TH, Press DJ, Tao L et al (2013) Impact of breast cancer subtypes on 3-year survival among adolescent and young adult women. Breast Cancer Res 15(5):R95

90. Liu P, Li X, Mittendorf EA et al (2013) Comparison of clinicopathologic features and survival in young American women aged 18-39 years in different ethnic groups with breast cancer. Br J Cancer 109(5):1302–1309

91. Ma H, Lu Y, Malone KE et al (2013) Mortality risk of black women and white women with invasive breast cancer by hormone receptors, HER2, and p53 status. BMC Cancer 13:225

92. Sachdev JC, Ahmed S, Mirza MM et al (2010) Does race affect outcomes in triple negative breast cancer? Breast Cancer 4:23–33

93. Albain KS, Unger JM, Crowley JJ et al (2009) Racial disparities in cancer survival among randomized clinical trials patients of the Southwest Oncology Group. J Natl Cancer Inst 101 (14):984–992

94. Hill DA, Nibbe A, Royce ME et al (2010) Method of detection and breast cancer survival disparities in Hispanic women. Cancer Epidemiol Biomarkers Prev 19(10):2453–2460

95. Ooi SL, Martinez ME, Li CI (2011) Disparities in breast cancer characteristics and outcomes by race/ethnicity. Breast Cancer Res Treat 127(3):729–738

96. Banegas MP, Li CI (2012) Breast cancer characteristics and outcomes among Hispanic Black and Hispanic White women. Breast Cancer Res Treat 134(3):1297–1304

97. Newman LA, Griffith KA, Jatoi I et al (2006) Meta-analysis of survival in African American and white American patients with breast cancer: ethnicity compared with socioeconomic status. J Clin Oncol 24(9):1342–1349

98. Wright JD, Fiorelli J, Schiff PB et al (2009) Racial disparities for uterine corpus tumors: changes in clinical characteristics and treatment over time. Cancer 115(6):1276–1285

99. Long B, Liu FW, Bristow RE (2013) Disparities in uterine cancer epidemiology, treatment, and survival among African Americans in the United States. Gynecol Oncol 130(3):652–659

100. Maxwell GL, Tian C, Risinger J et al (2006) Racial disparity in survival among patients with advanced/recurrent endometrial adenocarcinoma: a Gynecologic Oncology Group study. Cancer 107(9):2197–2205

101. Smotkin D, Nevadunsky NS, Harris K et al (2012) Histopathologic differences account for racial disparity in uterine cancer survival. Gynecol Oncol 127(3):616–619

102. Ortiz AP, Perez J, Otero-Dominguez Y et al (2010) Endometrial cancer in Puerto Rico: incidence, mortality and survival (1992-2003). BMC Cancer 10:31

103. Alexander DD, Waterbor J, Hughes T et al (2007) African-American and Caucasian disparities in colorectal cancer mortality and survival by data source: an epidemiologic review. Cancer Biomark 3(6):301–313

104. Phatak UR, Kao LS, Millas SG et al (2013) Interaction between age and race alters predicted survival in colorectal cancer. Ann Surg Oncol 20(11):3363–3369

105. Yothers G, Sargent DJ, Wolmark N et al (2011) Outcomes among black patients with stage II and III colon cancer receiving chemotherapy: an analysis of ACCENT adjuvant trials. J Natl Cancer Inst 103(20):1498–1506
106. White A, Coker AL, Du XL et al (2011) Racial/ethnic disparities in survival among men diagnosed with prostate cancer in Texas. Cancer 117(5):1080–1088
107. Jafri NS, Gould M, El-Serag HB et al (2013) Incidence and survival of colorectal cancer among Hispanics in the United States: a population-based study. Dig Dis Sci 58(7):2052–2060
108. Soto-Salgado M, Suarez E, Calo W et al (2009) Incidence and mortality rates for colorectal cancer in Puerto Rico and among Hispanics, non-Hispanic whites, and non-Hispanic blacks in the United States, 1998-2002. Cancer 115(13):3016–3023
109. Wong RJ (2010) Marked variations in proximal colon cancer survival by race/ethnicity within the United States. J Clin Gastroenterol 44(9):625–630
110. Wallace K, Hill EG, Lewin DN et al (2013) Racial disparities in advanced-stage colorectal cancer survival. Cancer Causes Control 24(3):463–471
111. Simon MS, Thomson CA, Pettijohn E et al (2011) Racial differences in colorectal cancer incidence and mortality in the Women's Health Initiative. Cancer Epidemiol Biomarkers Prev 20(7):1368–1378
112. Cohen JH, Schoenbach VJ, Kaufman JS et al (2006) Racial differences in clinical progression among Medicare recipients after treatment for localized prostate cancer (United States). Cancer Causes Control 17(6):803–811
113. Godley PA, Schenck AP, Amamoo MA et al (2003) Racial differences in mortality among Medicare recipients after treatment for localized prostate cancer. J Natl Cancer Inst 95 (22):1702–1710
114. Thompson I, Tangen C, Tolcher A et al (2001) Association of African-American ethnic background with survival in men with metastatic prostate cancer. J Natl Cancer Inst 93 (3):219–225
115. Evans S, Metcalfe C, Ibrahim F et al (2008) Investigating Black-White differences in prostate cancer prognosis: a systematic review and meta-analysis. Int J Cancer 123(2):430–435
116. Sridhar G, Masho SW, Adera T et al (2010) Do African American men have lower survival from prostate cancer compared with White men? A meta-analysis. Am J Mens Health 4 (3):189–206
117. Morehead-Gee AJ, Pfalzer L, Levy E et al (2012) Racial disparities in physical and functional domains in women with breast cancer. Support Care Cancer 20(8):1839–1847
118. Kwan ML, Darbinian J, Schmitz KH et al (2010) Risk factors for lymphedema in a prospective breast cancer survivorship study: the Pathways Study. Arch Surg 145(11):1055–1063
119. Eversley R, Estrin D, Dibble S et al (2005) Post-treatment symptoms among ethnic minority breast cancer survivors. Oncol Nurs Forum 32(2):250–256
120. Paskett ED, Naughton MJ, McCoy TP et al (2007) The epidemiology of arm and hand swelling in premenopausal breast cancer survivors. Cancer Epidemiol Biomarkers Prev 16 (4):775–782
121. Sayko O, Pezzin LE, Yen TW et al (2013) Diagnosis and treatment of lymphedema after breast cancer: a population-based study. PM R 5(11):915–923
122. Matthews AK, Tejeda S, Johnson TP et al (2012) Correlates of quality of life among African American and white cancer survivors. Cancer Nurs 35(5):355–364
123. Bowen DJ, Alfano CM, McGregor BA et al (2007) Possible socioeconomic and ethnic disparities in quality of life in a cohort of breast cancer survivors. Breast Cancer Res Treat 106(1):85–95
124. Rao D, Debb S, Blitz D et al (2008) Racial/ethnic differences in the health-related quality of life of cancer patients. J Pain Symptom Manage 36(5):488–496
125. Ashing-Giwa KT, Padilla G, Tejero J et al (2004) Understanding the breast cancer experience of women: a qualitative study of African American, Asian American, Latina and Caucasian cancer survivors. Psychooncology 13(6):408–428

126. Gill KM, Mishel M, Belyea M et al (2004) Triggers of uncertainty about recurrence and long-term treatment side effects in older African American and Caucasian breast cancer survivors. Oncol Nurs Forum 31(3):633–639

127. Shelby RA, Lamdan RM, Siegel JE et al (2006) Standardized versus open-ended assessment of psychosocial and medical concerns among African American breast cancer patients. Psychooncology 15(5):382–397

128. Fu OS, Crew KD, Jacobson JS et al (2009) Ethnicity and persistent symptom burden in breast cancer survivors. J Cancer Surviv 3(4):241–250

129. Ashing-Giwa KT, Tejero JS, Kim J et al (2007) Examining predictive models of HRQOL in a population-based, multiethnic sample of women with breast carcinoma. Qual Life Res 16 (3):413–428

130. Eton DT, Lepore SJ (2002) Prostate cancer and health-related quality of life: a review of the literature. Psychooncology 11(4):307–326

131. Palmer NR, Tooze JA, Turner AR et al (2013) African American prostate cancer survivors' treatment decision-making and quality of life. Patient Educ Couns 90(1):61–68

132. Penedo FJ, Dahn JR, Shen BJ et al (2006) Ethnicity and determinants of quality of life after prostate cancer treatment. Urology 67(5):1022–1027

133. Yanez B, Thompson EH, Stanton AL (2011) Quality of life among Latina breast cancer patients: a systematic review of the literature. J Cancer Surviv 5(2):191–207

134. Sammarco A, Konecny LM (2010) Quality of life, social support, and uncertainty among Latina and Caucasian breast cancer survivors: a comparative study. Oncol Nurs Forum 37 (1):93–99

135. Paxton RJ, Phillips KL, Jones LA et al (2012) Associations among physical activity, body mass index, and health-related quality of life by race/ethnicity in a diverse sample of breast cancer survivors. Cancer 118(16):4024–4031

136. Sobel-Fox RM, McSorley AM, Roesch SC et al (2013) Assessment of daily and weekly fatigue among African American cancer survivors. J Psychosoc Oncol 31(4):413–429

137. Von Ah DM, Russell KM, Carpenter J et al (2012) Health-related quality of life of African american breast cancer survivors compared with healthy African American women. Cancer Nurs 35(5):337–346

138. Gewandter JS, Fan L, Magnuson A et al (2013) Falls and functional impairments in cancer survivors with chemotherapy induced peripheral neuropathy (CIPN): a University of Rochester CCOP study. Support Care Cancer 21(7):2059–2066

139. Hasan S, Dinh K, Lombardo F et al (2004) Doxorubicin cardiotoxicity in African Americans. J Natl Med Assoc 96(2):196–199

140. Drafts BC, Twomley KM, D'Agostino R Jr et al (2013) Low to moderate dose anthracycline-based chemotherapy is associated with early noninvasive imaging evidence of subclinical cardiovascular disease. JACC Cardiovasc Imaging 6(8):877–885

141. Meropol NJ, Schrag D, Smith TJ et al (2009) American Society of Clinical Oncology guidance statement: the cost of cancer care. J Clin Oncol 27(23):3868–3874

142. Meneses K, Azuero A, Hassey L et al (2012) Does economic burden influence quality of life in breast cancer survivors? Gynecol Oncol 124(3):437–443

143. Freedman RA, Virgo KS, He Y et al (2011) The association of race/ethnicity, insurance status, and socioeconomic factors with breast cancer care. Cancer 117(1):180–189

144. Bristow RE, Powell MA, Al-Hammadi N et al (2013) Disparities in ovarian cancer care quality and survival according to race and socioeconomic status. J Natl Cancer Inst 105 (11):823–832

145. Pollack CE, Bekelman JE, Liao KJ et al (2011) Hospital racial composition and the treatment of localized prostate cancer. Cancer 117(24):5569–5578

146. Bouknight RR, Bradley CJ, Luo Z (2006) Correlates of return to work for breast cancer survivors. J Clin Oncol 24(3):345–353

147. Mujahid MS, Janz NK, Hawley ST et al (2011) Racial/ethnic differences in job loss for women with breast cancer. J Cancer Surviv 5(1):102–111

148. McGee SA, Durham DD, Tse CK et al (2013) Determinants of breast cancer treatment delay differ for African American and White women. Cancer Epidemiol Biomarkers Prev 22 (7):1227–1238
149. Pisu M, Azuero A, Meneses K et al (2011) Out of pocket cost comparison between Caucasian and minority breast cancer survivors in the Breast Cancer Education Intervention (BCEI). Breast Cancer Res Treat 127(2):521–529
150. Jayadevappa R, Malkowicz SB, Chhatre S et al (2010) Racial and ethnic variation in health resource use and cost for prostate cancer. BJU Int 106(6):801–808
151. Keegan TH, Quach T, Shema S et al (2010) The influence of nativity and neighborhoods on breast cancer stage at diagnosis and survival among California Hispanic women. BMC Cancer 10:603
152. Conroy SM, Maskarinec G, Wilkens LR et al (2011) Obesity and breast cancer survival in ethnically diverse postmenopausal women: the Multiethnic Cohort Study. Breast Cancer Res Treat 129(2):565–574
153. Braithwaite D, Tammemagi CM, Moore DH et al (2009) Hypertension is an independent predictor of survival disparity between African-American and white breast cancer patients. Int J Cancer 124(5):1213–1219
154. Yancik R, Wesley MN, Ries LA et al (2001) Effect of age and comorbidity in postmenopausal breast cancer patients aged 55 years and older. JAMA 285(7):885–892
155. Wu AH, Gomez SL, Vigen C et al (2013) The California Breast Cancer Survivorship Consortium (CBCSC): prognostic factors associated with racial/ethnic differences in breast cancer survival. Cancer Causes Control 24(10):1821–1836
156. Kwan ML, John EM, Caan BJ et al (2014) Obesity and mortality after breast cancer by race/ethnicity: the California breast cancer survivorship consortium. Am J Epidemiol 179(1):95–111
157. Henderson SO, Coetzee GA, Ross RK et al (2000) Elevated mortality rates from circulatory disease in African American men and women of Los Angeles County, California—a possible genetic susceptibility? Am J Med Sci 320(1):18–23
158. Henderson SO, Bretsky P, Henderson BE et al (2001) Risk factors for cardiovascular and cerebrovascular death among African Americans and Hispanics in Los Angeles, California. Acad Emerg Med 8(12):1163–1172
159. Henderson SO, Haiman CA, Wilkens LR et al (2007) Established risk factors account for most of the racial differences in cardiovascular disease mortality. PLoS One 2(4):e377
160. Luke DA (2005) Getting the big picture in community science: methods that capture context. Am J Community Psychol 35(3–4):185–200
161. DeNavas C, Proctor B, Smith JC (2011) Income, poverty, and health insurance coverage in the United States. In: Department USC (ed) U.S. Government Printing Office, Washington, DC, pp 60–239
162. Morland K, Wing S, Diez Roux A (2002) The contextual effect of the local food environment on residents' diets: the atherosclerosis risk in communities study. Am J Public Health 92 (11):1761–1767
163. Block JP, Scribner RA, DeSalvo KB (2004) Fast food, race/ethnicity, and income: a geographic analysis. Am J Prev Med 27(3):211–217
164. Lovasi GS, Hutson MA, Guerra M et al (2009) Built environments and obesity in disadvantaged populations. Epidemiol Rev 31:7–20
165. Williams CD, Stechuchak KM, Zullig LL et al (2013) Influence of comorbidity on racial differences in receipt of surgery among US veterans with early-stage non-small-cell lung cancer. J Clin Oncol 31(4):475–481

Chapter 5
The Biology of Aging: Role in Cancer, Metabolic Dysfunction, and Health Disparities

Nathan K. LeBrasseur, Derek M. Huffman, and Gerald V. Denis

Abstract Aging is the primary cause of the majority of chronic diseases and disabling conditions. However, some individuals depart from the recognized patterns. Certain "at-risk" or "frail" individuals demonstrate premature aging, with increased cellular senescence, inflammation, early onset of diabetes, cardiovascular disease and cancer, and reductions in the ability to perform activities of daily living; whereas other "protected" or "fit" individuals appear to undergo a protracted period of health despite increasing years, remaining physically active senior citizens without chronic pain, disability, or frailty. Significant effort is being expended to understand this spectrum of aging phenotypes, with the goal of identifying the most important interventions or preventive steps that will preserve a "healthy aging" that maximizes lifespan without pain and chronic problems. We discuss health disparities that influence unhealthy aging, with a focus on interacting mechanisms in cancer, inflammation, and obesity.

Keywords Aging • Disparities • Cancer • Energy balance • Inflammation • Insulin resistance • Senescence

N.K. LeBrasseur
Physical Medicine and Rehabilitation, Physiology and Biomedical Engineering, Robert and Arlene Kogod Center on Aging, Mayo Clinic, 200 First Street SW, Rochester, MN 55905, USA
e-mail: lebrasseur.nathan@mayo.edu

D.M. Huffman
Institute for Aging Research, Albert Einstein College of Medicine, 1300 Morris Park Avenue, Forchheimer Building, Room 236, Bronx, NY 10461, USA
e-mail: derek.huffman@einstein.yu.edu

G.V. Denis (✉)
Cancer Research Center and Department of Pharmacology and Experimental Therapeutics, Boston University School of Medicine, 72 East Concord Street, K520, Boston, MA 02118, USA
e-mail: gdenis@bu.edu

D.J. Bowen et al. (eds.), *Impact of Energy Balance on Cancer Disparities*,
Energy Balance and Cancer 9, DOI 10.1007/978-3-319-06103-0_5,
© Springer International Publishing Switzerland 2014

The Scope of Unhealthy Aging as a Public Health Issue

> When sapless age and weak unable limbs
> Should bring thy father to his drooping chair. *William Shakespeare. Henry VI, Part I*

The depredations and pains of aging have been apparent to all societies since humans developed agriculture. In the shift from hunter-gatherer cultures to agrarian, pastoral cultures, humans reduced their risk of early mortality from causes related to war, wounding during a nomadic hunt, and starvation or exposure to harsh weather. These developments allowed humans to live long enough to manifest the familiar health consequences of aging, including frailty, osteoporosis, dementia, cancer, and a variety of other chronic diseases. Frailty in the elderly has been defined as the presence of three or more of the following: reduced walking speed, self-reported exhaustion, poor grip strength, recent weight loss, and low levels of physical activity [66]. However, even in ancient societies, there are recorded differences in the ways in which people experienced these complications of old age, with some men and women avoiding long, gradual chronic declines in mobility and strength, maintaining active and healthy lives until just before a natural death. Attention has long centered on the question of how to "live well" and to preserve faculties for as long as possible. For example, much folk wisdom has accumulated about how to best preserve cognitive function during aging, but trustworthy recommendations and mechanism remain obscure. Without such mechanism, the development of rational therapy for cognitive decline will be difficult or impossible. Recent commentary has drawn a distinction between "lifespan" and "health span." In this chapter, we will explore new areas of research that seek to define the factors that preserve health span, and discuss preventive and treatment strategies for aging to maximize health in diverse activities of daily living among the elderly. After delineating the scope of the problem of unhealthy aging, we will discuss six topic areas: cancer as a disease of aging; age-related changes in body composition; inflammatory mechanisms and aging; obesity, inflammation, and cancer in unhealthy aging; social determinants of healthy and unhealthy aging; and therapeutic opportunities including several behavioral strategies.

The medical significance and urgency of this research is apparent when we consider the number of older Americans and their growing representation in the population. The US Centers for Disease Control, in its report The State of Aging and Health in America 2013, noted that the number of Americans aged 65 years or older will double during the next 25 years to about 72 million individuals. By 2030, older adults will account for 20 % of the US population [34]. Between 2000 and 2011, many Southern and Western US states experienced a significant increase in their population aged 65 years and older (Fig. 5.1). Among the Western states, Nevada experienced the largest percentage increase (53.1 %), followed by Arizona (37.3 %), Idaho (37.28 %), Colorado (37.23 %), Utah (35.28 %), New Mexico (32.61 %), Texas (30.21 %), and Washington (30.15 %). Nationally, most of these older Americans (81 %) live in cities, a geographic distribution that has implications for disparities in physical activity and interactions with the built environment.

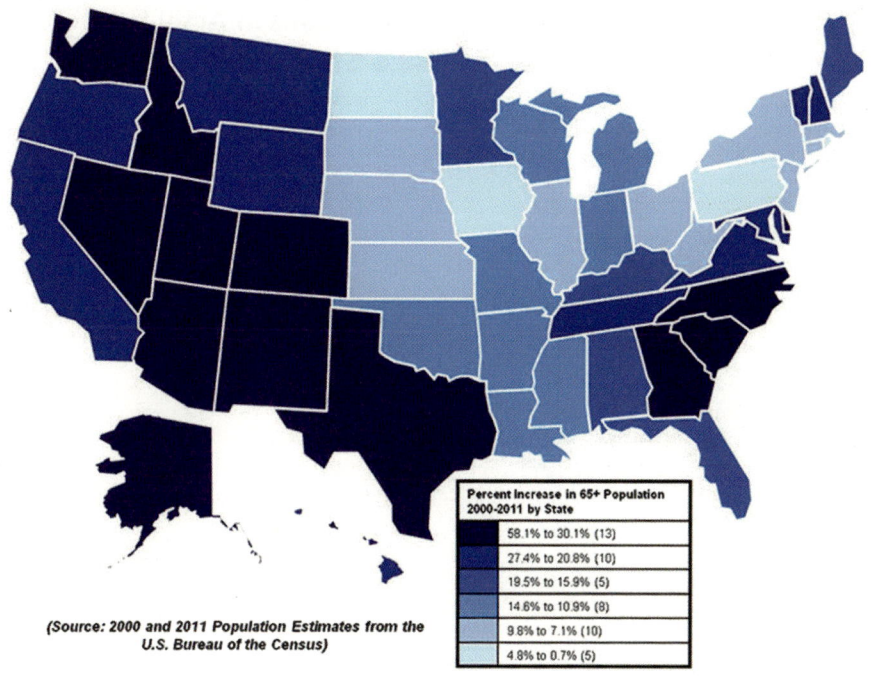

Fig. 5.1 Percent increase in number of Americans aged 65 years from 2000 to 2011. Data were obtained from: http://www.aoa.gov/Aging_Statistics/Profile/2012/8.aspx

Multiple chronic diseases are associated with aging, and the health care costs arising from this increased number of older adults will have a significant economic impact. Recent data from the Administration on Aging of the US Department of Health and Human Services has been used to compile a portrait of older Americans (aged 65 years and older) and issues of quality of life and health that affect them (Fig. 5.2). Disability of some kind was reported by 35 % of males and 38 % of females in 2011. Respondents that were institutionalized were excluded from the sample.

The historical lack of comprehensive health care for aging Americans and their expected chronic morbidities greatly complicates the economics and public policy issues. Without improvements in preventive medicine, such as reductions in obesity, smoking, alcohol abuse, and physical inactivity, it has been estimated that the major chronic diseases of "unhealthy aging" will accrue a cumulative economic impact of $1.6 trillion between 2007 and 2023 [47]. Multiple chronic conditions in older Americans include arthritis, asthma, chronic respiratory conditions, diabetes, heart disease, human immunodeficiency virus infection and hypertension [178], as well as some kinds of cancer. The prevalence of these conditions as comorbidities also increases with age [182, 191]. Significantly, some combinations of conditions, or clusters, of chronic conditions have synergistic interactions, but others do not

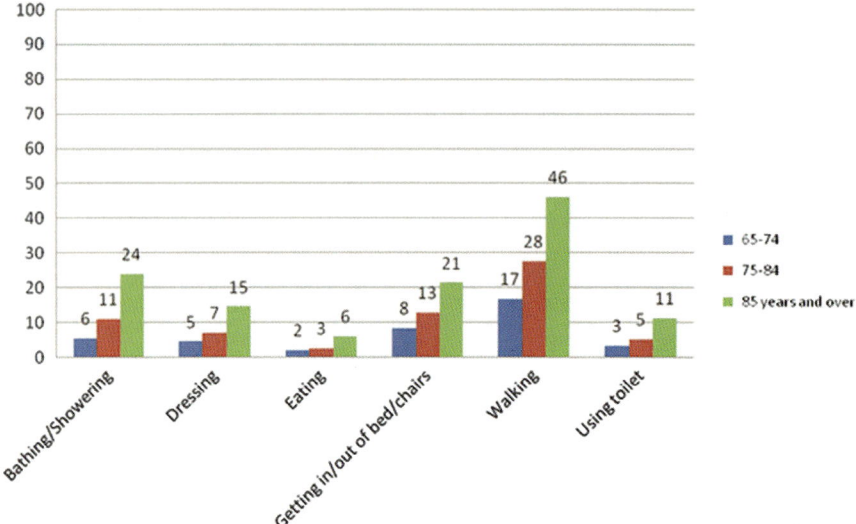

Fig. 5.2 Percent of Americans aged 65 years and older with limitations in activities of daily living, by age group in 2010. Data were obtained from: http://www.aoa.gov/Aging_Statistics/Profile/2012/16.aspx

[177]. It is also troubling that combinations of such chronic comorbidities are also characteristic of, and exacerbated by, unhealthy aging and frailty [178]. Consideration of underlying mechanism and the most effective intervention and prevention strategies is urgently required during this unfortunate era of declining public health expenditure and increasing disparities in health care delivery, because failures of political will and inadequate fiscal planning for anticipated increases in multiple chronic diseases in the aging US population will guarantee increased suffering and avoidable early mortality among older Americans. Discredited economic "austerity" policies in European, Latin American, and Asian economies that drastically curtail spending on health care for the elderly ensure that this pain will be global. No political excuse for inaction, or worse, the slashing of public health care spending, is acceptable when we have adequate scientific and medical knowledge to create cost-effective interventions now that will greatly improve this dire situation.

Cancer as a Disease of Aging

Molecular studies have provided a convincing link between the normal production of oxidative radicals (superoxide, hydroxyl radical and hydrogen peroxide), DNA damage in somatic cells, and gradually accumulating error in the genomic DNA of those cells that is improperly repaired. These accumulated errors have long been thought to be a primary cause of the molecular and cellular phenotypes we associate with unhealthy aging, including mutations linked to cancer [3]. These results have supported a rationale to recommend the consumption of dietary antioxidants to prevent aging and cancer. It is problematic that the postulated mechanisms are not sufficient to explain diverse cancers of childhood, which are in any case beyond the scope of this chapter.

Genetic and genomic instability is a well-recognized feature of cancer, and chromosomal abnormalities are thought to accumulate when mitotic checkpoints, as well as the machinery that enable proper chromosome segregation at mitosis, begin to fail. The resultant aneuploidy has been linked to aging in the case of increased risk of trisomy associated with female reproductive aging [145]. Instability in chromosome integrity and number has been suggested to be a general characteristic of aging [11], which may also help explain the rise in cancers with age in humans as mitotic regulation declines.

Additional research has developed links between mechanisms of senescence and aging. There is now a convincing body of evidence that cellular senescence is a fundamental mechanism of aging (reviewed in [148, 175]). Senescent cells have lost the ability to divide in response to stimuli that increase the risk of malignant transformation [117]. Senescence is a process that was selected through evolution to assure early life fitness by protection from cancer, but has unselected and undesirable consequences later in life. This evolutionary theme is referred to as *antagonistic pleiotropy*.

Senescence is sometimes considered to be a somatic defense mechanism against the oncogenic risks of mitosis [38]. Once a somatic cell has exited the cell cycle into a permanent post-mitotic state, the risks of genetic errors and mutations through DNA replication are ablated. Failures of function among cell cycle checkpoint proteins, such as p53, which has been famously termed the principal "gatekeeper" of genomic integrity [125], have implications for apoptosis and senescence. Deletion or functional inactivation of p53 and other tumor suppressor proteins is a hallmark of many cancers. The p53 protein has been shown to interact with a number of important transcription factors and co-regulators, including BRD7 [51], which possesses a bromodomain motif that recognizes motifs possessing acetylated lysine residues in proteins. The BRD7 protein also interacts with other tumor suppressor proteins such as BRCA1 [76]. The BRD family of proteins, a newly described group of chromatin-binding transcriptional co-regulators, has raised the profile of epigenetic mechanisms in the control of cancer [43, 17], senescence, apoptosis, and aging.

In addition to mitogenic signals, age-related DNA damage and mutations, telomere shortening, protein aggregation, and increased concentrations of reactive oxygen species also promote senescence [98, 116, 167]. As a result, the abundance of senescent cells increases in multiple tissues with advancing age [78, 154, 175, 183]. These cells secrete numerous biologically active molecules that are collectively referred to as the *senescence-associated secretory phenotype*, or SASP. Cytokines are an abundant component of the SASP and implicate senescent cells in the etiology of age-related inflammation. Cytokines and other components of the SASP, including growth factors and proteinases, disturb tissue architecture and perturb the functionality of neighboring cells [33]. Ultimately, these events culminate in tissue dysfunction and age-related pathologies.

Recent data suggest that elimination of senescent cells from a mouse model of accelerated aging delays the onset of multiple age-related phenotypes, including cataracts, muscle loss, lordokyphosis, and reduced physical performance [10]. The extent to which cellular senescence and the SASP underlie cancer, diabetes, atherosclerosis, frailty, and other age-related conditions in the context of conventional aging is an important area for investigation. It is tempting to speculate that interventions to prevent the accumulation of senescent cells or suppress the SASP would extend health span and compress late-life morbidity, and that interventions to augment the clearance of senescent cells would retard the progression of age-related conditions and associated disabilities. Disparities related to senescence-associated complications of aging have not been extensively studied. Potentially, lifestyle choices related to nutrition, physical activity, tobacco, and alcohol could affect the accumulation and clearance of senescent cells.

Age-Related Changes in Body Composition

In humans, aging is characterized by increased adiposity, which typically develops between the third and seventh decades of life. This phenotype is further exaggerated by the age-related decline in subcutaneous adipose tissue, leading to a redistribution of fat to the intra-abdominal region and ectopically in tissues such as skeletal muscle, bone marrow, and liver [61]. These unfavorable changes in body fat distribution, and particularly visceral fat accretion, have been found to predict more strongly all-cause [22, 94, 159] and disease-specific mortality risk [108] than measures of general obesity, such as body mass index (BMI). Indeed, studies in humans utilizing either imaging techniques (i.e., X-ray computed tomography and magnetic resonance imaging) or other anthropometric approaches to estimate intra-abdominal fat, such as waist-to-hip ratio, have found that estimates of abdominal obesity are a strong and independent predictor of several site-specific cancers, including colon [103, 140], esophageal [166], liver [153], and prostate [147]. For example, Moore et al. [140] found that waist circumference (a good measure of visceral fat accumulation) was a stronger predictor of colon cancer risk than BMI in both middle aged (30–54 years old) and older men and women (55–79 years old).

The underlying pathophysiology that links visceral adiposity to increased cancer risk and mortality is not entirely clear, but is likely to be multifactorial [90]. Visceral fat has been closely linked to the development of insulin resistance, dyslipidemias, and subsequent risk for type 2 diabetes, cardiovascular disease, and stroke [87]. Likewise, insulin resistance, which results in hyperinsulinemia, is also thought to promote cancer development due to the proliferative potential of insulin, and by suppressing circulating insulin-like growth factor (IGF) BP-1 and IGFBP-2 levels, leading to greater bioavailable circulating IGF-1. Visceral adipose tissue is also biologically distinct from other fat depots. In rodents, which demonstrate many of the metabolic manifestations of aging observed in humans, visceral fat has a more exaggerated gene expression profile and secretory capacity of cytokines and chemokines than other fat depots, including greater leptin, tumor necrosis factor-alpha (TNF-α), interleukin (IL)-6, IL-18, and plasminogen activation inhibitor-1 expression [9], which can be further provoked by nutrients [53–55].

Aged adipose tissue has also been shown to harbor a significant number of senescent cells [10], which can result in SASP as discussed above. Remarkably, selective clearance of p16Ink4a-expressing cells in mice, many of which were found in adipose tissue, delayed the onset of age-related pathologies [10]. Likewise, surgical removal of visceral fat in rodents has been shown to improve lifespan [143] and protect against the development of intestinal tumors [88]. Thus, given the close association between inflammation and diseases of aging (discussed in more detail below), the chronic, low-grade, pro-inflammatory state, which is associated with age-related visceral fat accretion and accumulation of senescent cells in fat, could be an important mechanism linking aging and obesity with aging and cancer risk.

Whereas the relationship between general and abdominal obesity and disease during middle and late-middle age has been clear, there has been some confusion regarding the role of excess adiposity in older adults with some studies showing a negative effect, no effect, or even a protective effect of fat mass. One important consideration when examining older adults is that many people at advanced ages are in fact abdominally obese, with greater ectopic fat stores, despite a seemingly normal BMI, a phenomena that can decrease the utility of BMI as a predictor of disease risk in older adults [165]. Furthermore, one must exercise caution when examining this relationship at advanced ages due to confounding, including the decline in fat mass that often coincides with late-life illness and disease. However, in weight-stable adults >75 years of age, higher BMI is predictive of greater mortality risk, and this relationship is stronger in males than females [165]. The nature and mechanisms of disparities in visceral fat deposition among older adults are insufficiently studied.

Sarcopenia, the age-related decline in muscle quantity and quality, represents a second major change in body composition commonly observed with aging [29]. The causes of sarcopenia are multifactorial, but the pro-inflammatory state associated with aging and obesity appears to contribute to the decline in skeletal muscle mass and function [29]. The loss of skeletal muscle mass with age has broad health implications, including a decline in physical function, onset of frailty, and increased disability among older adults [56]. Skeletal muscle is also a major site of glucose disposal in humans, and sarcopenia has been linked to insulin resistance [105], which can contribute to a more rapid onset of many other age-related diseases, including cancer. Furthermore, worsened insulin sensitivity in muscle is believed to contribute to the decline in skeletal muscle function, reduced protein synthesis rates, and accelerated skeletal muscle loss [120], placing insulin resistance in a vicious cycle of skeletal muscle decline and metabolic dysfunction.

Although it is not difficult to envision how age-related changes in skeletal muscle mass and function could contribute indirectly to cancer risk and other diseases, recent evidence has uncovered a possible direct role whereby skeletal muscle could modulate cancer risk. Specifically, similar to adipose tissue, skeletal muscle is now recognized as an endocrine organ, due to the secretion of numerous myokines, such as brain-derived neurotrophic factor, IL-6, and irisin, among others, in response to contraction [157, 179]. Indeed, Aoi et al. recently identified a novel myokine, secreted protein acidic and rich in cysteine (SPARC), which is secreted from skeletal muscle in response to exercise, has pro-apoptotic effects, and is necessary for the inhibition of intestinal tumorigenesis with exercise in mice [6]. Thus, sarcopenia along with reduced physical activity with age could lead to declines in myokines, such as SPARC [146], and embody an important link between skeletal muscle, aging, and cancer. Thus, visceral fat accrual and loss of skeletal muscle mass with aging represent two major phenotypic changes in humans that can predispose older adults to carcinogenesis. The mechanism (s) linking these shifts in body composition to cancer risk are not entirely known, but are likely multifaceted, encompassing both direct and indirect effects.

Inflammatory Mechanisms and Aging

It is widely appreciated that human aging is usually associated with an increase in the number and gravity of different chronic conditions that affect the geriatric individual. Similar patterns have been observed in rodent models of aging. Recent commentary has focused on prevention and treatment to reduce the number and severity of these conditions, thereby maximizing the length of time a healthy and active individual enjoys before death, and minimizing the end-stage conditions and interval that immediately precede natural death. Rapid, steep decline can be considered to be an aging ideal for humans. But which processes matter most? Which diseases are the most debilitating and what are the mechanisms that converge to create "comorbidities of aging"? What are the critical environmental and population factors that influence disparities among older adults? Do we have sufficient evidence to recommend priorities for geriatric patients that will have the greatest benefit to preserve daily functioning?

Preliminary and published evidence from our laboratories suggests that persistent, sterile, unresolved, chronic inflammation is one of the central causes of "unhealthy" aging that is associated with an increased number and severity of comorbidities, particularly cancer and type 2 diabetes. The immune system, in both its innate and adaptive arms, helps regulate all organ systems and controls responses to diverse exposures. We have discussed hypotheses that low-level, long-lasting inflammation may account for several comorbidities that affect apparently unrelated organ systems [43]. Low grade, sterile inflammation has been implicated in the specific features of frailty associated with aging in humans [58, 102, 123, 185]. Elevated levels in blood of pro-inflammatory cytokines, such as IL-6 and markers of monocyte activation, have been linked to frailty in geriatric patients [39, 99]. Furthermore, the premature aging phenotypes associated with persistent Human Immunodeficiency Virus (HIV) infection have been linked to chronic inflammation [8, 155, 192, 193]. As discussed above, the role of chronic, sterile inflammation in aging phenotypes has been reviewed recently and associated with senescence [148, 175].

Several animal models have also identified systemic inflammation as a culprit in pathologies associated with aging. For example, very recent work [28] has shown that inflammasome hyperactivity causes aging mice to develop systemic inflammation (neutrophilia, elevated leukocyte counts, splenomegaly, and leukocytosis of organs that are dependent on IL-1β and IL-18 for different stages of pathology). Additional animal models to probe these mechanisms are urgently needed.

Most seriously from the point of view of mortality risk, an aged immune system is less able to develop a robust response of adaptive immune to influenza immunization. During 1976–2007, estimates of annual influenza-associated deaths from respiratory and circulatory causes (including pneumonia and influenza causes) ranged from 3,349 in 1986–1987 to 48,614 in 2003–2004; approximately 90 % of influenza-associated deaths occur among adults aged ≥65 years [176]. Furthermore, seasonal influenza is implicated in excess mortality from cardiovascular diseases, stroke, diabetes, and pneumonia in older adults [134].

Diminished T cell function may offer a critical mechanism that links increased vulnerability to virus infection and decreased immune surveillance of cryptic cancers in geriatric patients. Specifically, thymic involution with age, dramatic decreases in naïve and memory T cell populations, and reduced numbers of $CD8^+$ T cells that infiltrate tumors, support a tolerogenic environment in which tumors grow well [128]. An increased ratio of anergic or functionally defective T cells to competent T cells in peripheral tissues could cause the tasks of tumor surveillance to fail in geriatric patients. Insufficient tumor surveillance in the aging individual opens the door to tumor growth and progression; thus cancers are indeed properly considered a disease of aging. In support of this idea, research in mouse models has shown that failure of immune surveillance of premalignant, senescent hepatocytes, which are normally cleared by $CD4^+$ T cells, leads to development of hepatocellular carcinomas [104]. Additional evidence points to declines in anti-inflammatory cytokines (IL-10 and transforming growth factor (TGF)-β) and adipokines (particularly adiponectin) associated with aging phenotypes.

The greatly increased prevalence of obesity, which in its metabolic complications particularly affects older Americans, represents a grave new threat to the health of hundreds of millions of people worldwide, although the impact will be felt greatest in the USA [60]. Among adults, overweight (BMI ≥ 25.0–29.9), obesity (BMI ≥ 30.0–39.9), and morbid obesity (BMI ≥ 40.0) manifest progressively serious complications, including insulin resistance, hypertension, cardiovascular disease, and type 2 diabetes [72, 180]. The interaction among obesity, chronic inflammation in insulin resistance, and inflammation-associated complications of aging define an important health disparity among older adults, as discussed below.

Many chronic inflammatory diseases of aging appear to be linked through abnormal metabolism. Some of the most significant examples are metabolic syndrome and type 2 diabetes, which are associated with glucose intolerance and insulin-resistant obesity that greatly increase in prevalence among older adults. Insulin resistance, metabolic syndrome, and type 2 diabetes are properly considered as diseases of unresolved chronic inflammation [149]. Obese, metabolically abnormal patients exhibit serum profiles characterized by elevated concentrations of pro-inflammatory cytokines (e.g., IL-1β, TNF-α, IL-6, IL-12, IL-18) [115], as well as acute phase proteins, such as C-reactive protein (CRP) [101], and decreased concentrations of anti-inflammatory, cardioprotective adipokines (e.g., high molecular weight adiponectin). This kind of chronic inflammation is both systemic [14, 27] and local, in white adipose tissue. Insulin-resistant adipose tissue, particularly in the visceral fat depots discussed above, is infiltrated by pro-inflammatory CD68$^+$ macrophages that produce TNF-α, which has long been known to induce insulin resistance in adipocytes directly [84]. These cellular and molecular features have been observed both in humans and in rodent models of obesity [188, 194], and affect T cell function [111].

Furthermore, type 2 diabetes in adult humans is associated with an increased ratio of pro- to anti-inflammatory T cells in peripheral blood [96]. B cells from adult type 2 diabetes patients, unlike B cells from nondiabetic donor controls, fail to secrete the generally anti-inflammatory cytokine IL-10 in response to stimulation through various Toll-Like Receptors [95]. Finally, new results have shown that B cells likely play a pathogenic role by promoting polarization of T cells towards increased production of pro-inflammatory cytokines [42]. These factors can improve over time after intentional weight loss and treatment with anti-inflammatory drugs [149]. We have discussed these mechanisms and interrelationships in detail [45, 149]. However, these specific immune mechanisms have never been studied in geriatric adults, geriatric obese adults, or geriatric type 2 diabetic adults. It is reasonable to speculate that failures of T cell function and tumor surveillance, B cell function, or failures of anti-inflammatory homeostasis in the immune system combine in unfortunate ways with declining metabolic health, increasing obesity, declining physical activity, and increasing chronic inflammation, to promote many comorbidities of aging such as insulin resistance, obesity-associated cancers, and frailty.

Finally, identification of at-risk older adults, particularly vulnerable individuals who experience stigma or race-based, class-based disparities, is important to public health goals. We propose some possible relationships among inflammatory, metabolic, and stress status and the comorbidities of obesity that may affect health disparities (Fig. 5.3). It will be essential to identify and to quantify the most important relationships in order to evaluate the success of short-term and long-term interventions in clinical trials. These interdisciplinary issues are not well studied, nor are optimal interventions defined that would reverse the hypothesized clustering effect of inflammation and other stressors on the underlying chronic conditions.

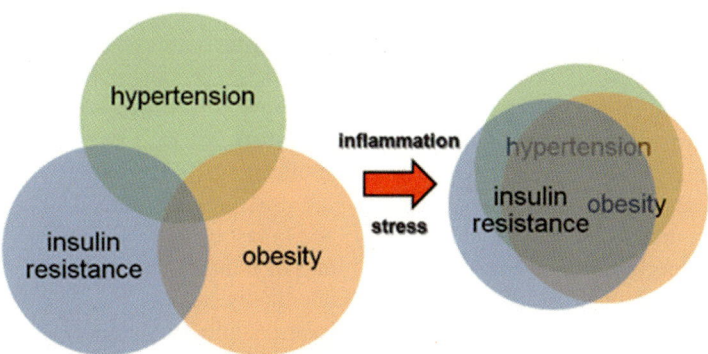

Fig. 5.3 Model for interactions among critical variables in aging: inflammatory, metabolic, and stress status. This scheme suggests testable hypotheses for how comorbidities might arise under conditions of "unhealthy aging" that compromise health span in older adults. Many conditions of aging do not occur in isolation; thus, single disease-focused therapies likely have limited utility to treat patients with complex and interacting diseases. The interacting problems can only be conceptualized properly from an interdisciplinary perspective

Obesity, Inflammation, and Cancer in Unhealthy Aging

As the obesity epidemic worsens, the incidence of cardiovascular disease, type 2 diabetes, and "obesity-associated cancer" is expected to increase. American Cancer Society epidemiologists first brought the problem of obesity-associated cancer to worldwide attention 10 years ago [30, 31]. Elevated levels of leptin, insulin, and IGF-1 found in obese, insulin-resistant patients have been linked to obesity-associated cancer [67, 70]. However, recent evidence suggests that, for certain obesity-associated cancers, metabolic status is important for risk, whereas insulin levels are not [100]. Abundant evidence also links unresolved, chronic inflammation to cancer [107], and more aggressive properties of tumor cells in insulin-resistant obesity [69, 91, 93]. Pro-inflammatory cytokines that are frequently elevated in "metabolically unhealthy" obesity, such as IL-6, are implicated in dangerous shifts in the properties of breast cancer cells [162, 172, 186]. Unresolved inflammation is independently associated with colon cancer in patients with inflammatory bowel diseases [114], which supports the idea that the inflammatory nature of insulin-resistant obesity is a critical mechanism that promotes certain obesity-associated cancers. Obesity-associated inflammation has been implicated in adverse outcomes in breast cancer in postmenopausal women, colon cancer in both men and women, and several other obesity-associated cancers. However, we do not yet know if specific cytokine and adipokine profiles are associated with increased risk for some obesity-associated cancers, but not others. Identification of these profiles is essential before targeted chemopreventive agents can be developed to supersede the broad-spectrum anti-inflammatory agents currently available.

Significantly, it appears that not all obesity conveys the same disease risks: recent data show that immunometabolic status stratifies cardiovascular disease risk in obesity [14, 74, 189]. For most subjects, as obesity increases, metabolic health declines. We have shown in animal models that inflammation [187] and the presence in adipose tissue of "crown-like structures" of CD68+, pro-inflammatory macrophages [36] are associated with metabolic status [44]. Similarly, in "metabolically abnormal" obese humans, crown-like structures in inflamed adipose tissue are associated with cardiovascular risk [7] and breast cancer risk [142].

Interestingly, there is an informative group of humans that appear to bend the rules of metabolism in obesity [46]. "Metabolically healthy but obese" persons [164], who account for about one quarter of obese adults, are overweight/obese but show relatively normal blood parameters [171] and are protected from cardiometabolic risk [20, 21]. They also lack crown-like structures in white adipose tissue [112]. The mechanistic relationship between cardiovascular risk and elevated insulin in obesity has been difficult to study because many factors, including inflammation, tend to co-vary in human subjects. The "metabolically healthy" obese phenotype may be useful to deconvolute some of the relevant variables. Cardiovascular risk in "metabolically healthy" obese women has been shown to be intermediate between lean and healthy women, and "metabolically abnormal" obese women [130]. Critically, "metabolically healthy" obese adults have diminished risks for cancer mortality compared to "metabolically abnormal" obese adults [32, 196]. Specific adipocyte [150, 151], adipose tissue distribution [184], and immunological features, particularly a reduced inflammatory profile [16, 44, 106, 113] and elevated adiponectin [1], separate "metabolically healthy" obese individuals from the obese and insulin-resistant general population, but the molecular mechanisms that uncouple obesity from cancer risks in this context are not well understood. We suspect that this cancer protection is primarily attributable to the attenuated inflammatory profile [16, 17, 44, 45].

The details of the cytokine signal transduction pathways critical for obesity-associated cancer progression are not well established for *any* of the obesity-associated cancers, nor is it understood why some obese adults become inflamed while others do not. Furthermore, it is not known whether the "metabolically healthy" obese state is stable among older obese adults, or whether there is a genetic or epigenetic component to the phenotype that influences cancer risk. Neither is it known whether there are unique qualities to unresolved, chronic inflammation among older obese adults that differ from younger obese adults, or whether the aforementioned multiple, chronic, inflammatory diseases that often accompany unhealthy aging influence risk for obesity-associated cancers in older adults. In many cases, the necessary observational work is incomplete or sample sizes are too small to enable statistically robust conclusions to be drawn about the role of such comorbidities in cancer risk among older adults. This area of investigation is urgently in need of further effort and better mechanistic understanding, in view of the prevalence of unhealthy aging and obesity among older humans and the potential role of health disparities in exacerbating these risks.

Social Determinants of Healthy and Unhealthy Aging

We don't stop playing because we grow old; we grow old because we stop playing. *George Bernard Shaw*

Groundbreaking studies in cultural anthropology have compared the social context in which senility and mental confusion arise in "Western" cultures and in India, suggesting that the role of supportive family structures must be considered. Elderly people in traditional Indian joint family arrangements were well cared for and asserted to be protected from development of dementia [37]. Forms of dementia in elderly humans can arise through an accumulation of neurofibrillary tangles, as in Alzheimer's disease, or an accumulation of small infarcts in vascular diseases of the brain, as in vascular dementia; but in neither case are these diseases thought of as normal processes of aging. The mechanisms that stratify risks for dementia and other chronic conditions of aging are just beginning to be studied in a comprehensive fashion. These anthropological observations suggest that family structures, measures of social interconnectedness, locus of control, and socioeconomic status play important roles as mediators of biological factors like stress and inflammation, as proposed in Fig. 5.3. Poverty and economic inequality in particular are widely acknowledged to critically affect mortality risks [50, 131].

There is now sufficient evidence to develop a hypothesis that social and environmental factors affect childhood development in ways that influence chronic disease risks, including cancer risks, when the affected children reach geriatric ages. For example, childhood obesity is a major public health problem in the USA that disproportionately affects African American and Hispanic children and children of low socioeconomic status [18, 35]. Cohorts of these racial/ethnic groups who live in environments characterized by high social stress (family and neighborhood) are associated with increased risk of obesity [23, 68, 73]. These stresses also strongly associate with cardiovascular disease and type 2 diabetes, mediated in part by severe or chronic stress on immune processes, which in turn influence physiological stress systems [52]. These stresses in childhood appear to affect health for many years into mature adulthood and beyond [133, 161, 163]. In particular, there is evidence that chronic comorbidities of aging are exacerbated among adults who experienced trauma as children [4, 49], including cancer risks [109]. Psychological trauma therefore also defines an important form of disparity that influences health risks among older adults.

These data considered together implicate social disruption, psychological stress, economic inequality, and early trauma as predisposing factors for chronic diseases that arise as comorbidities in geriatric patients. The rising incidence of obesity and obesity-associated cancer is therefore likely also be worsened by contemporary inequality and sharply diverging socioeconomic status among the wealthy and the poor in modern industrial cultures worldwide. Thus, it is not unreasonable to hypothesize that future cancer disparities will have some of their roots in current socioeconomic disparities. Current political movements and budgetary policies that will inevitably increase inequality and exacerbate social stress must therefore be vigorously opposed on public health grounds.

Therapeutic Opportunities

Aging is the major underlying risk factor for cancer risk in humans [127]. Identifying effective behavioral and pharmacologic strategies to help prevent cancer incidence and mortality requires a greater understanding of the mechanisms at the cancer-aging interface. Several of the known purported risk factors linking aging to carcinogenesis have been discussed in this chapter. The lifetime accumulation of DNA damage and genome instability induced by oxidative stress and replicative errors, as well as epigenomic changes as a result of environmental cues, have been linked to cancer [181]. Likewise, aging is associated with unfavorable changes in body composition, including increased visceral adiposity [61] and a decline in skeletal muscle mass [56]. These changes are associated with age-related metabolic dysfunction and a pro-inflammatory state [97], both of which can promote tumorigenesis. Aging is also characterized by a deterioration in innate and adaptive immunity, which can increase the likelihood of immune system evasion by a tumor cell [89, 144], as discussed above, and a decline in processes involved in cellular homeostasis [141, 195]. Thus, interventions are needed that can effectively target these processes in middle aged and older adults to break the aging-cancer link.

Behavioral Strategies

Caloric Restriction

It was first shown more than 100 years ago that a reduction in food intake could inhibit tumor formation in rats [173, 174]. Nearly a century later, caloric restriction (CR) remains the most robust intervention for preventing or delaying the onset of disease and extending lifespan [40]. CR results in a multitude of adaptive changes in humans and rodents, including a reduction in body weight, fat, and lean mass, and declines in insulin (and glucose), cytokines, chemokines, thyroid hormone, reproductive hormones, and circulating growth hormone/IGF-1 [13]. CR has also been shown to induce substantial shifts in the transcriptome, metabolome, and proteome, as well as increases in stress hormones such as cortisol (corticosterone in rodents). Collectively, these changes are believed to enhance stress resistance, minimize macromolecular and organelle damage, and improve cellular homeostasis.

Given the near universality of this intervention to improve health status and extend lifespan in animal models, several efforts in recent years have been undertaken to understand the feasibility and potential benefits of CR in humans. These studies have found that ~20–25 % CR is feasible and safe in adults, and reproduces many of the biological effects observed in rodents, including a reduction in body size, blood pressure, and body temperature, as well as several systemic markers including lower thyroid hormone, fasting insulin, low-density lipoprotein (LDL) cholesterol, and core body temperature [62, 64, 77, 122]. However, long-term CR surprisingly did not reduce circulating IGF-1 concentrations unless protein intake was also reduced, highlighting an important difference between humans and rodents regarding this important cancer risk factor [63]. Due to the limitations in performing long-term CR studies in humans to evaluate cancer outcomes, there is no direct evidence as yet that this intervention can effectively reduce cancer risk in humans. However, inference from the overwhelming evidence linking obesity to cancer risk [30] suggests that CR has promise as a therapeutic strategy for cancer prevention and control in humans.

Exercise

The beneficial effects of physical activity on health and function, including lower risk for developing cardiovascular disease, stroke, and cognitive decline, are well documented [59, 65, 129, 156]. Likewise, it has been shown that exercise can modestly improve mean lifespan in rodents [80–83] and life expectancy in humans [121], but unlike CR, does not extend maximum lifespan. Furthermore, compared to CR, the link between exercise and cancer has been less consistent [19]. This result is due in part to the variable nature of exercise interventions employed in preclinical and human studies, including differing modalities (i.e., resistance training, swimming, running), with varying degrees of intensity and duration. Second, most studies have tested the effect of exercise using a mixture of aggressive tumor-bearing mice and cell types (in rodents), as well as a limited number of studies on a range of advanced-stage cancers in humans, from glioma to breast cancer, with mixed results [19]. Furthermore, whereas some trials employ only an exercise intervention, others incorporate a combination strategy of exercise and diet to induce weight loss. Finally, most exercise studies are limited to testing its potential to prevent cancer recurrence, rather than its ability to prevent cancer development. Thus, data on the potential long-term protective effect of exercise in rodents and humans is not totally clear, but there is a growing body of evidence that exercise in humans can lead to improvements in factors associated with cancer risk, such as lowering circulating IGF-1, insulin, and cytokines [137, 139]. Thus, this observation coupled with the existing epidemiologic evidence showing that physical activity is typically associated with lower risk of breast, colon, and prostate cancer [26] suggests that efforts to pursue the optimal exercise prescription for each scenario (site and stage) should continue to be pursued.

Weight Loss

Given the known link between visceral obesity and cancer risk, weight (and abdominal fat) reduction, which is common side effect of CR, may be necessary in order to maximize the beneficial effects of exercise on cancer. Indeed, visceral fat accrual is a common hallmark of aging [61], and cancer occurrence also increases dramatically with age [127]. Thus, it is possible that reducing the amount of total and visceral fat with age via diet and/or exercise may be an important cancer prevention strategy as well as an adjuvant therapy for improving outcomes following a cancer diagnosis. Intervention studies in humans employing a regimen of 20 % CR versus an exercise regimen designed to induce similar reduction in body weight for 1 year reported similar improvements in cardiovascular disease risk factors [64] and reductions in oxidative damage to lymphocytes [79]. Along these lines, results from a Phase II study aimed at inducing >10 % weight loss by diet and exercise in obese postmenopausal women at risk for developing breast cancer were recently published [57]. In this study, the authors report that weight loss led to improvement in a multitude of endpoint markers, including several factors in breast tissue and serum predictive of lower breast cancer risk [57].

Dietary Strategies

The role of diet and dietary factors on cancer risk with aging has been an area of intense study for many years, and a detailed analysis of all the evidence is beyond the scope of this review. In general, diets high in fruit and vegetables, dietary fiber, and plant-based protein coupled with low intakes of saturated fats, red and processed meats, sugar-sweetened foods, and alcohol is the most grounded dietary prescription, based on the evidence, for reducing cancer risk or recurrence [110, 118]. Some evidence exists that specific micronutrient deficiencies with age can increase cancer risk, such as vitamin D, calcium [48], or folic acid [160]. However, over-supplementation of these factors can also lead to complications [132].

In addition, the use of multivitamins is not presently recommended as a strategy to prevent cancer due to a lack of evidence, and in some cases, harmful unintended consequences, such as evidence that long-term supplementation with beta carotene increased lung cancer risk in smokers [2]. Likewise, the issue can also be complicated, such as with omega-3 fatty acids, which have been tied to lower breast cancer risk [24], but increased risk for prostate cancer [25]. Green tea polyphenols, resveratrol, and soy isoflavones as a dietary supplement to reduce cancer risk also continue to be an active area of study with promising effects in the laboratory [85, 86, 136], but evidence that these factors are effective in humans is limiting for several reasons. Among these concerns, the challenge of performing long-term randomized trials for dietary factors that are well controlled and have cancer as an outcome is logistically and financially challenging. Second, it is difficult to control for confounding effects such as changes in physical activity or body weight with a dietary intervention. Third, it is often problematic to isolate the important effect(s) due to confounding, such as when one dietary component is removed (i.e., refined sugars) and replaced with another (i.e., fruits and vegetables) [158]. Finally, it is difficult to know the dose at which a bioactive component is most beneficial, a methodological concern that has plagued the resveratrol field [41], for example.

Pharmacologic Strategies

The modern era of aging research has ushered in a new genre of research focused on identifying pharmacologic interventions, also termed CR mimetics [15, 92], to treat some manifestations of aging and extend lifespan. Some compounds have produced mixed results, such as resveratrol, the polyphenol first reported to improve survival in mice fed a high-fat diet [5], an observation that could not be replicated on a low-fat diet [138]. Other agents with anti-inflammatory properties, including nordihydroguaiaretic acid and aspirin, led to a significant improvement in survival of male mice [170], but not female mice. Aspirin also reduced cancer incidence in Lynch syndrome patients [12] and extended lifespan in a mouse model of Lynch syndrome [135]. Furthermore, aspirin use has been linked to improved survival in colon cancer patients with *PI3KCA* mutations [152].

Interest is now focused on the potential of other drugs used to treat various age-related ailments, including statins, bisphosphonates, and metformin, as potential therapeutic agents for cancer treatment [71]. For example, the biguanide, metformin, has been shown to increase lifespan in yeast and mice [5], but not in rats [169] or *Drosophila* [168]. Interestingly, metformin, which is commonly prescribed to patients with type 2 diabetes and is well tolerated, has also emerged as an anticancer agent [126]. Specifically, metformin, which is believed to be an activator of $5'$-AMP-activated protein kinase (AMPK), has been associated with a ~30 % reduction in lifetime risk of cancer in type 2 diabetics [124]. Metformin is now under intense study in clinical trials, either alone or as an adjuvant therapy.

The most consistent and promising drug thus far, at least from the aging perspective, is the immunosuppressant, rapamycin, which among its many effects inhibits mammalian Target of Rapamycin (mTOR) and consistently extends lifespan in mice, regardless of whether it is started early or late in life [75, 138, 190]. However, rapamycin is unlikely to be used in humans due to numerous side effects, including hyperglycemia, dyslipidemia, immunosuppression, vasospasm, and renal failure [119]. However, the so-called rapalogs, which are under development to safely modulate mTOR activity, could represent an effective strategy in the future to treat diseases of aging, including cancer [119]. Therefore, while a healthy diet and exercise remain the cornerstone to an effective cancer prevention strategy, the prospect of one day developing drugs that can safely treat or even prevent diseases of aging, including cancer, has never been more realistic.

Acknowledgements The authors thank the National Institutes of Health for support with grants from the National Institute on Aging (AG041122—N.K.L. and AG037574—D.M.H.), the National Institute of Diabetes and Digestive and Kidney Diseases (DK090455—G.V.D.), and the National Cancer Institute (CA182898—G.V.D.); the Glenn Foundation for Medical Research (N.K.L.); the Prevent Cancer Foundation (D.M.H.); the Einstein Nathan Shock Center Healthy Aging Physiology Core (P30AG038072—D.M.H.); the Diabetes Research and Training Center (DK20541—D.M.H.) at the Albert Einstein College of Medicine; and the American Cancer Society (RSG0507201—G.V.D.). N.K.L. is a Principal Investigator at the Robert and Arlene Kogod Center on Aging at the Mayo Clinic. D.M.H. and G.V.D. are Chair and Chair-Elect, respectively, of the Obesity and Cancer Section of The Obesity Society. The authors report no conflicts of interest.

References

1. Aguilar-Salinas CA, Garcia EG, Robles L, Riaño D et al (2008) High adiponectin concentrations are associated with the metabolically healthy obese phenotype. J Clin Endocrinol Metab 93(10):4075–4079
2. Alpha-Tocopherol, Beta Carotene Cancer Prevention Study Group (1994) The effect of vitamin E and beta carotene on the incidence of lung cancer and other cancers in male smokers. N Engl J Med 330:1029–1035
3. Ames BN, Shigenaga MK, Hagen TM (1993) Oxidants, antioxidants, and the degenerative diseases of aging. Proc Natl Acad Sci U S A 90(17):7915–7922
4. Anda RF, Dong M, Brown DW, Felitti VJ, Giles WH, Perry GS et al (2009) The relationship of adverse childhood experiences to a history of premature death of family members. BMC Public Health 9:106. doi:10.1186/1471-2458-9-106
5. Anisimov VN, Berstein LM, Egormin PA, Piskunova TS, Popovich IG, Zabezhinski MA, Kovalenko IG, Poroshina TE, Semenchenko AV, Provinciali M, Re F, Franceschi C (2005) Effect of metformin on life span and on the development of spontaneous mammary tumors in HER-2/neu transgenic mice. Exp Gerontol 40:685–693
6. Aoi W, Naito Y, Takagi T, Tanimura Y, Takanami Y, Kawai Y, Sakuma K, Hang LP, Mizushima K, Hirai Y, Koyama R, Wada S, Higashi A, Kokura S, Ichikawa H, Yoshikawa T (2013) A novel myokine, secreted protein acidic and rich in cysteine (SPARC), suppresses colon tumorigenesis via regular exercise. Gut 62:882–889
7. Apovian CM, Bigornia S, Mott M, Meyers MR, Ulloor J, Gagua M, McDonnell M, Hess D, Joseph L, Gokce N (2008) Adipose macrophage infiltration is associated with insulin resistance and vascular endothelial dysfunction in obese subjects. Arterioscler Thromb Vasc Biol 28(9):1654–1659
8. Arnsten JH, Freeman R, Howard AA, Floris-Moore M, Lo Y, Klein RS (2007) Decreased bone mineral density and increased fracture risk in aging men with or at risk for HIV infection. AIDS 21:617–623
9. Atzmon G, Yang XM, Muzumdar R, Ma XH, Gabriely I, Barzilai N (2002) Differential gene expression between visceral and subcutaneous fat depots. Horm Metab Res 34:622–628
10. Baker DJ, Wijshake T, Tchkonia T, LeBrasseur NK, Childs BG, van de Sluis B, Kirkland JL, van Deursen JM (2011) Clearance of p16Ink4a-positive senescent cells delays ageing-associated disorders. Nature 479:232–236
11. Baker DJ, Dawlaty MM, Wijshake T, Jeganathan KB, Malureanu L, van Ree JH, Crespo-Diaz R, Reyes S, Seaburg L, Shapiro V et al (2013) Increased expression of BubR1 protects against aneuploidy and cancer and extends healthy lifespan. Nat Cell Biol 15:96–102
12. Barton MK (2012) Daily aspirin reduces colorectal cancer incidence in patients with Lynch syndrome. CA Cancer J Clin 62:143–144

13. Barzilai N, Huffman DM, Muzumdar RH, Bartke A (2012) The critical role of metabolic pathways in aging. Diabetes 61:1315–1322
14. Bastard JP, Maachi M, Lagathu C, Kim MJ, Caron M, Vidal H, Capeau J, Feve B (2006) Recent advances in the relationship between obesity, inflammation, and insulin resistance. Eur Cytokine Netw 17(1):4–12
15. Baur JA, Pearson KJ, Price NL, Jamieson HA, Lerin C, Kalra A, Prabhu VV, Allard JS, Lopez-Lluch G, Lewis K, Pistell PJ, Poosala S, Becker KG, Boss O, Gwinn D, Wang M, Ramaswamy S, Fishbein KW, Spencer RG, Lakatta EG, Le Couteur D, Shaw RJ, Navas P, Puigserver P, Ingram DK, de Cabo R, Sinclair DA (2006) Resveratrol improves health and survival of mice on a high-calorie diet. Nature 444:337–342
16. Belkina AC, Denis GV (2010) Obesity genes and insulin resistance. Curr Opin Endocrinol Diabetes Obes 17(5):472–477
17. Belkina AC, Denis GV (2012) BET domain co-regulators in obesity, inflammation and cancer. Nat Rev Cancer 12(7):465–477
18. Bethell C, Simpson L, Stumbo S et al (2010) National, state, and local disparities in childhood obesity. Health Aff (Millwood) 29:347–356
19. Betof AS, Dewhirst MW, Jones LW (2013) Effects and potential mechanisms of exercise training on cancer progression: a translational perspective. Brain Behav Immun 30 Suppl: S75–S87
20. Blüher M (2010) The distinction of metabolically 'healthy' from 'unhealthy' obese individuals. Curr Opin Lipidol 21(1):38–43
21. Blüher M (2012) Are there still healthy obese patients? Curr Opin Endocrinol Diabetes Obes 19(5):341–346
22. Boggs DA, Rosenberg L, Cozier YC, Wise LA, Coogan PF, Ruiz-Narvaez EA, Palmer JR (2011) General and abdominal obesity and risk of death among black women. N Engl J Med 365:901–908
23. Boynton-Jarrett R, Fargnoli J, Suglia SF et al (2010) Association between maternal intimate partner violence and incident obesity in preschool-aged children: results from the fragile families and child well-being study. Arch Pediatr Adolesc Med 164:540–546
24. Brasky TM, Lampe JW, Potter JD, Patterson RE, White E (2010) Specialty supplements and breast cancer risk in the VITamins And Lifestyle (VITAL) Cohort. Cancer Epidemiol Biomarkers Prev 19:1696–1708
25. Brasky TM, Darke AK, Song X, Tangen CM, Goodman PJ, Thompson IM, Meyskens FL Jr, Goodman GE, Minasian LM, Parnes HL, Klein EA, Kristal AR (2013) Plasma phospholipid fatty acids and prostate cancer risk in the SELECT trial. J Natl Cancer Inst 105:1132–1141
26. Brown JC, Winters-Stone K, Lee A, Schmitz KH (2012) Cancer, physical activity, and exercise. Compr Physiol 2:2775–2809
27. Browning LM, Krebs JD, Magee EC, Frühbeck G, Jebb SA (2008) Circulating markers of inflammation and their link to indices of adiposity. Obes Facts 1(5):259–265
28. Brydges SD, Broderick L, McGeough MD, Pena CA, Mueller JL, Hoffman HM (2013) Divergence of IL-1, IL-18, and cell death in NLRP3 inflammasomopathies. J Clin Invest 123 (11):4695–4705. doi:10.1172/JCI71543
29. Buford TW, Anton SD, Judge AR, Marzetti E, Wohlgemuth SE, Carter CS, Leeuwenburgh C, Pahor M, Manini TM (2010) Models of accelerated sarcopenia: critical pieces for solving the puzzle of age-related muscle atrophy. Ageing Res Rev 9:369–383
30. Calle EE, Rodriguez C, Walker-Thurmond K, Thun MJ (2003) Overweight, obesity, and mortality from cancer in a prospectively studied cohort of U.S. adults. N Engl J Med 348 (17):1625–1638
31. Calle EE, Kaaks R (2004) Overweight, obesity and cancer: epidemiological evidence and proposed mechanisms. Nat Rev Cancer 4(8):579–591
32. Calori G, Lattuada G, Piemonti L, Garancini MP, Ragogna F, Villa M, Mannino S, Crosignani P, Bosi E, Luzi L, Ruotolo G, Perseghin G (2011) Prevalence, metabolic features,

and prognosis of metabolically healthy obese Italian individuals: the Cremona Study. Diabetes Care 34(1):210–215

33. Campisi J (2010) Cellular senescence: putting the paradoxes in perspective. Curr Opin Genet Dev 21(1):107–112

34. Centers for Disease Control and Prevention (2013) The state of aging and health in America 2013. Centers for Disease Control and Prevention, US Dept of Health and Human Services, Atlanta, GA, www.cdc.gov/aging

35. Centers for Disease Control and Prevention (2009) Obesity prevalence among low-income, preschool-aged children—United States, 1998-2008. MMWR Morb Mortal Wkly Rep 58:769–773

36. Cinti S, Mitchell G, Barbatelli G, Murano I, Ceresi E, Faloia E, Wang S, Greenberg AS, Obin MS (2005) Adipocyte death defines macrophage localization and function in adipose tissue of obese mice and humans. J Lipid Res 46:2347–2355

37. Cohen L (1998) No aging in India: Alzheimer's, the bad family and other modern things. University of California Press, Berkeley, CA

38. Collado M, Gil J, Efeyan A, Guerra C, Schuhmacher AJ, Barradas M et al (2005) Tumour biology: senescence in premalignant tumours. Nature 436:642

39. Collerton J, Martin-Ruiz C, Davies K et al (2012) Frailty and the role of inflammation, immunosenescence and cellular ageing in the very old: cross-sectional findings from the Newcastle 85+ study. Mech Ageing Dev 133:456–466

40. Colman RJ, Anderson RM, Johnson SC, Kastman EK, Kosmatka KJ, Beasley TM, Allison DB, Cruzen C, Simmons HA, Kemnitz JW, Weindruch R (2009) Caloric restriction delays disease onset and mortality in rhesus monkeys. Science 325:201–204

41. Crandall JP, Barzilai N (2013) Exploring the promise of resveratrol: where do we go from here? Diabetes 62:1022–1023

42. DeFuria J, Belkina AC, Jagannathan-Bogdan M, Snyder-Cappione J, Carr JD, Nersesova Y, Markham D, Strissel KJ, Watkins A, Allen J, Bouchard J, Toraldo G, Jasuja R, Obin MS, McDonnell ME, Apovian C, Denis GV, Nikolajczyk BS (2013) B cells promote inflammation in obesity and type 2 diabetes through regulation of T-cell function and an inflammatory cytokine profile. Proc Natl Acad Sci U S A 110:5133–5138

43. Denis GV (2010) Bromodomain coactivators in cancer, obesity, type 2 diabetes, and inflammation. Discov Med 10(55):489–499

44. Denis GV, Obin M (2012) 'Metabolically healthy obesity': origins and implications. Mol Aspects Med 34(1):59–70, 2012.10.004 DOI:10.1016/j.mam.%202012;%2034(1):%2059-70.%202012.10.004 http://dx.doi.org/10.1016/j.mam

45. Denis GV, Bowen DJ (2013) Uncoupling obesity from cancer: bromodomain co-regulators that control networks of inflammatory genes. In: Dannenberg AJ, Berger NA (eds) Energy balance and cancer, chapter 3, vol 8: obesity, inflammation and cancer. Springer, New York, pp 61–82

46. Després JP (2012) What is "metabolically healthy obesity"?: from epidemiology to pathophysiological insights. J Clin Endocrinol Metab 97(7):2283–2285

47. DeVol R, Bedroussian A (2007) An unhealthy America: the economic burden of chronic disease. Charting a new course to save lives and increase productivity and economic growth. . Milken Institute, Santa Monica, CA

48. Di Rosa M, Malaguarnera M, Zanghì A, Passaniti A, Malaguarnera L (2013) Vitamin D3 insufficiency and colorectal cancer. Crit Rev Oncol Hematol 88:594–612

49. Dong M, Anda RF, Dube SR, Giles WH, Felitti VJ (2003) The relationship of exposure to childhood sexual abuse to other forms of abuse, neglect, and household dysfunction during childhood. Child Abuse Negl 27(6):625–639. doi:10.1016/S0145-2134(03)00105-4

50. Donkin A, Goldblatt P, Lynch K (2002) Inequalities in life expectancy by social class 1972–1999. Health Stat Q 15:5–15

51. Drost J, Mantovani F, Tocco F, Elkon R, Comel A, Holstege H, Kerkhoven R, Jonkers J, Voorhoeve PM, Agami R, Del Sal G (2010) BRD7 is a candidate tumour suppressor gene required for p53 function. Nat Cell Biol 12(4):380–389

52. Ehlert U (2013) Enduring psychobiological effects of childhood adversity. Psychoneuroendocrinology 38:1850–1857
53. Einstein FH, Atzmon G, Yang XM, Ma XH, Rincon M, Rudin E, Muzumdar R, Barzilai N (2005) Differential responses of visceral and subcutaneous fat depots to nutrients. Diabetes 54:672–678
54. Einstein FH, Fishman S, Bauman J, Thompson RF, Huffman DM, Atzmon G, Barzilai N, Muzumdar RH (2008) Enhanced activation of a "nutrient-sensing" pathway with age contributes to insulin resistance. FASEB J 22:3450–3457
55. Einstein FH, Huffman DM, Fishman S, Jerschow E, Heo HJ, Atzmon G, Schechter C, Barzilai N, Muzumdar RH (2010) Aging per se increases the susceptibility to free fatty acid-induced insulin resistance. J Gerontol A Biol Sci Med Sci 65:800–808
56. Evans WJ, Paolisso G, Abbatecola AM, Corsonello A, Bustacchini S, Strollo F, Lattanzio F (2010) Frailty and muscle metabolism dysregulation in the elderly. Biogerontology 11:527–536
57. Fabian CJ, Kimler BF, Donnelly JE, Sullivan DK, Klemp JR, Petroff BK, Phillips TA, Metheny T, Aversman S, Yeh HW, Zalles CM, Mills GB, Hursting SD (2013) Favorable modulation of benign breast tissue and serum risk biomarkers is associated with >10 % weight loss in postmenopausal women. Breast Cancer Res Treat 142:119–132
58. Ferrucci L, Harris TB, Guralnik JM, Tracy RP, Corti MC, Cohen HJ, Penninx B, Pahor M, Wallace R, Havlik RJ (1999) Serum IL-6 level and the development of disability in older persons. J Am Geriatr Soc 47(6):639–646
59. Finley CE, LaMonte MJ, Waslien CI, Barlow CE, Blair SN, Nichaman MZ (2006) Cardiorespiratory fitness, macronutrient intake, and the metabolic syndrome: the Aerobics Center Longitudinal Study. J Am Diet Assoc 106:673–679
60. Finucane MM, Stevens GA, Cowan MJ, Danaei G, Lin JK, Paciorek CJ, Singh GM, Gutierrez HR, Lu Y, Bahalim AN, Farzadfar F, Riley LM, Ezzati M, Global Burden of Metabolic Risk Factors of Chronic Diseases Collaborating Group (Body Mass Index) (2011) National, regional, and global trends in body-mass index since 1980: systematic analysis of health examination surveys and epidemiological studies with 960 country-years and 9.1 million participants. Lancet 377(9765):557–567
61. Folsom AR, Kaye SA, Sellers TA, Hong CP, Cerhan JR, Potter JD, Prineas RJ (1993) Body fat distribution and 5-year risk of death in older women. JAMA 269:483–487
62. Fontana L, Meyer TE, Klein S, Holloszy JO (2004) Long-term calorie restriction is highly effective in reducing the risk for atherosclerosis in humans. Proc Natl Acad Sci U S A 101:6659–6663
63. Fontana L, Klein S, Holloszy JO (2006) Long-term low-protein, low-calorie diet and endurance exercise modulate metabolic factors associated with cancer risk. Am J Clin Nutr 84:1456–1462
64. Fontana L, Villareal DT, Weiss EP, Racette SB, Steger-May K, Klein S, Holloszy JO (2007) Calorie restriction or exercise: effects on coronary heart disease risk factors. A randomized controlled trial. Am J Physiol Endocrinol Metab 293(1):E197–E202
65. Franco-Martin M, Parra-Vidales E, Gonzalez-Palau F, Bernate-Navarro M, Solis A (2013) The influence of physical exercise in the prevention of cognitive deterioration in the elderly: a systematic review. Rev Neurol 56:545–554
66. Fried LP, Tangen CM, Walston J, Newman AB, Hirsch C, Gottdiener J et al (2001) Frailty in older adults: evidence for a phenotype. J Gerontol 56(3):M146–M156
67. Gallagher EJ, LeRoith D (2011) Minireview: IGF, insulin, and cancer. Endocrinology 152:2546–2551
68. Garasky S, Stewart SD, Gundersen C et al (2009) Family stressors and child obesity. Soc Sci Res 38:755–766
69. Geng Y, Chandrasekaran S, Hsu JW, Gidwani M, Hughes AD, King MR (2013) Phenotypic switch in blood: effects of pro-inflammatory cytokines on breast cancer cell aggregation and adhesion. PLoS One 8(1):e54959

70. Giovannucci E (2001) Insulin, insulin-like growth factors and colon cancer: a review of the evidence. J Nutr 131:3109S–3120S
71. Gronich N, Rennert G (2013) Beyond aspirin-cancer prevention with statins, metformin and bisphosphonates. Nat Rev Clin Oncol 10:625–642
72. Guh DP, Zhang W, Bansback N, Amarsi Z, Birmingham CL, Anis AH (2009) The incidence of co-morbidities related to obesity and overweight: a systematic review and meta-analysis. BMC Public Health 9:88
73. Gundersen C, Mahatmya D, Garasky S et al (2011) Linking psychosocial stressors and childhood obesity. Obes Rev 12:e54–e63
74. Hamer M, Stamatakis E (2012) Metabolically healthy obesity and risk of all-cause and cardiovascular disease mortality. J Clin Endocrinol Metab 97(7):2482–2488
75. Harrison DE, Strong R, Sharp ZD, Nelson JF, Astle CM, Flurkey K, Nadon NL, Wilkinson JE, Frenkel K, Carter CS, Pahor M, Javors MA, Fernandez E, Miller RA (2009) Rapamycin fed late in life extends lifespan in genetically heterogeneous mice. Nature 460:392–395
76. Harte MT, O'Brien GJ, Ryan NM, Gorski JJ, Savage KI, Crawford NT, Mullan PB, Harkin DP (2010) BRD7, a subunit of SWI/SNF complexes, binds directly to BRCA1 and regulates BRCA1-dependent transcription. Cancer Res 70(6):2538–2547
77. Heilbronn LK, de Jonge L, Frisard MI, DeLany JP, Larson-Meyer DE, Rood J, Nguyen T, Martin CK, Volaufova J, Most MM, Greenway FL, Smith SR, Deutsch WA, Williamson DA, Ravussin E (2006) Effect of 6-month calorie restriction on biomarkers of longevity, metabolic adaptation, and oxidative stress in overweight individuals: a randomized controlled trial. JAMA 295:1539–1548
78. Herbig U, Ferreira M, Condel L, Carey D, Sedivy JM (2006) Cellular senescence in aging primates. Science 311:1257
79. Hofer T, Fontana L, Anton SD, Weiss EP, Villareal D, Malayappan B, Leeuwenburgh C (2008) Long-term effects of caloric restriction or exercise on DNA and RNA oxidation levels in white blood cells and urine in humans. Rejuvenation Res 11:793–799
80. Holloszy JO, Smith EK, Vining M, Adams S (1985) Effect of voluntary exercise on longevity of rats. J Appl Physiol 59:826–831
81. Holloszy JO, Smith EK (1987) Effects of exercise on longevity of rats. Fed Proc 46:1850–1853
82. Holloszy JO, Schechtman KB (1991) Interaction between exercise and food restriction: effects on longevity of male rats. J Appl Physiol 70:1529–1535
83. Holloszy JO (1997) Mortality rate and longevity of food-restricted exercising male rats: a reevaluation. J Appl Physiol 82:399–403
84. Hotamisligil GS, Shargill NS, Spiegelman BM (1993) Adipose expression of tumor necrosis factor-α: direct role in obesity-linked insulin resistance. Science 259:87–91
85. Hsu A, Bray TM, Ho E (2010) Anti-inflammatory activity of soy and tea in prostate cancer prevention. Exp Biol Med (Maywood) 235:659–667
86. Hsu A, Bruno RS, Lohr CV, Taylor AW, Dashwood RH, Bray TM, Ho E (2011) Dietary soy and tea mitigate chronic inflammation and prostate cancer via NFkappaB pathway in the Noble rat model. J Nutr Biochem 22:502–510
87. Huffman DM, Barzilai N (2010) Contribution of adipose tissue to health span and longevity. Interdiscip Top Gerontol 37:1–19
88. Huffman DM, Augenlicht LH, Zhang X, Lofrese JJ, Atzmon G, Chamberland JP, Mantzoros CS (2013) Abdominal obesity, independent from caloric intake, accounts for the development of intestinal tumors in Apc(1638N/+) female mice. Cancer Prev Res (Phila) 6:177–187
89. Hurez V, Daniel BJ, Sun L, Liu AJ, Ludwig SM, Kious MJ, Thibodeaux SR, Pandeswara S, Murthy K, Livi CB, Wall S, Brumlik MJ, Shin T, Zhang B, Curiel TJ (2012) Mitigating age-related immune dysfunction heightens the efficacy of tumor immunotherapy in aged mice. Cancer Res 72:2089–2099
90. Hursting SD, Dunlap SM, Ford NA, Hursting MJ, Lashinger LM (2013) Calorie restriction and cancer prevention: a mechanistic perspective. Cancer Metab 1:10

91. Iliopoulos D, Hirsch HA, Wang G, Struhl K (2011) Inducible formation of breast cancer stem cells and their dynamic equilibrium with non-stem cancer cells via IL6 secretion. Proc Natl Acad Sci U S A 108(4):1397–1402

92. Ingram DK, Zhu M, Mamczarz J, Zou S, Lane MA, Roth GS, deCabo R (2006) Calorie restriction mimetics: an emerging research field. Aging Cell 5:97–108

93. Iyengar P, Combs TP, Shah SJ, Gouon-Evans V, Pollard JW, Albanese C, Flanagan L, Tenniswood MP, Guha C, Lisanti MP, Pestell RG, Scherer PE (2003) Adipocyte-secreted factors synergistically promote mammary tumorigenesis through induction of anti-apoptotic transcriptional programs and proto-oncogene stabilization. Oncogene 22(41):6408–6423

94. Jacobs EJ, Newton CC, Wang Y, Patel AV, McCullough ML, Campbell PT, Thun MJ, Gapstur SM (2010) Waist circumference and all-cause mortality in a large US cohort. Arch Intern Med 170:1293–1301

95. Jagannathan M et al (2010) Toll-like receptors regulate B cell cytokine production in patients with diabetes. Diabetologia 53:1461–1471

96. Jagannathan-Bogdan M, McDonnell ME, Shin H, Rehman Q, Hasturk H, Apovian CM, Nikolajczyk BS (2011) Elevated proinflammatory cytokine production by a skewed T cell compartment requires monocytes and promotes inflammation in type 2 diabetes. J Immunol 186(2):1162–1172. doi:10.4049/jimmunol.1002615

97. Jensen GL (2008) Inflammation: roles in aging and sarcopenia. JPEN J Parenter Enteral Nutr 32:656–659

98. Jeyapalan JC, Sedivy JM (2008) Cellular senescence and organismal aging. Mech Ageing Dev 129:467–474

99. Justice AC, Freiberg MS, Tracy R et al (2012) Does an index composed of clinical data reflect effects of inflammation, coagulation, and monocyte activation on mortality among those aging with HIV? Clin Infect Dis 54:984–994

100. Kabat GC, Kim MY, Strickler HD, Shikany JM, Lane D, Luo J, Ning Y, Gunter MJ, Rohan TE (2012) A longitudinal study of serum insulin and glucose levels in relation to colorectal cancer risk among postmenopausal women. Br J Cancer 106(1):227–232

101. Kahn SE, Zinman B, Haffner SM, O'Neill MC, Kravitz BG, Yu D, Freed MI, Herman WH, Holman RR, Jones NP, Lachin JM, Viberti GC, ADOPT Study Group (2006) Obesity is a major determinant of the association of C-reactive protein levels and the metabolic syndrome in type 2 diabetes. Diabetes 55(8):2357–2364

102. Kanapuru B, Ershler WB (2009) Inflammation, coagulation, and the pathway to frailty. Am J Med 122(7):605–613

103. Kang HW, Kim D, Kim HJ, Kim CH, Kim YS, Park MJ, Kim JS, Cho SH, Sung MW, Jung HC, Lee HS, Song IS (2010) Visceral obesity and insulin resistance as risk factors for colorectal adenoma: a cross-sectional, case-control study. Am J Gastroenterol 105:178–187

104. Kang TW, Yevsa T, Woller N, Hoenicke L, Wuestefeld T, Dauch D, Hohmeyer A, Gereke M, Rudalska R, Potapova A, Iken M, Vucur M, Weiss S, Heikenwalder M, Khan S, Gil J, Bruder D, Manns M, Schirmacher P, Tacke F, Ott M, Luedde T, Longerich T, Kubicka S, Zender L (2011) Senescence surveillance of pre-malignant hepatocytes limits liver cancer development. Nature 479(7374):547–551. doi:10.1038/nature10599

105. Karakelides H, Nair KS (2005) Sarcopenia of aging and its metabolic impact. Curr Top Dev Biol 68:123–148

106. Karelis AD, Faraj M, Bastard JP, St-Pierre DH, Brochu M, Prud'homme D, Rabasa-Lhoret R (2005) The metabolically healthy but obese individual presents a favorable inflammation profile. J Clin Endocrinol Metab 90(7):4145–4150

107. Karin M, Greten FR (2005) NF-kappaB: linking inflammation and immunity to cancer development and progression. Nat Rev Immunol 5:749–759

108. Kartheuser AH, Leonard DF, Penninckx F, Paterson HM, Brandt D, Remue C, Bugli C, Dozois E, Mortensen N, Ris F, Tiret E (2013) Waist circumference and waist/hip ratio are better predictive risk factors for mortality and morbidity after colorectal surgery than body mass index and body surface area. Ann Surg 258:722–730

109. Keinan-Boker L, Vin-Raviv N, Liphshitz I, Linn S, Barchana M (2009) Cancer incidence in Israeli Jewish survivors of World War II. J Natl Cancer Inst 101(21):1489–1500. doi:10.1093/jnci/djp327

110. Key TJ (2011) Fruit and vegetables and cancer risk. Br J Cancer 104:6–11

111. Kintscher U, Hartge M, Hess K, Foryst-Ludwig A, Clemenz M, Wabitsch M, Fischer-Posovszky P, Barth TF, Dragun D, Skurk T, Hauner H, Blüher M, Unger T, Wolf AM, Knippschild U, Hombach V, Marx N (2008) T-lymphocyte infiltration in visceral adipose tissue: a primary event in adipose tissue inflammation and the development of obesity-mediated insulin resistance. Arterioscler Thromb Vasc Biol 28:1304–1310

112. Klöting N, Fasshauer M, Dietrich A, Kovacs P, Schön MR, Kern M, Stumvoll M, Blüher M (2010) Insulin-sensitive obesity. Am J Physiol Endocrinol Metab 299(3):E506–E515

113. Koster A, Stenholm S, Alley DE, Kim LJ, Simonsick EM, Kanaya AM, Visser M, Houston DK, Nicklas BJ, Tylavsky FA, Satterfield S, Goodpaster BH, Ferrucci L, Harris TB, Health ABC Study (2010) Body fat distribution and inflammation among obese older adults with and without metabolic syndrome. Obesity (Silver Spring) 18(12):2354–2361

114. Kraus S, Arber N (2009) Inflammation and colorectal cancer. Curr Opin Pharmacol 9 (4):405–410

115. Kristiansen OP, Mandrup-Poulsen T (2005) Interleukin-6 and diabetes: the good, the bad, or the indifferent? Diabetes 54(Suppl.):S114–S124

116. Ksiazek K, Mikula-Pietrasik J, Olijslagers S, Jorres A, von Zglinicki T, Witowski J (2009) Vulnerability to oxidative stress and different patterns of senescence in human peritoneal mesothelial cell strains. Am J Physiol Regul Integr Comp Physiol 296:R374–R382

117. Kuilman T, Michaloglou C, Mooi WJ, Peeper DS (2010) The essence of senescence. Genes Dev 24:2463–2479

118. Kushi LH, Byers T, Doyle C, Bandera EV, McCullough M, McTiernan A, Gansler T, Andrews KS, Thun MJ (2006) American Cancer Society Guidelines on Nutrition and Physical Activity for cancer prevention: reducing the risk of cancer with healthy food choices and physical activity. CA Cancer J Clin 56:254–281, quiz 313-4

119. Lamming DW, Ye L, Sabatini DM, Baur JA (2013) Rapalogs and mTOR inhibitors as anti-aging therapeutics. J Clin Invest 123:980–989

120. Lee CG, Boyko EJ, Strotmeyer ES, Lewis CE, Cawthon PM, Hoffman AR, Everson-Rose SA, Barrett-Connor E, Orwoll ES (2011) Association between insulin resistance and lean mass loss and fat mass gain in older men without diabetes mellitus. J Am Geriatr Soc 59:1217–1224

121. Lee IM, Hsieh CC, Paffenbarger RS Jr (1995) Exercise intensity and longevity in men. The Harvard Alumni Health Study. JAMA 273:1179–1184

122. Lefevre M, Redman LM, Heilbronn LK, Smith JV, Martin CK, Rood JC, Greenway FL, Williamson DA, Smith SR, Ravussin E (2009) Caloric restriction alone and with exercise improves CVD risk in healthy non-obese individuals. Atherosclerosis 203:206–213

123. Leng SX, Xue QL, Tian J, Walston JD, Fried LP (2007) Inflammation and frailty in older women. J Am Geriatr Soc 55(6):864–871

124. Leone A, Di Gennaro E, Bruzzese F, Avallone A, Budillon A (2014) New perspective for an old antidiabetic drug: metformin as anticancer agent. Cancer Treat Res 159:355–376

125. Levine AJ (1997) p53, the cellular gatekeeper for growth and division. Cell 88(3):323–331

126. Libby G, Donnelly LA, Donnan PT, Alessi DR, Morris AD, Evans JM (2009) New users of metformin are at low risk of incident cancer: a cohort study among people with type 2 diabetes. Diabetes Care 32:1620–1625

127. Lopez-Otin C, Blasco MA, Partridge L, Serrano M, Kroemer G (2013) The hallmarks of aging. Cell 153:1194–1217

128. Lu B, Chen L, Liu L, Zhu Y, Wu C, Jiang J, Zhang X (2011) T-cell-mediated tumor immune surveillance and expression of B7 co-inhibitory molecules in cancers of the upper gastrointestinal tract. Immunol Res 50(2–3):269–275. doi:10.1007/s12026-011-8227-9

129. Manini TM, Everhart JE, Patel KV, Schoeller DA, Colbert LH, Visser M, Tylavsky F, Bauer DC, Goodpaster BH, Harris TB (2006) Daily activity energy expenditure and mortality among older adults. JAMA 296:171–179

130. Marini MA, Succurro E, Frontoni S, Hribal ML, Andreozzi F, Lauro R, Perticone F, Sesti G (2007) Metabolically healthy but obese women have an intermediate cardiovascular risk profile between healthy nonobese women and obese insulin-resistant women. Diabetes Care 30(8):2145–2147

131. Marmot M (2005) Social determinants of health inequalities. Lancet 365:1099–1104

132. Martinez ME, Jacobs ET, Baron JA, Marshall JR, Byers T (2012) Dietary supplements and cancer prevention: balancing potential benefits against proven harms. J Natl Cancer Inst 104:732–739

133. Matthews KA (2005) Psychological perspectives on the development of coronary heart disease. Am Psychol 60(8):783–796. doi:10.1037/0003-066X.60.8.783

134. McElhaney JE, Zhou X, Talbot HK, Soethout E, Bleackley RC, Granville DJ, Pawelec G (2012) The unmet need in the elderly: how immunosenescence, CMV infection, co-morbidities and frailty are a challenge for the development of more effective influenza vaccines. Vaccine 30(12):2060–2067. doi:10.1016/j.vaccine.2012.01.015, Epub 2012 Jan 27

135. McIlhatton MA, Tyler J, Kerepesi LA, Bocker-Edmonston T, Kucherlapati MH, Edelmann W, Kucherlapati R, Kopelovich L, Fishel R (2011) Aspirin and low-dose nitric oxide-donating aspirin increase life span in a Lynch syndrome mouse model. Cancer Prev Res (Phila) 4:684–693

136. Meeran SM, Katiyar SK (2008) Cell cycle control as a basis for cancer chemoprevention through dietary agents. Front Biosci 13:2191–2202

137. Mikkelsen UR, Couppe C, Karlsen A, Grosset JF, Schjerling P, Mackey AL, Klausen HH, Magnusson SP, Kjaer M (2013) Life-long endurance exercise in humans: circulating levels of inflammatory markers and leg muscle size. Mech Ageing Dev 134:531–540

138. Miller RA, Harrison DE, Astle CM, Baur JA, Boyd AR, de Cabo R, Fernandez E, Flurkey K, Javors MA, Nelson JF, Orihuela CJ, Pletcher S, Sharp ZD, Sinclair D, Starnes JW, Wilkinson JE, Nadon NL, Strong R (2011) Rapamycin, but not resveratrol or simvastatin, extends life span of genetically heterogeneous mice. J Gerontol A Biol Sci Med Sci 66:191–201

139. Mina DS, Connor MK, Alibhai SM, Toren P, Guglietti C, Matthew AG, Trachtenberg J, Ritvo P (2013) Exercise effects on adipokines and the IGF axis in men with prostate cancer treated with androgen deprivation: a randomized study. Can Urol Assoc J 7:E692–E698

140. Moore LL, Bradlee ML, Singer MR, Splansky GL, Proctor MH, Ellison RC, Kreger BE (2004) BMI and waist circumference as predictors of lifetime colon cancer risk in Framingham Study adults. Int J Obes Relat Metab Disord 28:559–567

141. Morimoto RI, Cuervo AM (2009) Protein homeostasis and aging: taking care of proteins from the cradle to the grave. J Gerontol A Biol Sci Med Sci 64:167–170

142. Morris PG, Hudis CA, Giri D, Morrow M, Falcone DJ, Zhou XK, Du B, Brogi E, Crawford CB, Kopelovich L, Subbaramaiah K, Dannenberg AJ (2011) Inflammation and increased aromatase expression occur in the breast tissue of obese women with breast cancer. Cancer Prev Res (Phila) 4(7):1021–1029

143. Muzumdar R, Allison DB, Huffman DM, Ma X, Atzmon G, Einstein FH, Fishman S, Poduval AD, McVei T, Keith SW, Barzilai N (2008) Visceral adipose tissue modulates mammalian longevity. Aging Cell 7:438–440

144. Myers CE, Mirza NN, Lustgarten J (2011) Immunity, cancer and aging: lessons from mouse models. Aging Dis 2:512–523

145. Nagaoka SI, Hassold TJ, Hunt PA (2012) Human aneuploidy: mechanisms and new insights into an age-old problem. Nat Rev Genet 13:493–504. doi:10.1038/nrg3245

146. Nakamura K, Yamanouchi K, Nishihara M (2013) Secreted protein acidic and rich in cysteine internalization and its age-related alterations in skeletal muscle progenitor cells. Aging Cell 13:175–184

147. Nemesure B, Wu SY, Hennis A, Leske MC (2012) Central adiposity and prostate cancer in a black population. Cancer Epidemiol Biomarkers Prev 21:851–858

148. Newgard CB, Sharpless NE (2013) Coming of age: molecular drivers of aging and therapeutic opportunities. J Clin Invest 123:946–950

149. Nikolajczyk BS, Jagannathan-Bogdan M, Denis GV (2012) The outliers become a stampede as immunometabolism reaches a tipping point. Immunol Rev 249(1):253–275

150. O'Connell J, Lynch L, Cawood TJ, Kwasnik A, Nolan N, Geoghegan J, McCormick A, O'Farrelly C, O'Shea D (2010) The relationship of omental and subcutaneous adipocyte size to metabolic disease in severe obesity. PLoS One 5(4):e9997

151. O'Connell J, Lynch L, Hogan A, Cawood TJ, O'Shea D (2011) Preadipocyte factor-1 is associated with metabolic profile in severe obesity. J Clin Endocrinol Metab 96(4):E680–E684

152. Ogino S, Liao X, Chan AT (2013) Aspirin, PIK3CA mutation, and colorectal-cancer survival. N Engl J Med 368:289–290

153. Ohki T, Tateishi R, Shiina S, Goto E, Sato T, Nakagawa H, Masuzaki R, Goto T, Hamamura K, Kanai F, Yoshida H, Kawabe T, Omata M (2009) Visceral fat accumulation is an independent risk factor for hepatocellular carcinoma recurrence after curative treatment in patients with suspected NASH. Gut 58:839–844

154. Ohtani N, Yamakoshi K, Takahashi A, Hara E (2010) Real-time in vivo imaging of p16gene expression: a new approach to study senescence stress signaling in living animals. Cell Div 5:1

155. Oursler KK, Goulet JL, Crystal S et al (2011) Association of age and comorbidity with physical function in HIV-infected and uninfected patients: results from the Veterans Aging Cohort Study. AIDS Patient Care STDS 25:13–20

156. Paffenbarger RS Jr, Lee IM (1997) Intensity of physical activity related to incidence of hypertension and all-cause mortality: an epidemiological view. Blood Press Monit 2:115–123

157. Pedersen BK (2013) Muscle as a secretory organ. Compr Physiol 3:1337–1362

158. Pericleous M, Mandair D, Caplin ME (2013) Diet and supplements and their impact on colorectal cancer. J Gastrointest Oncol 4:409–423

159. Pischon T, Boeing H, Hoffmann K, Bergmann M, Schulze MB, Overvad K, van der Schouw YT, Spencer E, Moons KG, Tjonneland A, Halkjaer J, Jensen MK, Stegger J, Clavel-Chapelon F, Boutron-Ruault MC, Chajes V, Linseisen J, Kaaks R, Trichopoulou A, Trichopoulos D, Bamia C, Sieri S, Palli D, Tumino R, Vineis P, Panico S, Peeters PH, May AM, Bueno-de-Mesquita HB, van Duijnhoven FJ, Hallmans G, Weinehall L, Manjer J, Hedblad B, Lund E, Agudo A, Arriola L, Barricarte A, Navarro C, Martinez C, Quiros JR, Key T, Bingham S, Khaw KT, Boffetta P, Jenab M, Ferrari P, Riboli E (2008) General and abdominal adiposity and risk of death in Europe. N Engl J Med 359:2105–2120

160. Piyathilake CJ, Macaluso M, Alvarez RD, Bell WC, Heimburger DC, Partridge EE (2009) Lower risk of cervical intraepithelial neoplasia in women with high plasma folate and sufficient vitamin B12 in the post-folic acid fortification era. Cancer Prev Res (Phila) 2:658–664

161. Repetti RL, Taylor SE, Seeman TE (2002) Risky families: family social environments and the mental and physical health of offspring. Psychol Bull 128(2):330–366. doi:10.1037/0033-2909.128.2.330

162. Sasser AK, Sullivan NJ, Studebaker AW, Hendey LF, Axel AE, Hall BM (2007) Interleukin-6 is a potent growth factor for ER-α-positive human breast cancer. FASEB J 21 (13):3763–3770

163. Shonkoff JP, Boyce WT, McEwen BS (2009) Neuroscience, molecular biology, and the childhood roots of health disparities: building a new framework for health promotion and disease prevention. JAMA 301(21):2252–2259. doi:10.1001/jama.2009.754

164. Sims EA (2001) Are there persons who are obese, but metabolically healthy? Metabolism 50 (12):1499–1504

165. Singh PN, Haddad E, Tonstad S, Fraser GE (2011) Does excess body fat maintained after the seventh decade decrease life expectancy? J Am Geriatr Soc 59:1003–1011
166. Singh S, Sharma AN, Murad MH, Buttar NS, El-Serag HB, Katzka DA, Iyer PG (2013) Central adiposity is associated with increased risk of esophageal inflammation, metaplasia, and adenocarcinoma: a systematic review and meta-analysis. Clin Gastroenterol Hepatol 11:1399.e1397–1412.e1397
167. Sitte N, Merker K, Von Zglinicki T, Davies KJ, Grune T (2000) Protein oxidation and degradation during cellular senescence of human BJ fibroblasts: part II—aging of nondividing cells. FASEB J 14:2503–2510
168. Slack C, Foley A, Partridge L (2012) Activation of AMPK by the putative dietary restriction mimetic metformin is insufficient to extend lifespan in Drosophila. PLoS One 7:e47699
169. Smith DL Jr, Elam CF Jr, Mattison JA, Lane MA, Roth GS, Ingram DK, Allison DB (2010) Metformin supplementation and life span in Fischer-344 rats. J Gerontol A Biol Sci Med Sci 65:468–474
170. Strong R, Miller RA, Astle CM, Floyd RA, Flurkey K, Hensley KL, Javors MA, Leeuwenburgh C, Nelson JF, Ongini E, Nadon NL, Warner HR, Harrison DE (2008) Nordihydroguaiaretic acid and aspirin increase lifespan of genetically heterogeneous male mice. Aging Cell 7:641–650
171. Succurro E, Marini MA, Frontoni S, Hribal ML, Andreozzi F, Lauro R, Perticone F, Sesti G (2008) Insulin secretion in metabolically obese, but normal weight, and in metabolically healthy but obese individuals. Obesity (Silver Spring) 16(8):1881–1886
172. Sullivan NJ, Sasser AK, Axel AE, Vesuna F, Raman V, Ramirez N, Oberyszyn TM, Hall BM (2009) Interleukin-6 induces an epithelial-mesenchymal transition phenotype in human breast cancer cells. Oncogene 28(33):2940–2947
173. Tannenbaum A (1942) The genesis and growth of tumors. II. Effects of caloric restriction per se. Cancer Res 2:460–464
174. Tannenbaum A (1944) The dependence of the genesis of induced skin tumors on the caloric intake during different stages of carcinogenesis. Cancer Res 4:673–679
175. Tchkonia T, Zhu Y, van Deursen J, Campisi J, Kirkland JL (2013) Cellular senescence and the senescent secretory phenotype: therapeutic opportunities. J Clin Invest 123(3):966–972
176. Thompson MG et al (2010) Updated estimates of mortality associated with seasonal influenza through the 2006-2007 influenza season. MMWR 59(33):1057–1062
177. Thorpe KE, Ogden LL, Galactionova K (2010) Chronic conditions account for rise in Medicare spending from 1987 to 2006. Health Aff 29(4):1–7
178. U.S. Department of Health and Human Services. Multiple chronic conditions—a strategic framework: optimum health and quality of life for individuals with multiple chronic conditions. Washington, DC, December 2010
179. Vamvini MT, Aronis KN, Panagiotou G, Huh JY, Chamberland JP, Brinkoetter MT, Petrou M, Christophi CA, Kales SN, Christiani DC, Mantzoros CS (2013) Irisin mRNA and circulating levels in relation to other myokines in healthy and morbidly obese humans. Eur J Endocrinol 169:829–834
180. Van Gaal LF, Mertens IL, De Block CE (2006) Mechanisms linking obesity with cardiovascular disease. Nature 444(7121):875–880
181. Vijg J, Suh Y (2013) Genome instability and aging. Annu Rev Physiol 75:645–668
182. Vogeli C, Shields AE, Lee TA, Gibson TB, Marder WD, Weiss KB, Blumenthal D (2007) Multiple chronic conditions: prevalence, health consequences, and implications for quality, care management, and costs. J Gen Intern Med 22(Suppl 3):391–395
183. Waaijer ME, Parish WE, Strongitharm BH, van Heemst D, Slagboom PE, de Craen AJ, Sedivy JM, Westendorp RG, Gunn DA, Maier AB (2012) The number of p16INK4a positive cells in human skin reflects biological age. Aging Cell 11:722–725
184. Wajchenberg BL (2000) Subcutaneous and visceral adipose tissue: their relation to the metabolic syndrome. Endocr Rev 21(6):697–738

185. Walston J, McBurnie MA, Newman A, Tracy RP, Kop WJ, Hirsch CH, Gottdiener J, Fried LP, Cardiovascular Health Study (2002) Frailty and activation of the inflammation and coagulation systems with and without clinical comorbidities: results from the Cardiovascular Health Study. Arch Intern Med 162(20):2333–2341

186. Walter M, Liang S, Ghosh S, Hornsby PJ, Li R (2009) Interleukin 6 secreted from adipose stromal cells promotes migration and invasion of breast cancer cells. Oncogene 28:2745–2755

187. Wang F, Liu H, Blanton WP, Belkina AC, Lebrasseur NK, Denis GV (2009) *Brd2* disruption in mice causes severe obesity without type 2 diabetes. Biochem J 425(1):71–83

188. Weisberg SP, McCann D, Desai M, Rosenbaum M, Leibel RL, Ferrante AW Jr (2003) Obesity is associated with macrophage accumulation in adipose tissue. J Clin Invest 112:1796–1808

189. Wildman RP, Muntner P, Reynolds K, McGinn AP, Rajpathak S, Wylie-Rosett J, Sowers MR (2008) The obese without cardiometabolic risk factor clustering and the normal weight with cardiometabolic risk factor clustering: prevalence and correlates of 2 phenotypes among the US population (NHANES 1999-2004). Arch Intern Med 168(15):1617–1624

190. Wilkinson JE, Burmeister L, Brooks SV, Chan CC, Friedline S, Harrison DE, Hejtmancik JF, Nadon N, Strong R, Wood LK, Woodward MA, Miller RA (2012) Rapamycin slows aging in mice. Aging Cell 11:675–682

191. Wolff JL, Starfield B, Anderson G (2002) Prevalence, expenditures, and complications of multiple chronic conditions in the elderly. Arch Intern Med 162(20):2269–2276

192. Womack JA, Goulet JL, Gibert C et al (2011) Increased risk of fragility fractures among HIV infected compared to uninfected male veterans. PLoS One 6:e17217

193. Womack JA, Goulet JL, Gibert C, Brandt CA, Skanderson M, Gulanski B, Rimland D, Rodriguez-Barradas MC, Tate J, Yin MT, Justice AC, Veterans Aging Cohort Study Project Team (2013) Physiologic frailty and fragility fracture in HIV-infected male veterans. Clin Infect Dis 56(10):1498–1504. doi:10.1093/cid/cit056. Epub 2013 Feb 1

194. Xu H, Barnes GT, Yang Q, Tan G, Yang D, Chou CJ, Sole J, Nichols A, Ross JS, Tartaglia LA, Chen H (2003) Chronic inflammation in fat plays a crucial role in the development of obesity-related insulin resistance. J Clin Invest 112(12):1821–1830

195. Zhang C, Cuervo AM (2008) Restoration of chaperone-mediated autophagy in aging liver improves cellular maintenance and hepatic function. Nat Med 14:959–965

196. Moore LL, Chadid S, Singer MR, Denis GV (2014) Metabolic health reduces risk of obesity-related cancer in Framingham study adults. The Framingham Study. Cancer Epidemiol Biomarkers Prevent (in press)

Chapter 6
Energy Balance and Multiple Myeloma in African Americans

Graham A. Colditz, Kari Bohlke, Su-Hsin Chang, and Kenneth Carson

Abstract Multiple myeloma (MM) is a plasma B-cell malignancy that is characterized by clonal proliferation of malignant plasma cells in the bone marrow, monoclonal (M) protein in blood or urine, and organ dysfunction. The incidence of multiple myeloma (MM) is more than twice as high among African Americans as among whites in the USA (DeSantis et al., CA Cancer J Clin 63:151–166, 2013), but the reasons for this are still not well understood (Benjamin et al., Cancer Metastasis Rev 22:87–93, 2003; Greenberg et al., Leukemia 26:609–614, 2012). Obesity—a risk factor for MM that is more prevalent in African Americans than in whites—has been hypothesized to contribute to racial disparities in this disease. This chapter reviews the evidence regarding energy balance, race, and MM.

Keywords Multiple myeloma • Monoclonal gammopathy of unknown significance (MGUS) • Smoldering multiple myeloma • M protein • Waldenström macroglobulinemia • Obesity • Vitamin D

The Burden of Multiple Myeloma

With an estimated 22,350 new diagnoses in 2013, multiple myeloma accounts for roughly 15 % of hematologic malignancies and approximately 1 % of all cancers in the USA [1]. The median age at diagnosis is 69 years [2].

G.A. Colditz (✉) • K. Bohlke • S.-H. Chang
Alvin J. Siteman Cancer Center, Washington University School of Medicine, Campus Box 8109, 660S. Euclid Avenue, St. Louis, MO 63110, USA
e-mail: colditzg@wudosis.wustl.edu; kari.b@alumni.stanford.edu; changsh@wudosis.wustl.edu

K. Carson
Division of Public Health Sciences, Department of Surgery, Washington University School of Medicine, Campus Box 8109, 660S. Euclid Avenue, St. Louis, MO 63110, USA
e-mail: kecarson@wustl.edu

D.J. Bowen et al. (eds.), *Impact of Energy Balance on Cancer Disparities*, Energy Balance and Cancer 9, DOI 10.1007/978-3-319-06103-0_6, © Springer International Publishing Switzerland 2014

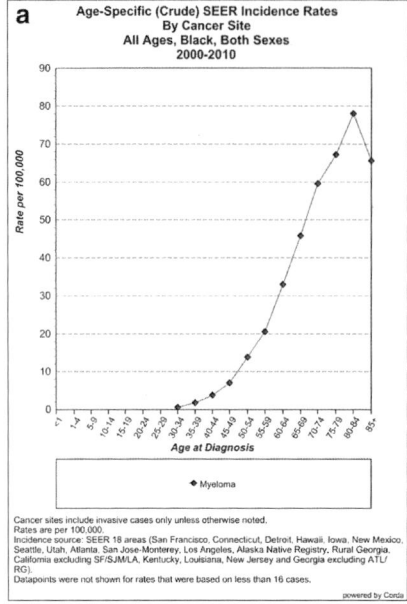

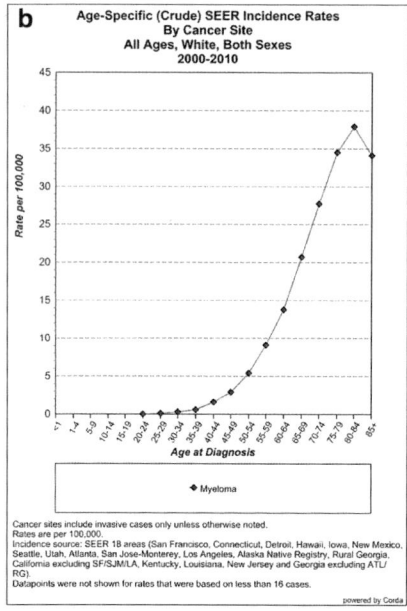

Fig. 6.1 Age-specific incidence rates of multiple myeloma for blacks (**a**) and whites (**b**), SEER 2000–2010. (Note different scales)

Well-established risk factors for MM are increasing age, male gender, African American race [3], and family history of a hematologic malignancy [4, 6]. The reasons for this excess risk remain unknown [4, 5]. In SEER data from 2006 to 2010, age-adjusted incidence of MM (per 100,000 people) was 7.1 for white men, 4.2 for white women, 14.4 for black men, and 10.2 for black women [5]. Age- and race-specific incidence rates of multiple myeloma are provided in Fig. 6.1. Similar patterns are found for mortality data, with mortality rates (per 100,000 people) of 4.0 for white men, 2.5 for white women, 7.9 for black men, and 5.4 for black women [5]. Trends in incidence and mortality over time are illustrated in Fig. 6.2. Overall, incidence rates increased between 1975 and 2010, but mortality rates have declined since the mid-1990s [5].

Survival with multiple myeloma has improved with the introduction of newer approaches to treatment [7], and racial differences in survival with multiple myeloma tend to be smaller than racial differences in incidence. In SEER 18 data from 2003 to 2009, 5-year relative survival with multiple myeloma was 44.7 % for white men, 41.2 % for white women, 41.9 % for black men, and 43.6 % for black women [5]. Improvements in relative survival over time have been larger for whites than for blacks [8].

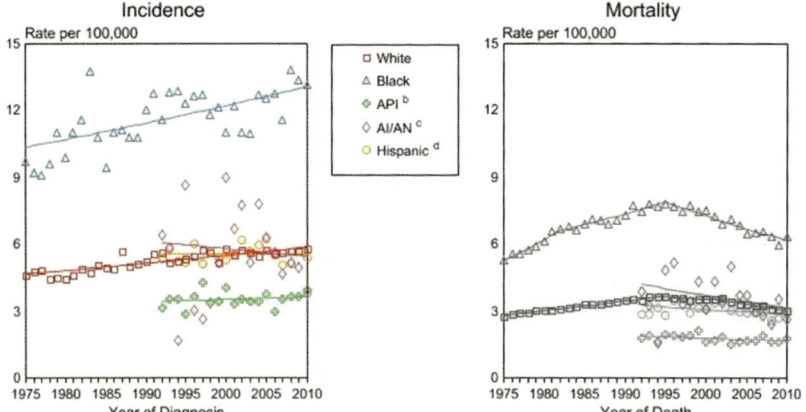

SEER Incidence and US Death Rates[a]
Myeloma, Both Sexes
Joinpoint Analyses for Whites and Blacks from 1975-2010
and for Asian/Pacific Islanders, American Indians/Alaska Natives and Hispanics from 1992-2010

Source: Incidence data for whites and blacks are from the SEER 9 areas (San Francisco, Connecticut, Detroit, Hawaii, Iowa, New Mexico, Seattle, Utah, Atlanta).
Incidence data for Asian/Pacific Islanders, American Indians/Alaska Natives and Hispanics are from the SEER 13 Areas (SEER 9 Areas, San Jose-Monterey,
Los Angeles, Alaska Native Registry and Rural Georgia). Mortality data are from US Mortality Files, National Center for Health Statistics, CDC.
[a] Rates are age-adjusted to the 2000 US Std Population (19 age groups - Census P25-1103).
Regression lines are calculated using the Joinpoint Regression Program Version 4.0.3, April 2013, National Cancer Institute. Joinpoint analyses for Whites and
Blacks during the 1975-2010 period allow a maximum of 5 joinpoints. Analyses for other ethnic groups during the period 1992-2010 allow a maximum of 3 joinpoints.
[b] API = Asian/Pacific Islander.
[c] AI/AN = American Indian/Alaska Native. Rates for American Indian/Alaska Native are based on the CHSDA(Contract Health Service Delivery Area) counties.
[d] Hispanic is not mutually exclusive from whites, blacks, Asian/Pacific Islanders, and American Indians/Alaska Natives. Incidence data for Hispanics are based on
NHIA and exclude cases from the Alaska Native Registry. Mortality data for Hispanics exclude cases from Connecticut, the District of Columbia, Maine,
Maryland, Minnesota, New Hampshire, New York, North Dakota, South Carolina, Oklahoma, and Vermont.

Fig. 6.2 SEER incidence and US death rates[a] myeloma, both sexes. Joinpoint analyses for Whites and Blacks from 1975 to 2010 and for Asian/Pacific Islanders, American Indians/Alaska natives, and Hispanics from 1992 to 2010

Precursors to Multiple Myeloma

Monoclonal gammopathy of undetermined significance (MGUS) and smoldering multiple myeloma are the known precursors of MM [9]. In both conditions, patients are asymptomatic [10]. MGUS—which is present in more than 3 % of whites over the age of 50 [11]—is characterized by the clonal proliferation of plasma cells in the bone marrow (<10 %) and the presence of M protein in the serum (<3 g/dL) and the absence of end-organ damage, such as hypercalcemia, renal insufficiency, anemia, and lytic bone lesions (CRAB features) that can be attributed to the plasma cell proliferative disorder [10]. CRAB features are also absent in smoldering multiple myeloma. Smoldering multiple myeloma is distinguished from MGUS by a higher percentage of clonal bone marrow cells (≥10 %) and/or higher serum levels of M protein (≥3 g/dL) [10], and has a higher rate of progression to MM [12].

MGUS progresses to MM or another related condition at a rate of roughly 1 % per year [13]. Three distinct types of MGUS have been described: non-IgM, IgM, and light chain [14]. IgM MGUS tends to progress to Waldenström macroglobulinemia rather than to multiple myeloma [14]. In the predominantly white population of Olmstead County, Minnesota, IgM MGUS accounted for 17 % of all MGUS cases [11]. IgM MGUS accounts for a smaller proportion of MGUS among Africans and African Americans [15, 16]. Exclusion of IgM MGUS would be appropriate when considering the epidemiology of multiple myeloma precursors, but many studies conducted to date report on MGUS as a whole.

In the USA, patterns of MGUS by age, sex, and race are similar to those of multiple myeloma: the prevalence of MGUS is higher at older ages, among men, and among African Americans [17]. In a study of 1,000 black women and 996 white women with a similar prevalence of obesity and similar socioeconomic status, the prevalence of MGUS was 3.9 % among the black women and 2.1 % among the white women [18]. After further accounting for factors such as age, education, obesity, and household income, black race was associated with an 80 % increase in the likelihood of MGUS [18]. In a study of US Veterans Affairs (VA) hospital discharge diagnoses, the age-adjusted prevalence of MGUS was three times higher (95 % confidence interval (CI) 2.7–3.3) among African Americans than among whites [19]. Among those with MGUS, however, 10-year risk of progression to MM was similar in the two groups: 17 % for African Americans and 15 % for whites. This suggests that the higher risk of MM in African Americans stems from an increase in the risk of MGUS, rather than from more frequent progression of MGUS to MM. Other, earlier studies also reported a higher prevalence of MGUS among African Americans [20, 21].

In a study conducted outside of the USA, the age-adjusted prevalence of MGUS among Ghanaian men between the ages of 50 and 74 was 5.84 % [15], which the authors note is higher than the reported prevalence among white men in Olmstead County, MN [11]. The finding that both African Americans and Africans have a higher a higher prevalence of MGUS than US whites raises the possibility of race-related genetic susceptibility, though environmental factors may also play a role.

Together, these studies suggest that the higher incidence of MM among African Americans stems from the more frequent occurrence of MGUS. Prevention strategies that begin early in life—prior to the development of MGUS—are likely to be important in reducing racial disparities in multiple myeloma.

Energy Balance and Cancer

Excess body weight and the absence of regular physical activity each contribute to cancer incidence. In a 2007 report titled *Food, Nutrition, Physical Activity, and the Prevention of Cancer: a Global Perspective*, the World Cancer Research Fund and the American Institute of Cancer research concluded that there is convincing or probable evidence for a relationship between body fatness and cancers of the esophagus (adenocarcinoma), pancreas, colorectum, breast (postmenopausal), endometrium, kidney, and gallbladder [22]. A 2003 report from the Cancer Prevention Study II study estimated that overweight and obesity account for an estimated 14 % of cancer deaths in men and 20 % of cancer deaths in women in the USA [23]. There is also convincing or probable evidence that regular physical activity reduces the risk of the colon cancer, postmenopausal breast cancer, and endometrial cancer [22].

The evidence linking energy balance and cancer incidence and mortality is most extensive for solid tumors, but a growing number of studies suggest that it is also contributes to hematologic malignancies. Meta-analyses of prospective studies have reported statistically significant, positive associations between obesity and the risk of leukemia [24], lymphoma [25], and multiple myeloma [26]. Studies of physical activity and the incidence of hematologic malignancies have produced less consistent results [27, 28].

Obesity and Multiple Myeloma

Data from several prospective studies support an association between overweight and obesity and the risk of multiple myeloma. A meta-analysis [26] included information from 15 cohort studies of multiple myeloma incidence and five cohort studies of multiple myeloma mortality. Compared with normal-weight individuals, risk of multiple myeloma was elevated among those who were overweight (relative risk (RR) = 1.12, 95 % CI 1.07–1.18) or obese (RR = 1.21, 95 % CI 1.08–1.35). Multiple myeloma mortality was also increased among those who were overweight (RR = 1.15, 95 % CI 1.04–1.27) or obese (RR = 1.54, 95 % CI 1.35–1.76) [26].

A positive association between obesity and MGUS was reported in a study of black and white women: obesity increased the risk of MGUS by 80 % (p-value = 0.04) after adjustment for race, age, education, household income, and diabetes [18].

These results suggest that overweight and obesity increase MGUS prevalence, multiple myeloma incidence, and multiple myeloma mortality. The mechanisms that link obesity with MGUS and multiple myeloma are still not well understood; several potential mechanisms have been proposed, including alterations in circulating levels of adipokines (polypeptide hormones produced by adipose tissue) [29, 30].

Physical Activity and Multiple Myeloma

In contrast to the generally consistent relationship between obesity and incidence of multiple myeloma, there is little evidence thus far that physical activity affects the incidence of multiple myeloma and other plasma cells neoplasms. The prospective VITamins And Lifestyle (VITAL) study collected information about 666 incident cases of hematologic malignancies, 80 of which involved plasma cell disorders. Regular physical activity reduced the risk of certain hematologic malignancies, particularly myeloid neoplasms, but did not reduce the risk of plasma cell disorders [27]. A combined analysis of data from the Nurses' Health Study and the Health

Professionals Follow-Up Study collected information about 215 incident cases of multiple myeloma; physical activity was not statistically significantly related to risk [31]. A lack of association between physical activity and risk of multiple myeloma was also reported in the European Prospective Investigation into Cancer and Nutrition (EPIC) study, which collected information about 165 cases [32]. Physical activity at different ages was assessed in the NIH-AARP Diet and Health Study; results at each age were null for both men and women [33]. In the Cancer Prevention Study II cohort, physical activity had a borderline-significant, inverse association with risk of multiple myeloma in women (hazard ratio for highest level of activity versus no activity: 0.52, 95 % CI 0.27–1.00), but was not associated with risk of multiple myeloma in men [34]. Time spent sitting followed a similar pattern: it increased the risk of multiple myeloma in women but not in men [34]. An inverse association with physical activity was reported in a study conducted in Japan: those who walked the least had an increased risk of multiple myeloma [35].

In summary, the evidence for an effect of physical activity on risk of multiple myeloma is far weaker than the evidence for an effect of adiposity, and likely null. Several large prospective studies have failed to find an association between physical activity and risk of multiple myeloma.

Obesity and Multiple Myeloma in African Americans

The growing evidence for an association between obesity and risk of multiple myeloma raises questions about whether this modifiable risk factor contributes to racial disparities in multiple myeloma incidence. The prevalence of obesity in the USA has reached alarming levels, and is especially high among African American women (Fig. 6.3). In 2009–2010 NHANES data, age-adjusted prevalence of obesity among women was 58.5 % among non-Hispanic blacks (95 % CI 52.4–64.3 %), 41.4 % (37.4–45.6 %) among Hispanics, and 32.2 % (95 % CI 29.2–35.3 %) among non-Hispanic whites [36]. Among men, prevalence of obesity was 38.8 % (95 % CI 33.9–43.9 %) among non-Hispanic blacks, 37.0 % (32.5–41.7 %) among all Hispanics, and 36.2 % (95 % CI 31.8–40.8 %) among non-Hispanic whites [36]. Differences by race among men are more apparent at higher levels of obesity. Grade 2 or 3 obesity (a body mass index (BMI) of ≥ 35 kg/m^2) occurs in 20 % of non-Hispanic black men, 11.9 % of Hispanic men, and 12.1 % for non-Hispanic white men [36]. The comparable numbers for women are 30.7 % for non-Hispanic black women, 18.1 % for Hispanic women, and 16.6 % for non-Hispanic white women [36].

Differences in BMI across racial and ethnic groups begin at an early age. For children between the ages of 2 and 19 years, obesity is commonly defined as being at or above the 95 % percentile of sex- and age-specific BMI [37]. In 2009–2010 NHANES data, the prevalence of obesity among children and adolescents this age range was higher among non-Hispanic blacks than among non-Hispanic whites for both girls and boys [38]. Among boys, the prevalence of obesity was 24.3 % among

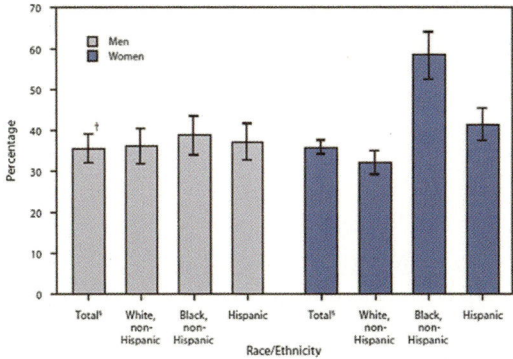

* Defined as a body mass index (weight [kg] / height [m]^2) ≥30.

† 95% confidence interval.

§ Includes other races (i.e., Asians and American Indians/Alaska Natives) not shown separately because of small sample sizes, which affect reliability of estimates.

Among adults aged ≥20 years during 2009–2010, 35.5% of men and 35.8% of women were obese. Among men, 38.8% of non-Hispanic blacks, 37.0% of Hispanics, and 36.2% of non-Hispanic whites were obese. Among women, 58.5% of non-Hispanic blacks, 41.4% of Hispanics, and 32.2% of non-Hispanic whites were obese.

Sources: Ogden CL, Carroll MD, Kit BK, Flegal KM. Prevalence of obesity in the United States, 2009–2010. NCHS data brief no. 82. Hyattsville, MD: US Department of Health and Human Services, CDC, National Center for Health Statistics; 2012.

Flegal KM, Carroll MD, Kit BK, Ogden CL. Prevalence of obesity and trends in body mass index among US adults, 1999–2010. JAMA 2012;307:491–7.

National Health and Nutrition Examination Survey, 2009–2010. Available at http://www.cdc.gov/nchs/nhanes.htm.

Fig. 6.3 Prevalence of obesity* among adults aged ≥20 years, by race/ethnicity and sex— National Health and Nutrition Examination Survey, USA, 2009–2010. (From http://www.cdc.gov/mmwr/preview/mmwrhtml/mm6107a5.htm)

non-Hispanic blacks, 23.4 % among Hispanics, and 16.1 % among non-Hispanic whites. Among girls, these numbers were 24.3 % for non-Hispanic blacks, 18.9 % among Hispanics, and 11.7 % among non-Hispanic whites. A difference between non-Hispanic blacks and non-Hispanic whites was less apparent among very young children (from birth to 2 years); during these ages, high weight-for-length affects 8.7 % of non-Hispanic blacks, 14.8 % of Hispanics, and 8.4 % of non-Hispanic whites.

Several studies have provided race-specific estimates of the effect of obesity on risk of multiple myeloma. Samanic et al. evaluated the relationship between obesity and a broad range of cancer types among black and white US veterans. The effect of obesity on risk of multiple myeloma did not vary significantly by race: RR = 1.22 (95 % CI 1.05–1.40) for white men; RR = 1.26 (95 % CI 1.02–1.56) for black men [39]. An effect of obesity among both blacks and whites was also found in a case–control study by Brown et al.: odds ratio (OR) = 1.9 (95 % CI 1.2–3.1) for whites; OR = 1.5 (95 % CI 0.9–2.4) for blacks [40]. In analyses by race and gender, obesity increased the risk of multiple myeloma among white men, white women, and black women, but not black men [40]. Friedman and Herrinton conducted one of the early studies of race, obesity, and multiple myeloma; in a large managed care organization, increasing BMI was statistically significantly associated with an increased risk of multiple myeloma only among white men [41].

Though studies that stratified by both race and gender produced inconsistent results, analyses by race alone suggest that obesity increases the risk of multiple myeloma in both African Americans and whites. Patterns of obesity, however, do not clearly explain the higher risk of multiple myeloma in African Americans. African American men have the highest risk of both multiple myeloma and MGUS, yet the overall prevalence of obesity in African American men is similar to that of white men, and much lower than the prevalence in African American women.

Other Factors That May Contribute to Higher Rates of Multiple Myeloma in African Americans

The patterns of obesity described above, coupled with an effect of race on MGUS even after adjusting for obesity [18], suggest that factors other than (or in addition to) obesity contribute to the higher rates of multiple myeloma among African Americans. The other factors remain poorly understood, but ongoing research into the biology of MGUS and multiple myeloma may provide clues. Among people with MGUS, African Americans tend to have lower levels of M protein, a lower prevalence of IgM gammopathy, and a higher frequency of abnormal serum free light chain ratios than whites [3, 16]. Genomic differences in multiple myeloma among African Americans and whites are also being evaluated [42].

Variation in cancer risk by race raises questions about a potential role of vitamin D. Serum levels of 25-hydroxyvitamin D (25(OH)D) tend to be lower in African Americans than in whites, and low 25(OH)D has been linked with an increased risk of cancer [43]. Evidence for an association between vitamin D or sun exposure and risk of multiple myeloma, however, is limited and inconsistent [44–47].

As studies continue to evaluate racial variability in multiple myeloma, it will be important to focus on the etiologically relevant time period of exposure. Characteristics or behaviors that are present prior to the development of MGUS (many years before a diagnosis of multiple myeloma) are likely to be those that drive the higher rates of multiple myeloma among African Americans.

Conclusions

Though obesity does not fully explain the racial disparities in multiple myeloma incidence, it does appear to modestly increase the risk of both MGUS and multiple myeloma, and is one of the only modifiable risk factors identified thus far for these conditions. Maintenance of a healthy body weight reduces the risk of several types of cancer regardless of race and must be a cornerstone of cancer prevention efforts [48].

Ongoing investigation into the other factors that contribute to the higher risk of multiple myeloma among African Americans may provide clues to the etiology of this disease as well as to potential new approaches to prevention and treatment.

Funding Sources   GAC is supported in part by an American Cancer Society Clinical Research Professorship and by the Breast Cancer Research Foundation.

References

1. American Cancer Society (2013) Cancer facts & figures 2013. American Cancer Society, Atlanta
2. Howlader N, Noone AM, Krapcho M, et al (2012) SEER Cancer Statistics Review, 1975-2010. National Cancer Institute. Based on November 2012 SEER data submission, posted to the SEER web site, April 2013. http://seer.cancer.gov/csr/1975_2010/
3. DeSantis C, Naishadham D, Jemal A (2013) Cancer statistics for African Americans, 2013. CA Cancer J Clin 63:151–166
4. Benjamin M, Reddy S, Brawley OW (2003) Myeloma and race: a review of the literature. Cancer Metastasis Rev 22:87–93
5. Greenberg AJ, Vachon CM, Rajkumar SV (2012) Disparities in the prevalence, pathogenesis and progression of monoclonal gammopathy of undetermined significance and multiple myeloma between blacks and whites. Leukemia 26:609–614
6. Alexander DD, Mink PJ, Adami HO et al (2007) Multiple myeloma: a review of the epidemiologic literature. Int J Cancer 120(Suppl 12):40–61
7. Kumar SK, Rajkumar SV, Dispenzieri A et al (2008) Improved survival in multiple myeloma and the impact of novel therapies. Blood 111:2516–2520
8. Waxman AJ, Mink PJ, Devesa SS et al (2010) Racial disparities in incidence and outcome in multiple myeloma: a population-based study. Blood 116:5501–5506
9. Landgren O, Waxman AJ (2010) Multiple myeloma precursor disease. JAMA 304:2397–2404
10. Kyle RA, Durie BG, Rajkumar SV et al (2010) Monoclonal gammopathy of undetermined significance (MGUS) and smoldering (asymptomatic) multiple myeloma: IMWG consensus perspectives risk factors for progression and guidelines for monitoring and management. Leukemia 24:1121–1127
11. Kyle RA, Therneau TM, Rajkumar SV ct al (2006) Prevalence of monoclonal gammopathy of undetermined significance. N Engl J Med 354:1362–1369
12. Kyle RA, Remstein ED, Therneau TM et al (2007) Clinical course and prognosis of smoldering (asymptomatic) multiple myeloma. N Engl J Med 356:2582–2590
13. Kyle RA, Therneau TM, Rajkumar SV et al (2002) A long-term study of prognosis in monoclonal gammopathy of undetermined significance. N Engl J Med 346:564–569
14. Rajkumar SV, Kyle RA, Buadi FK (2010) Advances in the diagnosis, classification, risk stratification, and management of monoclonal gammopathy of undetermined significance: implications for recategorizing disease entities in the presence of evolving scientific evidence. Mayo Clin Proc 85:945–948
15. Landgren O, Katzmann JA, Hsing AW et al (2007) Prevalence of monoclonal gammopathy of undetermined significance among men in Ghana. Mayo Clin Proc 82:1468–1473
16. Weiss BM, Minter A, Abadie J et al (2011) Patterns of monoclonal immunoglobulins and serum free light chains are significantly different in black compared to white monoclonal gammopathy of undetermined significance (MGUS) patients. Am J Hematol 86:475–478
17. Wadhera RK, Rajkumar SV (2010) Prevalence of monoclonal gammopathy of undetermined significance: a systematic review. Mayo Clin Proc 85:933–942
18. Landgren O, Rajkumar SV, Pfeiffer RM et al (2010) Obesity is associated with an increased risk of monoclonal gammopathy of undetermined significance among black and white women. Blood 116:1056–1059

19. Landgren O, Gridley G, Turesson I et al (2006) Risk of monoclonal gammopathy of undetermined significance (MGUS) and subsequent multiple myeloma among African American and white veterans in the United States. Blood 107:904–906
20. Cohen HJ, Crawford J, Rao MK, Pieper CF, Currie MS (1998) Racial differences in the prevalence of monoclonal gammopathy in a community-based sample of the elderly. Am J Med 104:439–444
21. Singh J, Dudley AW Jr, Kulig KA (1990) Increased incidence of monoclonal gammopathy of undetermined significance in blacks and its age-related differences with whites on the basis of a study of 397 men and one woman in a hospital setting. J Lab Clin Med 116:785–789
22. World Cancer Research Fund/American Institute for Cancer Research (2007) Food, nutrition, physical activity, and the prevention of cancer: a global perspective. AICR, Washington DC
23. Calle EE, Rodriguez C, Walker-Thurmond K, Thun MJ (2003) Overweight, obesity, and mortality from cancer in a prospectively studied cohort of U.S. adults. N Engl J Med 348:1625–1638
24. Castillo JJ, Reagan JL, Ingham RR et al (2012) Obesity but not overweight increases the incidence and mortality of leukemia in adults: a meta-analysis of prospective cohort studies. Leukemia Res 36:868–875
25. Larsson SC, Wolk A (2011) Body mass index and risk of non-Hodgkin's and Hodgkin's lymphoma: a meta-analysis of prospective studies. Eur J Cancer 47:2422–2430
26. Wallin A, Larsson SC (2011) Body mass index and risk of multiple myeloma: a meta-analysis of prospective studies. Eur J Cancer 47:1606–1615
27. Walter RB, Buckley SA, White E (2013) Regular recreational physical activity and risk of hematologic malignancies: results from the prospective VITamins And lifestyle (VITAL) study. Ann Oncol 24:1370–1377
28. Vermaete NV, Wolter P, Verhoef GE et al (2013) Physical activity and risk of lymphoma: a meta-analysis. Cancer Epidemiol Biomarkers Prev 22:1173–1184
29. Beason T, Colditz GA (2012) Obesity and multiple myeloma. In: Mittelman SD, Berger NA (eds) Energy balance and hematologic malignancies. Energy balance and cancer, vol 5. Springer, New York
30. Hofmann JN, Liao LM, Pollak MN et al (2012) A prospective study of circulating adipokine levels and risk of multiple myeloma. Blood 120:4418–4420
31. Birmann BM, Giovannucci E, Rosner B, Anderson KC, Colditz GA (2007) Body mass index, physical activity, and risk of multiple myeloma. Cancer Epidemiol Biomarkers Prev 16: 1474–1478
32. van Veldhoven CM, Khan AE, Teucher B et al (2011) Physical activity and lymphoid neoplasms in the European Prospective Investigation into Cancer and nutrition (EPIC). Eur J Cancer 47:748–760
33. Hofmann JN, Moore SC, Lim U et al (2013) Body mass index and physical activity at different ages and risk of multiple myeloma in the NIH-AARP diet and health study. Am J Epidemiol 177:776–786
34. Teras LR, Gapstur SM, Diver WR, Birmann BM, Patel AV (2012) Recreational physical activity, leisure sitting time and risk of non-Hodgkin lymphoid neoplasms in the American Cancer Society Cancer Prevention Study II Cohort. Int J Cancer 131:1912–1920
35. Khan MM, Mori M, Sakauchi F et al (2006) Risk factors for multiple myeloma: evidence from the Japan Collaborative Cohort (JACC) study. Asian Pac J Cancer Prev 7:575–581
36. Flegal KM, Carroll MD, Kit BK, Ogden CL (2012) Prevalence of obesity and trends in the distribution of body mass index among US adults, 1999-2010. JAMA 307:491–497
37. Ogden CL, Flegal KM (2010) Changes in terminology for childhood overweight and obesity. Natl Health Stat Report 25:1–5
38. Ogden CL, Carroll MD, Kit BK, Flegal KM (2012) Prevalence of obesity and trends in body mass index among US children and adolescents, 1999-2010. JAMA 307:483–490
39. Samanic C, Gridley G, Chow WH, Lubin J, Hoover RN, Fraumeni JF Jr (2004) Obesity and cancer risk among white and black United States veterans. Cancer Causes Control 15:35–43

40. Brown LM, Gridley G, Pottern LM et al (2001) Diet and nutrition as risk factors for multiple myeloma among blacks and whites in the United States. Cancer Causes Control 12:117–125
41. Friedman GD, Herrinton LJ (1994) Obesity and multiple myeloma. Cancer Causes Control 5:479–483
42. Baker A, Braggio E, Jacobus S et al (2013) Uncovering the biology of multiple myeloma among African Americans: a comprehensive genomics approach. Blood 121:3147–3152
43. Giovannucci E, Liu Y, Rimm EB et al (2006) Prospective study of predictors of vitamin D status and cancer incidence and mortality in men. J Natl Cancer Inst 98:451–459
44. Boffetta P, van der Hel O, Kricker A et al (2008) Exposure to ultraviolet radiation and risk of malignant lymphoma and multiple myeloma—a multicentre European case-control study. Int J Epidemiol 37:1080–1094
45. Chang ET, Canchola AJ, Cockburn M et al (2011) Adulthood residential ultraviolet radiation, sun sensitivity, dietary vitamin D, and risk of lymphoid malignancies in the California Teachers Study. Blood 118:1591–1599
46. Lim U, Freedman DM, Hollis BW et al (2009) A prospective investigation of serum 25-hydroxyvitamin D and risk of lymphoid cancers. Int J Cancer 124:979–986
47. van Leeuwen MT, Turner JJ, Falster MO et al (2013) Latitude gradients for lymphoid neoplasm subtypes in Australia support an association with ultraviolet radiation exposure. Int J Cancer 133:944–951
48. Dart H, Wolin KY, Colditz GA (2012) Commentary: eight ways to prevent cancer: a framework for effective prevention messages for the public. Cancer Causes Control 23:601–608

Chapter 7
Single Nucleotide Polymorphisms in Obesity and Inflammatory Genes in African Americans with Colorectal Cancer

Melissa Kang and Temitope O. Keku

Abstract Colorectal cancer (CRC) is one of the most commonly diagnosed cancers in the world and has one of the highest mortality rates among all cancers. In the USA, a racial disparity exists in CRC with the highest incidence and worst survival in African Americans compared to other races. This disparity persists even after taking into account the stage of CRC at diagnosis, treatment differences, or socioeconomic status. Some argue that African Americans may have more risk factors for CRC such as higher rates of obesity and insulin resistance. However, when adjusted for diet, physical inactivity, or central obesity, increased risk factors present in African Americans are not fully explained. These observations suggest that there may be genetic or biological differences between races that confer worse CRC outcomes in African Americans. Single nucleotide polymorphisms (SNPs), which occur in a race-specific manner, influence development of obesity, insulin resistance, chronic inflammation, and CRC. SNPs alter levels of circulating inflammatory cytokines and growth promoting hormone peptides while modifying response to environmental stimuli. SNPs that occur distinctly in African American populations may provide important insights into the racial disparities observed in CRC incidence and survival. This chapter provides an overview of racial disparities in CRC and the potential contribution of SNPs in obesity and inflammatory related genes.

Keywords Colorectal cancer • Single nucleotide polymorphism • SNP influence on adipokines • Insulin and inflammatory cytokines • Racial disparities in obesity • Insulin resistance • Genetic susceptibility studies • Genome-wide association studies

M. Kang (✉) • T.O. Keku (✉)
Division of Gastroenterology and Hepatology, Department of Medicine, University of North Carolina, 103 Mason Farm Road, 7340-C Medical Biomolecular Research Building, CB # 7032, Chapel Hill, NC 27599-7032, USA
e-mail: melissa_kang@med.unc.edu; tokeku@med.unc.edu

D.J. Bowen et al. (eds.), *Impact of Energy Balance on Cancer Disparities*,
Energy Balance and Cancer 9, DOI 10.1007/978-3-319-06103-0_7,
© Springer International Publishing Switzerland 2014

Colorectal Cancer (CRC) Disparity and Racial Differences in Risk Factors

Colorectal cancer (CRC) is the third most commonly diagnosed cancer in both men and women in the USA with an estimated 102,480 new cases of colon and 40,340 new cases of rectal cancers in 2013 (American Cancer Society (ACS), SEER data). It is also the third leading cause of cancer-related deaths in the USA and is expected to account for 50,830 deaths in 2013 (ACS, SEER data). There are many risk factors associated with sporadic CRC including increasing age, obesity [1–8], a diet high in red or processed meats [9–12], alcohol [13, 14], smoking [12, 15], a personal or family history of CRC or polyps [12, 16–18], and type 2 diabetes [19–22]. Physical activity [12, 23, 24], consumption of calcium, and higher levels of vitamin D [25–27] as well as regular use of nonsteroidal anti-inflammatory drugs (NSAIDs) have been observed to reduce risk [28–33].

The 1-, 5-, and 10-year survival rates for CRC depend on the stage at diagnosis (ACS, Cancer Statistics, 2013). When CRC is detected at an early, localized stage, the 5-year survival is 90 %, but only 39 % of CRCs are diagnosed at this stage. If the cancer has spread to distant organs, the 5-year survival decreases to 12 % (ACS, Cancer facts and figures, 2013). However, there is a significant racial variability in the incidence and mortality from CRC. African Americans are typically diagnosed at a younger age, at a more advanced stage, and have worse survival from CRC than other ethnic groups in the USA (Table 7.1). This disparity persists even after adjustments for stage, treatment, and socioeconomic status [34–39], suggesting that there may be a biological or genetic difference or predisposition depending on ethnicity. This chapter presents racial differences in the risk factors between African Americans and Caucasians that may help explain the worse outcomes observed in African American patients with CRC. The connection between obesity, insulin resistance, inflammation, and CRC is discussed in relation to the single nucleotide polymorphisms (SNPs) that influence the levels of circulating adipokines, insulin, and inflammatory cytokines. We will also demonstrate how these genetic polymorphisms can modulate susceptibility to obesity, diabetes, and CRC in a race-specific manner. Lastly, we will address some of the limitations of genetic susceptibility studies and possible future directions in this area of study.

Obesity and Insulin Resistance

Obesity is an important CRC risk factor that could contribute to racial disparities in CRC outcomes. The consumption of diets high in saturated fats and low in fruits and vegetables, as well as sedentary lifestyles, are thought to contribute significantly to the increased prevalence of obesity in Western countries [40]. Furthermore, the higher incidence of CRC in westernized nations is believed to be partly contingent upon the increased consumption of Western diets and physical

Table 7.1 Colorectal cancer incidence and mortality rates, by race/ethnicity, USA, 2005–2009 (American Cancer Society, Surveillance Research, 2013)[a]

	Incidence		Mortality	
	Men	Women	Men	Women
Caucasian	52.8	39.2	19.5	13.6
African American	65.1	48.0	29.8	19.8
Asian American/Pacific Islander	41.4	32.1	13.1	9.6
American Indian/Alaska Native[b]	50.7	41.1	18.8	14.6
Hispanic/Latino	46.9	33.3	15.3	10.2

[a]Per 100,000, age adjusted to the 2000 US standard population
[b]Data based on Contract Health Service Delivery Area counties

inactivity. Data from the 1999 to 2000 National Health and Nutrition Examination Survey (NHANES) showed that 31 % of Americans were obese, and this rate increased to 36 % between 2009 and 2010 [41, 42]. Obesity is defined by body mass index (BMI) >30 kg/m^2 and overweight by BMI 25–29.9 kg/m^2. While BMI is positively correlated with fat mass, it does not distinguish between lean mass and fat, nor does it take into account the body fat distribution (i.e., subcutaneous and central (visceral) fat). Men tend to have more visceral fat while women usually have greater amounts of subcutaneous fat. It has been suggested that subcutaneous and visceral fat deposits are metabolically distinct from each other with visceral fat secreting higher levels of cytokines and hormones [43]. Therefore, several studies have suggested using measures of central obesity as a predictor of disease (waist to hip ratio (WHR)), with suggested cutoffs for men and women: WHR > 0.90 in men or >0.85 in women or waist circumference >102 cm in men or >88 cm in women [44].

Regardless of which measure of obesity is used, studies have demonstrated that African Americans have higher rates of obesity than Caucasians [41, 42]. NHANES data from 2009 to 2010 demonstrated that among men, age-adjusted obesity prevalence was lowest in non-Hispanic Caucasians (36.2 %) and highest among African Americans (38.8 %). For women, the overall age-adjusted obesity prevalence was 35.8 % with the lowest prevalence among Caucasian women (32.2 %); the highest rate was seen among African Americans (58.5 %) [41]. Furthermore, African Americans have more difficulty losing weight than Caucasians after a weight loss surgery or randomized diet and exercise intervention [45–51].

Obesity is believed to predispose to CRC via increased secretion of hormones and cytokines, such as tumor necrosis factor-α (TNF-α), C-reactive protein (CRP), and interleukin (IL)-6, which have been implicated in chronic low grade inflammation, cellular proliferation, cell survival, and invasion (Fig. 7.1). Moreover, obesity is also a risk factor for the development of diabetes mellitus. Type 2 diabetes as well as exogenous insulin administration have been associated with increased CRC risk, which is thought to be due to the growth-promoting effects of insulin and its related growth factors such as insulin-like growth factor (IGF-1) [19–22, 52–56]. For example in animal studies, it insulin supplementation resulted in increased colonic epithelial proliferation and aberrant colonic crypt foci formation via

Fig. 7.1 Mechanisms underlying the relationship between obesity and cancer. Abbreviations: *CRP* C-reactive protein, *TNF* tumor necrosis factor, *IL* interleukin, *ROS* reactive oxygen species, *IGF* insulin-like growth factor, *IGFBP* insulin-like growth factor binding protein

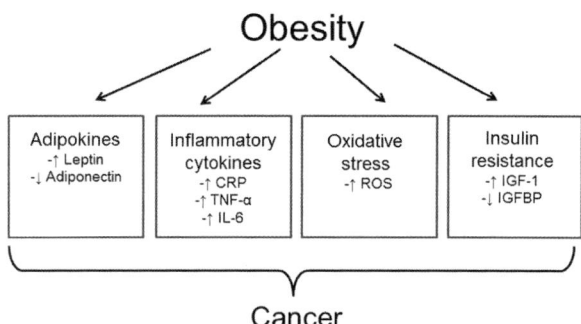

activation of pathways such as ERK and AKT that are involved in cellular proliferation [54–56].

Existing literature suggests that African Americans are at a greater risk for insulin resistance than Caucasians. Brancati et al. found that diabetes incidence was 1.85-fold greater in African American women than in Caucasian women even after adjustments for BMI, WHR, physical activity, and dietary intake [57]. Cavicchia et al. found that the association between type 2 diabetes and colon cancer was stronger in African Americans (OR = 1.72, 95 % CI 1.21–2.46) than among Caucasians (OR = 1.24, 95 % CI 0.73–2.11) [58].

Hyperinsulinemia may modify cancer risk not only through the direct effects of insulin but also indirectly by increased production of IGF-1. In the presence of insulin, growth hormone receptor expression is increased in the liver, leading to elevated production of IGF-1. Chronic hyperinsulinemia has been associated with elevated circulating levels of IGF-1 [59]. Also, adoption of a Western diet is associated with increases in serum IGF-1 levels [60]. Extensive evidence, both in vitro and in vivo, suggests that IGF-1 promotes CRC growth, prevents apoptosis, and increases metastasis [53, 60]. IGF binding proteins (IGFBP), especially IGFBP-3, sequester and inhibit the actions of circulating IGFs [53]. Research has shown that circulating IGF levels may differ by race, where African American women had significantly higher IGF-1 levels than Caucasian women [61, 62]. Furthermore, recent studies suggest that African Americans have higher hyperinsulinemia and insulin resistance even with less visceral fat mass or similar adiposity to Caucasians [63–71]. These observations indicate that higher insulin resistance and diabetes in African Americans are not fully explained by higher rates of obesity or diet and suggest underlying biological or genetic differences as an explanation for ethnic variance in insulin resistance.

Adipokines

Adipokines are hormones secreted by fat cells. Leptin is an adipokine that is involved in food intake, energy expenditure and balance. Leptin also affects cell proliferation, migration, and angiogenesis [72–75]. Leptin has been shown to be positively associated with obesity, insulin resistance, and CRC [76–82]. Studies have demonstrated that African Americans have higher levels of leptin independent of fat mass compared to Caucasians, although this may be partially due to leptin's positive association with insulin [83–86]. Adiponectin is another cytokine that is secreted by adipocytes. It influences energy metabolism, however, contrary to the action of leptin, adiponectin is an insulin sensitizer and exhibits anti-inflammatory and anti-atherogenic properties [87]. Thus, adiponectin has been inversely related to obesity and CRC risk [88–92]. Animal studies demonstrate that adiponectin knockout mice have significantly greater number and larger tumors and inflammation than mice with intact adiponectin [93, 94]. When mice on a high-fat diet were fed adiponectin, tumor growth was suppressed, and serum insulin levels were decreased [95]. In humans, African American children and women have lower levels of adiponectin than their Caucasian counterparts, even after accounting for adiposity [84, 86, 96, 97]. To support this, Schutte et al. found that ethnicity was a significant contributor to variances in adiponectin levels [98]. These studies on leptin and adiponectin suggest that plasma levels of these adipokines could be influenced by genetic variation and race, and as such may differentially impact CRC risk.

Inflammatory Markers and Cytokines

In addition to insulin dysregulation, visceral adipose tissue has also been shown to play a role in systemic inflammation. Obesity has been established as a chronic low-grade inflammatory state with elevated circulating levels of pro-inflammatory cytokines, such as IL-6 and TNF-α, and reduced anti-inflammatory mediators. These pro-inflammatory cytokines, produced by adipose tissue, are involved in a positive feedback loop in the liver to produce acute phase reactants such as C-Reactive Protein (CRP) that could further contribute to inflammation. All of these cytokines and resulting inflammation have been associated with tumorigenesis and CRC.

TNF-α, since its first description, has been found to have multiple functions in the areas of inflammation, apoptosis, production of other cytokines, and insulin resistance. TNF-α also has a role in tumor development [99, 100]. TNF-α has been shown to be positively associated with lymph node metastasis, CRC recurrence, or presence of colorectal adenomas, a precursor to CRC [101, 102]. Because adipocytes also produce TNF-α, obese subjects tend to have higher levels of TNF-α. Resection of visceral fat or weight loss lowers TNF-α levels in both animals and humans [103–110]. Reduction in TNF-α leads to decreased activation of JUN, a transcription factor involved in cellular proliferation and anti-apoptotic signaling [111, 112]. In rodent models, exposure to TNF-α induced insulin resistance, whereas neutralization or knockout of TNF-α increased insulin sensitivity, suggesting that TNF-α is a key player in insulin resistance [108]. In humans it was observed that weight loss affected overweight African American and Caucasian women differently [68, 113]. Intra-abdominal adipose tissue, serum concentrations of TNF-α, CRP, IL-6, soluble TNF receptors all decreased with weight loss in Caucasian women, but only IL-6 and CRP decreased in African American women. Further, though Caucasian women had greater visceral adipose tissue and TNF-α than African American women at baseline, they had better insulin sensitivity. Olson et al. also supported race-dependent levels of TNF-α by showing that African Americans have the lowest TNF-α level compared to Caucasians or Hispanics, even after adjustment for age, gender and BMI [114]. In this study population, serum glucose and insulin levels were moderately correlated with TNF-α in Caucasians, but not in African Americans. These studies suggest that circulating TNF-α and its effects could be varied by race and genetic background.

IL-6 is a pro-inflammatory cytokine that is also produced in visceral adipose tissue. It has been observed to promote cancer proliferation and inhibit apoptosis, where an increased IL-6 expression was positively associated with increasing BMI, adenomas, and advanced stage of CRC [102, 115–120]. CRP, an acute phase reactant produced in response to factors released by macrophages and adipocytes, is a measure of inflammation. Similarly, CRP has been positively associated with CRC, and elevated CRP level has been demonstrated to be a poor prognostic factor in patients with CRC [121–124]. Supplementation with vitamin D and calcium, which are suggested to exert anti-inflammatory and protective effects from CRC via binding of free fatty and bile acids, decreased TNF-α, IL-6, and CRP in patients with adenomas [125]. Multiple studies found that African Americans had higher levels of CRP and IL-6 than Caucasians [126–128], and even after controlling for socioeconomic status, exercise, and BMI, IL-6 continued to remain higher in African Americans [128]. Similar to that in TNF-α, genetic alterations have been proposed as one of the reasons why the levels of IL-6 and CRP and associated conditions may differ between races.

In the beginning of this chapter, we mentioned that inhibitors of the cyclooxygenase (COX) enzymes, such as aspirin and other NSAIDs decrease the risk of CRC. There are two types of COX. COX-1 is expressed in most normal cells and is involved in the protection of the gastric mucosa and the regulation of platelet aggregation. COX-2, on the other hand, is expressed in inflammatory cells or cancer cells, regulates production of prostaglandins, and is involved in tumorigenesis [129]. Consistent with these findings, COX-2 expression was shown to be elevated in 55–80 % of human colon cancers and adenomas compared with normal mucosa in multiple studies [130–133]. More recently, COX-2 expression has been revealed as a poor prognostic marker in CRC [131, 134–136].

As COX-2 expression and circulating levels of above mentioned cytokines have all been linked to CRC, the study of these markers have elucidated mechanisms and pathways involved in pathogenesis of cancer. It turns out that several of the inflammation and insulin markers discussed above are genetically regulated. SNPs in genes that encode these markers can influence their production and levels in circulation. SNPs are single nucleotide sequence variations in the genome that affect transcription and translation of a gene, thus, providing a unique method of studying how genetic mutations influence CRC risk and outcomes. In the following section, we will describe how these SNPs influence COX-2 gene expression, circulating levels of insulin-IGF markers, adipokines and inflammatory cytokines, and how they contribute to racial disparities in CRC.

Single Nucleotide Polymorphisms (SNPs)

SNPs are variations in nucleotide sequences in the DNA that occur with greater than 1 % frequency in at least one population and are inherited from generation to generation [137]. A set of associated SNP alleles in a region of a chromosome is called a haplotype. If two or more SNPs have a greater than random chance of being inherited together, they are said to be in a linkage disequilibrium. About ten million SNPs have been discovered and characterized in humans, however, only a fraction of these are known to confer a biological effect [138]. Because SNPs occur in any region of the genome, both in coding and noncoding regions, SNPs can lead to changes in amino acid sequences, which result in changes in the function or amount of encoded protein, alter enzyme stability, or cause no phenotypically visible change. Data indicate that many diseases are not directly triggered by SNPs, but are likely a result of complex interactions among multiple genes and environment. SNPs are believed to function by modifying responses to environmental exposures of health hazards or medications. For example, depending on a presence of a SNP, a patient may be a responder to a certain drug, while others are non-responders or even develop toxicities. Over the last few years with advances in technology, the field of genetic epidemiology has evolved from studying SNPs in candidate genes to genome-wide association studies (GWAS). Earlier studies evaluated candidate genes chosen for biological relevance to CRC. More recent studies have used GWAS to discover hundreds of thousands of SNPs simultaneously to assess disease associations. We will focus on some of the more well-known SNPs in insulin, diabetes, and inflammation in the next section and examine their contribution to racial disparities in CRC.

SNPs and CRC

To date, multiple CRC susceptibility alleles have been identified across chromosomes using GWAS. These loci include 8q24.21 (rs6983267), 8q23.3 (rs16892766), 10p14 (rs10795668), 11q23.1 (rs3802842), 15q13.3 (rs4779584), 18q21 (rs4939827), 14q22.2 (rs4444235), 16q22.1 (rs9929218), 19q13.1 (rs10411210), and 20p12.3 (rs961253), and range in effect sizes of odds ratio (OR) = 1.10–1.70 [139–145]. Given that there is limited information regarding how these genotypes affect phenotypes, we will focus on 8q24.21 (rs6983267), a locus that has been extensively studied. Several studies have demonstrated that rs6983267 is associated with increased CRC risk [139–145]. Information regarding how this polymorphism affects the function of the gene as well as the relationship to the risk of CRC is being discovered in more detail. Rs6983267 has been mapped nearby to the MYC gene, which is mutated in many forms of cancer, leading to dysregulation of genes involved in cell proliferation [140]. Takatsuno et al. found that MYC gene expression was highest in cells with the GG genotype (homozygous risk allele) and lowest in cells with the TT wild type genotype [146]. Abuli et al. showed that the G risk allele was more common in advanced stages, highly differentiated tumors, and in patients with a family history of CRC in first-degree relatives [139]. Dai et al. also observed that a polymorphism at rs6983267 was significantly associated with survival for patients with stage III disease on 5-fluorouracil therapy, and there was a cumulative effect when multiple unfavorable genotypes occurred concurrently [147]. Moreover, there are racial differences in the frequencies of rs6983267 alleles. The G risk allele frequency for rs6983267 is 50 % in Europeans and 80–100 % in Africans [140]. Although most SNP studies have been performed mostly in Caucasians, this is one of the few SNPs that have been studied across ethnicities. It was found to confer risk of CRC in African Americans (OR = 1.52, 95 % confidence interval (CI) 1.11–2.07) as well as in Caucasians of European descent (OR = 1.12, 95 % CI 1.01–1.25) [148]. However, not all studies have demonstrated consistent results. For example, Kupfer et al. found a trend of association with CRC in Caucasians but not in African Americans [149]. Provided that the G risk allele occurs more frequently and has greater odds of being associated with CRC in African Americans, it is possible that this SNP could partially contribute to the racial disparity observed in CRC outcomes.

SNPs in Diabetes, IGFs, and Adipokines

Through GWAS many SNPs related to obesity, diabetes, and insulin resistance have been identified. For example, SNPs in the *fat mass and obesity-associated gene (FTO)* were discovered to associate with the risk of type 2 diabetes or obesity in Caucasians, and in particular strong associations were seen at locus rs9939609 [150–154]. Association of polymorphisms at this locus with BMI and diabetes has since been reproduced in Asian populations, as well [155–158]. However, the minor A allele at locus rs9939609, though occurring at similar frequencies in African Americans and Caucasians, was not found to be associated with obesity or diabetes in African Americans contrary to that observed in Caucasians [153, 154, 159, 160]. Rather, SNPs at other loci on the *FTO* gene were associated with elevated BMI in African Americans [153, 159], suggesting that African Americans and Caucasians have race-specific polymorphisms that could confer disease risk differentially. When assessing associations with colorectal adenomas, Nock et al. observed that having one variant allele polymorphisms on the *FTO* gene at rs9939609 or r8050136 was associated with adenomas in African Americans

[161]. However, these associations were not observed in Caucasians. Cheng et al. evaluated 19 established type 2 diabetes risk variants for their associations with CRC by race [162]. The T variant allele at rs7578597 on the *THADA* gene was protective against CRC in Caucasians, Latinos, and Japanese Americans, but not in African Americans, and this association was strongest among those without diabetes or with normal BMIs. These results further provide a possible genetic explanation for the ethnic differences in the risk of diabetes and obesity in relation to colon cancer while demonstrating the complexities involved in interpreting effects of SNPs.

SNPs play an important physiological role because allelic variation results in differential expression of peptides in plasma. Increased frequency of minor alleles at different loci of the IGF-1 gene has been observed to affect levels of growth factors, even independent of BMI [163–169]. The IGF-1 gene has a polymorphism, a microsatellite $(CA)_n$ in the promoter region, that has been shown to influence IGF-1 levels. The$(CA)_{19}$ repeat is the most common allele reported in Caucasian (62–68 %), Japanese (41 %), Indian Pakistani (56 %), and African American (38 %) populations [170]. Keku et al. observed that Caucasians with homozygous IGF-1 19/19 genotype had an increased risk of colon cancer but not African Americans [168]. Wong et al. did not observe any changes in the risk among those who carried one or two copies of the $(CA)_{19}$ allele in Singapore Chinese, but $(CA)_{21}$ homozygosity was associated with approximately half of the risk for CRC in this population [170]. Morimoto et al. detected that in a mostly Caucasian population in the USA, having an IGF-1 genotype other than the homozygous 19-repeat allele was associated with a modestly increased risk of CRC (OR $= 1.3, 95$ % CI 1.0–1.6) [171]. This finding was supported by Pechlivanis et al. in Germans [172]. These results provide support to the idea that SNPs differ distinctly by race and could serve to explain the racial disparities in CRC.

A number of studies have previously linked common SNPs in the leptin and leptin receptor genes such as A+19G (rs2167270), Gln223Arg (rs1137101), and G-2548A (rs7799039) to obesity [173–179]. Friedlander et al. evaluated SNPs in leptin and leptin receptor genes by race and observed that African Americans and Caucasians have different SNPs in leptin receptor genes (rs3828033 and rs1137101 in African Americans and rs3828033 and rs6696954 in Caucasians) that were associated with weight changes over time [180]. More recent studies evaluated the association of SNPs in these genes to the risk of CRC. Slattery et al. observed a reduction in colon cancer risk with variant AA genotype at rs2167270 in the leptin gene in a mostly Caucasian cohort and showed that vitamin D receptor and IGF-1 polymorphisms interact with leptin gene polymorphisms to modify the risk [181]. In a different Caucasian cohort, Vasku et al. observed a difference in the genotype distribution of Gln223Arg polymorphisms in the leptin receptor gene among CRC patients with low (I–II) and high stages (III–IV). Wild-type AA genotype was associated with an elevated risk to patients in III–IV stages [182] suggesting that SNPs could influence not only the risk of CRC but also the prognosis. In a Chinese cohort, Liu et al. demonstrated a different group of SNPs in the leptin receptor gene modifying CRC risk [183]. Two common polymorphisms in the adiponectin gene, T45G and G276T, have been associated with obesity and insulin resistance in European whites and Asians [184–191]. More recently, variant SNPs at −11,391 G > A and −11,377 C > G have also been found to affect serum adiponectin levels, obesity, and insulin resistance in Caucasians [192–196]. Additionally, Woo et al. confirmed that A and G haplotypes of SNPs at −11,391 and −10,068, respectively, were associated with elevated adiponectin levels in African Americans [197]. In a Chinese population, adiponectin (*ADIPOQ*) rs1063538 variant CC genotype was associated with increased CRC risk (OR = 1.94, 95 % CI 1.48–2.54) compared with TT [198]. This CC genotype demonstrated a further increased CRC risk in those with a family history of cancer (OR = 3.18, 95 % CI 1.73–5.82) or smoking history (OR = 4.52, 95 % CI 2.78–7.34), demonstrating that SNPs can modify response to environmental agents. Keku et al. studied a SNP at rs1501299 of the adiponectin gene (*APM1*) between African Americans and Caucasians and did not find any increased risk of CRC with the variant genotype in either race [168], which could suggest that either the sample size was too small to detect a significant difference or that a SNP at this locus does not influence CRC risk. All these studies demonstrate evidence as to how various ethnicities have differential disease risk while delineating the limitations and difficulties in studying genetic variations related to CRC.

SNPs in Inflammatory Markers and CRC

As inflammatory cytokines also play a pivotal role in insulin resistance, studies have evaluated SNPs in the TNF-α gene. Multiple SNPs including −1,031 (T → C), −863 (C → A), −857 (C → A), −851 (C → T), −419 (G → C), −376 (G → A), −308 (G → A) (rs1800629), −238 (G → A), −162 (G → A), and −49 (G → A) have been reported [199]. Most extensively studied are the polymorphisms at −308 and −238. Dalziel et al. observed that those with a homozygous variant −308 AA had higher insulin resistance, systolic blood pressure and lower high density lipoprotein than those with wild type GG profile [200]. Sookoian et al. determined that those with −308 A variant alleles were at an increased risk of obesity (OR = 1.23, 95 % CI 1.05–1.45) [201]. Fontaine-Bisson et al. similarly observed an association between higher insulin resistance and the variant A allele, however, this relationship only occurred in obese individuals [202]. De Luis et al. found that on a diet lower in fats or carbohydrates, those with wild type −308 GG genotype improved in many aspects of metabolic profile including BMI, fat mass, blood pressure, insulin, and cholesterol, but those with the variant A allele only improved in BMI and fat mass without other added benefits [203]. Of note, these studies were conducted in a Caucasian population. Conclusions from these studies suggest that the variant A allele at −308 in the TNF-α gene confers susceptibility and higher risk for insulin resistance and obesity, and those with the wild-type allele at this locus may benefit more from a healthier diet. However, contrary to these findings, Rasmussen et al. did not see any associations between genetic variants in −308 and −238 with altered insulin sensitivity, BMI, WHR, fat mass, or insulin concentrations [204]. In the only study available among African

Table 7.2 Association of individual or combined inflammation related genes and colon cancer among African Americans and Caucasians (Keku et al., unpublished)

		African American		Caucasian	
		Case/control	OR (95 % CI)	Case/control	OR (95 % CI)
IL-6	GG	177/234	1.0 (Ref)	82/149	1.0 (Ref)
	GC	35/34	1.5 (0.9, 2.6)	136/231	1.1 (0.8, 1.5)
	CC	1/0	–	47/82	1.0 (0.7, 1.6)
TNF-α	GG	158/219	1.0 (Ref)	182/308	1.0 (Ref)
	AG	53/44	**1.8 (1.1, 2.8)**	79/133	1.0 (0.7, 1.4)
	AA	2/5	0.5 (0.08, 2.5)	4/21	**0.3 (0.1, 0.9)**
CRP	GG	136/175	1.0 (Ref)	116/230	1.0 (Ref)
	AG	66/81	1.0 (0.7, 1.5)	122/190	1.2 (0.9, 1.7)
	AA	11/12	1.3 (0.6, 3.1)	27/42	1.3 (0.7, 2.2)
# Var Alleles	0–1	46/62	1.0 (Ref)	9/36	1.0 (Ref)
	2–3	116/159	1.0 (0.6, 1.6)	130/200	**2.7 (1.2, 5.8)**

Cases subjects with colon cancer, *Controls* subjects without colon cancer
Values in bold demonstrate that while having a AG genotype in TNF-A locus confers increased risk to African Americans, it does not in Caucasians. Similarly, having an AA genotype demonstrate decreased risk in Caucasians, it had no effect in African Americans

women, the variant allele frequency at −308 did not differ between those with normal and obese weights [205]. However, when dietary fat intake was 30 % of the total energy intake, those with the variant A allele had reduced odds of being obese, but with increasing fat content of the diet, obesity risk increased at a faster rate. Reports on associations between CRC risk and TNF-α (at −308 or −238) demonstrate a null relationship in Caucasians and Asians [206–211]. In our work, we observed that African Americans with AG genotype at −308 had an 80 % increased risk (OR = 1.8, 95 % CI 1.1–2.8) of colon cancer compared to African Americans with wild-type GG genotype (Table 7.2). Further, increased risk was only present in those with GG (OR = 1.9, 95 % CI 1.2–2.9) or AG (OR = 3.5, 95 % CI 1.9–6.5) genotypes who had not used NSAIDs regularly compared to those with wild type GG genotype who reported taking NSAIDs regularly (Table 7.3). Similar protective effects from NSAIDs were observed for Caucasians. Our results support the previous conclusions regarding the preventative role of regular NSAID use in CRC. Further, if confirmed in larger studies and as SNP testing becomes more available, it is possible that in the future the determination of SNPS at TNF-α −308 could be useful in individualizing medical care to reduce colon cancer risk, especially among those who use NSAIDs.

SNPs on the IL-6 gene have also been evaluated for associations with obesity and insulin resistance. The majority of the studies on Caucasian populations demonstrated a positive association with increasing BMI, glucose, or insulin resistance with an increasing number of variant C allele at IL-6 −174 [212–217]. Further, this in turn influences the response to diet or exercise [212, 214, 218]. Clinical studies demonstrated that circulating IL-6 level was correlated with CRC stage, survival, or hepatic metastasis [219, 220], and that serum concentrations could be altered by SNPs in the IL-6 gene promoter region [221, 222]. However, the

Table 7.3 Joint effects of individual or combined cytokine genes and NSAIDs use on colon cancer risk among African Americans and Caucasians (Keku et al., unpublished)

Gene		African Americans NSAID use				Caucasians NSAID use			
		Yes		No		Yes		No	
		Cases/controls	OR (95 % CI)	Cases/controls	OR (95 % CI)	Cases/controls	OR (95 % CI)	Cases/controls	OR (95 % CI)
IL-6	GG	60/117	1.0 (Ref)	116/117	**1.8 (1.2, 2.8)**	30/87	1.0 (Ref)	52/62	**2.4 (1.3, 4.1)**
	GC	7/16	0.9 (0.4, 2.5)	28/18	**3.2 (1.6, 6.4)**	62/138	1.3 (0.8, 2.2)	74/93	**2.2 (1.3, 3.8)**
	CC	0/0	–	1/0	–	20/34	1.7 (0.8, 3.4)	27/48	1.6 (0.8, 3.0)
TNF-α	GG	53/110	1.0 (Ref)	104/109	**1.9 (1.2, 2.9)**	74/168	1.0 (Ref)	108/140	**1.7 (1.2, 2.5)**
	AG	14/20	1.5 (0.7, 3.2)	39/24	**3.5 (1.9, 6.5)**	37/78	1.1 (0.6, 1.7)	42/55	1.7 (1.0, 2.7)
	AA	0/3	–	2/2	1.4 (0.2, 10.7)	1/13	0.2 (0.02, 1.4)	3/8	0.8 (0.2, 3.2)
CRP	GG	41/87	1.0 (Ref)	94/88	**2.1 (1.3, 3.5)**	50/126	1.0 (Ref)	66/104	1.6 (1.0, 2.5)
	AG	21/39	1.1 (0.6, 2.1)	45/42	**2.1 (1.2, 3.8)**	52/112	1.1 (0.7, 1.8)	70/78	**2.1 (1.3, 3.4)**
	AA	5/7	1.7 (0.5, 5.8)	6/5	2.6 (0.7, 9.2)	10/21	1.2 (0.5, 2.8)	17/21	1.9 (0.9, 4.0)
# Variant alleles	0–1	14/27	1.0 (Ref)	32/35	1.5 (0.7, 3.5)	4/19	1.0 (Ref)	5/17	1.3 (0.3, 5.7)
	2–3	38/82	0.9 (0.4, 1.8)	77/77	1.8 (0.9, 3.7)	47/115	1.9 (0.6, 6.1)	83/85	**4.6 (1.5, 14.3)**

Cases subjects with colon cancer, *Controls* subjects without colon cancer

Values in bold demonstrate that while having GG or GC genotype in IL-6 locus were associated with increased risk to both African Americans and Caucasians, TNF-alph and CRP, there was a differential risk depending on ethnicity literature examining the relationship between SNPs and CRC risk at this locus is inconclusive. Some endorse the variant C allele to be a risk for CRC [208, 223, 224] while others show that the G allele is the risk [182, 210]. A possible explanation for this discrepancy is that there is a significant interaction between NSAID use and SNPs at IL-6 −174. When individuals with the C polymorphism regularly or currently took NSAIDs, they had a reduced CRC risk, and those who were alcohol drinkers had increased risk [223, 225]. Modification of risk by NSAID use was also supported by our work (Table 7.3). Unlike TNF-α, where allelic variations are reported to occur in similar frequencies between Caucasians and African Americans, minor allele frequencies at IL-6 −174 are different depending on race. Whereas in Caucasians minor C allele is relatively common (38 %), in African Americans and Asians it is relatively rare (10 % and 3 %, respectively), which would imply that African Americans with a risk allele in IL-6 −174 as a whole would derive less benefit with medication and lifestyle interventions in reducing CRC risk [226, 227].

Circulating CRP levels are also affected by SNPs in genes coding for CRP as well as in other inflammatory genes such as in IL-6, IL-1γ, and TNF-α [228–236]. Racial differences in CRP gene polymorphisms that alter CRP levels have been best studied thus far in cardiovascular disease. Variant alleles at 1,919 A > T and 2,667 G > C were associated with higher and lower levels of CRP, respectively, in Caucasian participants, while in African Americans, the variant T allele at 790 A > T was associated with higher CRP levels [237]. Additionally, at rs3093058 having a variant T allele was associated with increased levels of CRP while having a variant A allele at rs1205 and variant G allele at rs2808630 were associated with decreased levels of CRP compared with the homozygous referent genotypes in African Americans [238]. SNPs and haplotypes in the CRP gene have been associated with obesity such that if positive for the minor allele at rs1205, as BMI increased, CRP levels in men also increased [239]. Similar results with minor alleles altering CRP levels in obese individuals were observed at rs2794521, rs1800947, and rs1130864 [240, 241]. Regarding CRC risk, Tsilidis et al. found that in Caucasians who are variant C allele carriers at the −717 T > C promoter (rs2794521) or 2,407 T > C (rs2808630), there was an approximately 50 % increased risk [221] while Asian patients with CRC were observed to have a higher percentage of homogeneous wild-type TT genotype at −757 T > C (rs3093059) than those without CRC [242]. Further, in this Asian population SNPs affected the prognosis of CRC. Homogeneous wild type TT genotype at −757 T > C was associated with a longer disease-free interval than those with TC and CC genotypes, and homogenous wild-type AA genotype at +2,147 A > G was associated with a shorter cancer-specific survival. Whether these or those SNPs at other loci in the CRP gene will influence prognosis of African American patients with CRC is yet to be determined.

As chronic inflammation is a risk factor for CRC, polymorphisms in one of the inflammatory mediators, called mannose-binding lectin 2 (MBL2), also have been shown to modify risks to various cancers. There are specific haplotypes that have been demonstrated to correlate with MBL serum levels [243–245]. Further, frequencies of SNPs and haplotypes that occur in this gene also differ between Caucasians and African Americans [244, 246]. Zanetti et al. demonstrated that compared to the highly secreting haplotype (HYPA), haplotypes that produced low (LYQC) and moderate (LYPA) plasma levels of MBL had increased susceptibility to colon cancer (LYQC OR = 2.28, 95 % CI 1.20–4.30; LYPA OR = 2.6, 95 % CI 1.33–5.08), but only in African Americans and not in Caucasians [246]. Accordingly, African Americans with CRC had lower plasma MBL levels than Caucasians with CRC. As it has been observed that MBL can bind to ligands on colorectal tumor cells and inhibit tumor progression [247, 248], it is possible that genetic mutations that lead to lower plasma levels of MBL in African Americans predispose them to CRC. This represents exciting new evidence that could aid in explaining why African Americans have higher incidence of CRC.

Lastly, COX-2 gene polymorphisms in relation to CRC and adenoma risk have been evaluated. In mostly Caucasian populations having a variant C allele at −765 G > C was protective [249–251] while in Asian populations it appeared to increase the risk of CRC [252–254]. Further, the risks conferred by the SNP were dependent on the use of NSAIDs, smoking, or obesity. Siemes et al. found that carriers of the C allele who used NSAIDs in the 5 years prior to the diagnosis of CRC had survivals that were twice as long as those who were homozygous for G allele who did not use NSAIDs [250]. Xing et al. observed that smokers with homozygous wild type GG genotype had 2.68-fold increased risk of CRC compared to nonsmokers, and those who were overweight had twice the risk of CRC compared to those with normal BMIs [253]. In African Americans, the polymorphism at Val511Ala has been studied. While the SNP that changes valine to alanine at 511 is present in 4–7 % of African Americans, it is nearly absent in Caucasians [255, 256], and although not statistically significant, Ala COX-2 genotype has been shown to decrease the risk of CRC and colorectal adenomas [256, 257]. Polymorphisms at other loci in African Americans demonstrated that 6,064 T > C, 10,848 G > A, 10,935 A > G, 5,229 G > T, and 10,935 A > G may affect adenoma risk; however, these studies demonstrated contradicting risk associations regarding the SNP at 10,935 [258, 259].

Limitations of Genetic Susceptibility Studies, Future Directions, and Conclusion

Throughout this chapter, we mentioned examples of SNPs at same genetic loci resulting in conflicting associations. This showcases one of the current limitations of gene susceptibility studies using polymorphisms. Many of the association studies evaluating the relationship between SNPs and disease have been plagued by the inability to produce results that are consistent. Possible explanations for these findings could be inadequately powered studies with small sample sizes, false positivity due to execution of large number of tests, inability to determine linkage disequilibrium, and admixture of study populations [260]. Throughout this chapter, examples of differential effects of SNPs depending on ethnicity have been presented, which suggest that a polymorphism as a disease marker from one population cannot be applied to a different group of patients. Furthermore, though statistically significant associations have been identified through GWAS, an enormous gap exists in providing biological mechanisms for how these SNPs are related to disease pathogenesis. Similarly, the majority of these genomic markers explain only a small magnitude of susceptibility risk at roughly 10 %, and with slightly higher risk for homozygotes, which points to the possibility that if higher powered GWAS could be performed, effect sizes may increase [261]. Lastly, there are limitations in these studies of identifying gene–gene or gene–environment interactions, which have had profound impacts in some cases [261].

Despite these limitations, a number of GWAS have been performed that identified and replicated genetic variants associated with CRC risk and prognosis. Programs such as the International HapMap project, the National Institute of Health's dbSNP database, and the National Human Genome Research Institute's catalog of published GWAS studies work to bring together available knowledge. Results from these studies have confirmed key pathways as well as lifestyle factors that could affect primary and secondary prevention of CRC. New findings can also lead to targeted therapies as well as identify high-risk populations that could benefit from early intervention. Future studies will reveal if currently discovered risk alleles will endure to have real clinical applications.

In conclusion, the higher incidence and worse survival in CRC observed in African Americans compared to Caucasians may be partly due to higher obesity and insulin resistance present in African Americans. While diet and exercise likely impact the incidence of these conditions, literature has shown that African Americans have a disproportionately higher insulin resistance while having lower central obesity than Caucasians. Further, currently available evidence demonstrates that levels of cytokines and peptide hormones that contribute to obesity, insulin resistance, chronic inflammation, and CRC are influenced by genetic polymorphisms that are present in a race-specific manner. These polymorphisms have been associated with altered CRC risk while modifying response to environmental agents such as NSAIDs, smoking, and obesity. Given that there are millions of SNPs, most of which have yet to be discovered and their biological significance elucidated, the extent to which SNPs or combinations of SNPs actually contribute to the worse prognosis observed in African Americans with CRC is yet to be determined. However, as GWAS and genetic analyses become more widespread and available, it becomes increasingly possible to be able to explain why African Americans or other populations (i.e., those with family history of sporadic cancer) may have increased CRC risk and understand the biological mechanisms behind disease pathogenesis. Accordingly, in the future we may be able to offer interventions and treatments tailored to specific genotypes and high-risk populations.

References

1. Calle EE, Rodriguez C, Walker-Thurmond K, Thun MJ (2003) Overweight, obesity, and mortality from cancer in a prospectively studied cohort of U.S. adults. N Engl J Med 348 (17):1625–1638. doi:10.1056/NEJMoa021423
2. Dehal A, Garrett T, Tedders SH, Arroyo C, Afriyie-Gyawu E, Zhang J (2011) Body mass index and death rate of colorectal cancer among a national cohort of U.S. adults. Nutr Cancer 63(8):1218–1225. doi:10.1080/01635581.2011.607539
3. Larsson SC, Wolk A (2007) Obesity and colon and rectal cancer risk: a meta-analysis of prospective studies. Am J Clin Nutr 86(3):556–565
4. Murphy TK, Calle EE, Rodriguez C, Kahn HS, Thun MJ (2000) Body mass index and colon cancer mortality in a large prospective study. Am J Epidemiol 152(9):847–854
5. Parr CL, Batty GD, Lam TH, Barzi F, Fang X, Ho SC, Jee SH, Ansary-Moghaddam A, Jamrozik K, Ueshima H, Woodward M, Huxley RR (2010) Body-mass index and cancer mortality in the Asia-Pacific Cohort Studies Collaboration: pooled analyses of 424,519 participants. Lancet Oncol 11(8):741–752. doi:10.1016/S1470-2045(10)70141-8
6. Renehan AG, Tyson M, Egger M, Heller RF, Zwahlen M (2008) Body-mass index and incidence of cancer: a systematic review and meta-analysis of prospective observational studies. Lancet 371(9612):569–578. doi:10.1016/S0140-6736(08)60269-X
7. Wang Y, Jacobs EJ, Patel AV, Rodriguez C, McCullough ML, Thun MJ, Calle EE (2008) A prospective study of waist circumference and body mass index in relation to colorectal cancer incidence. Cancer Causes Control 19(7):783–792. doi:10.1007/s10552-008-9141-x
8. Whitlock G, Lewington S, Sherliker P, Clarke R, Emberson J, Halsey J, Qizilbash N, Collins R, Peto R (2009) Body-mass index and cause-specific mortality in 900 000 adults: collaborative analyses of 57 prospective studies. Lancet 373(9669):1083–1096. doi:10.1016/S0140-6736(09)60318-4
9. Chan DS, Lau R, Aune D, Vieira R, Greenwood DC, Kampman E, Norat T (2011) Red and processed meat and colorectal cancer incidence: meta-analysis of prospective studies. PLoS One 6(6):e20456. doi:10.1371/journal.pone.0020456
10. Chao A, Thun MJ, Connell CJ, McCullough ML, Jacobs EJ, Flanders WD, Rodriguez C, Sinha R, Calle EE (2005) Meat consumption and risk of colorectal cancer. JAMA 293 (2):172–182. doi:10.1001/jama.293.2.172
11. Cross AJ, Ferrucci LM, Risch A, Graubard BI, Ward MH, Park Y, Hollenbeck AR, Schatzkin A, Sinha R (2010) A large prospective study of meat consumption and colorectal cancer risk: an investigation of potential mechanisms underlying this association. Cancer Res 70(6):2406–2414. doi:10.1158/0008-5472.CAN-09-3929
12. Johnson CM, Wei C, Ensor JE, Smolenski DJ, Amos CI, Levin B, Berry DA (2013) Meta-analyses of colorectal cancer risk factors. Cancer Causes Control. doi:10.1007/s10552-013-0201-5
13. Cho E, Smith-Warner SA, Ritz J, van den Brandt PA, Colditz GA, Folsom AR, Freudenheim JL, Giovannucci E, Goldbohm RA, Graham S, Holmberg L, Kim DH, Malila N, Miller AB, Pietinen P, Rohan TE, Sellers TA, Speizer FE, Willett WC, Wolk A, Hunter DJ (2004) Alcohol intake and colorectal cancer: a pooled analysis of 8 cohort studies. Ann Intern Med 140(8):603–613
14. Fedirko V, Tramacere I, Bagnardi V, Rota M, Scotti L, Islami F, Negri E, Straif K, Romieu I, La Vecchia C, Boffetta P, Jenab M (2011) Alcohol drinking and colorectal cancer risk: an overall and dose-response meta-analysis of published studies. Ann Oncol 22(9):1958–1972. doi:10.1093/annonc/mdq653
15. Botteri E, Iodice S, Bagnardi V, Raimondi S, Lowenfels AB, Maisonneuve P (2008) Smoking and colorectal cancer: a meta-analysis. JAMA 300(23):2765–2778. doi:10.1001/jama.2008.839

16. Atkin WS, Morson BC, Cuzick J (1992) Long-term risk of colorectal cancer after excision of rectosigmoid adenomas. N Engl J Med 326(10):658–662. doi:10.1056/NEJM199203053261002

17. Cottet V, Pariente A, Nalet B, Lafon J, Milan C, Olschwang S, Bonaiti-Pellie C, Faivre J, Bonithon-Kopp C (2007) Colonoscopic screening of first-degree relatives of patients with large adenomas: increased risk of colorectal tumors. Gastroenterology 133(4):1086–1092. doi:10.1053/j.gastro.2007.07.023

18. Winawer SJ, Zauber AG, Gerdes H, O'Brien MJ, Gottlieb LS, Sternberg SS, Bond JH, Waye JD, Schapiro M, Panish JF et al (1996) Risk of colorectal cancer in the families of patients with adenomatous polyps. National Polyp Study Workgroup. N Engl J Med 334(2):82–87. doi:10.1056/NEJM199601113340204

19. Deng L, Gui Z, Zhao L, Wang J, Shen L (2012) Diabetes mellitus and the incidence of colorectal cancer: an updated systematic review and meta-analysis. Dig Dis Sci 57(6):1576–1585. doi:10.1007/s10620-012-2055-1

20. He J, Stram DO, Kolonel LN, Henderson BE, Le Marchand L, Haiman CA (2010) The association of diabetes with colorectal cancer risk: the Multiethnic Cohort. Br J Cancer 103(1):120–126. doi:10.1038/sj.bjc.6605721

21. Larsson SC, Orsini N, Wolk A (2005) Diabetes mellitus and risk of colorectal cancer: a meta-analysis. J Natl Cancer Inst 97(22):1679–1687. doi:10.1093/jnci/dji375

22. Yang YX, Hennessy S, Lewis JD (2005) Type 2 diabetes mellitus and the risk of colorectal cancer. Clin Gastroenterol Hepatol 3(6):587–594

23. Boyle T, Keegel T, Bull F, Heyworth J, Fritschi L (2012) Physical activity and risks of proximal and distal colon cancers: a systematic review and meta-analysis. J Natl Cancer Inst 104(20):1548–1561. doi:10.1093/jnci/djs354

24. Wolin KY, Yan Y, Colditz GA, Lee IM (2009) Physical activity and colon cancer prevention: a meta-analysis. Br J Cancer 100(4):611–616. doi:10.1038/sj.bjc.6604917

25. Aune D, Lau R, Chan DS, Vieira R, Greenwood DC, Kampman E, Norat T (2012) Dairy products and colorectal cancer risk: a systematic review and meta-analysis of cohort studies. Ann Oncol 23(1):37–45. doi:10.1093/annonc/mdr269

26. Cho E, Smith-Warner SA, Spiegelman D, Beeson WL, van den Brandt PA, Colditz GA, Folsom AR, Fraser GE, Freudenheim JL, Giovannucci E, Goldbohm RA, Graham S, Miller AB, Pietinen P, Potter JD, Rohan TE, Terry P, Toniolo P, Virtanen MJ, Willett WC, Wolk A, Wu K, Yaun SS, Zeleniuch-Jacquotte A, Hunter DJ (2004) Dairy foods, calcium, and colorectal cancer: a pooled analysis of 10 cohort studies. J Natl Cancer Inst 96(13):1015–1022

27. Shaukat A, Scouras N, Schunemann HJ (2005) Role of supplemental calcium in the recurrence of colorectal adenomas: a metaanalysis of randomized controlled trials. Am J Gastroenterol 100(2):390–394. doi:10.1111/j.1572-0241.2005.41220.x

28. Vargas AJ, Thompson PA (2012) Diet and nutrient factors in colorectal cancer risk. Nutr Clin Pract 27(5):613–623. doi:10.1177/0884533612454885

29. Baron JA, Cole BF, Sandler RS, Haile RW, Ahnen D, Bresalier R, McKeown-Eyssen G, Summers RW, Rothstein R, Burke CA, Snover DC, Church TR, Allen JI, Beach M, Beck GJ, Bond JH, Byers T, Greenberg ER, Mandel JS, Marcon N, Mott LA, Pearson L, Saibil F, van Stolk RU (2003) A randomized trial of aspirin to prevent colorectal adenomas. N Engl J Med 348(10):891–899. doi:10.1056/NEJMoa021735

30. Johnson CC, Hayes RB, Schoen RE, Gunter MJ, Huang WY (2010) Non-steroidal anti-inflammatory drug use and colorectal polyps in the prostate, lung, colorectal, and ovarian cancer screening trial. Am J Gastroenterol 105(12):2646–2655. doi:10.1038/ajg.2010.349

31. Rothwell PM, Wilson M, Elwin CE, Norrving B, Algra A, Warlow CP, Meade TW (2010) Long-term effect of aspirin on colorectal cancer incidence and mortality: 20-year follow-up of five randomised trials. Lancet 376(9754):1741–1750. doi:10.1016/S0140-6736(10)61543-7

32. Ruder EH, Laiyemo AO, Graubard BI, Hollenbeck AR, Schatzkin A, Cross AJ (2011) Non-steroidal anti-inflammatory drugs and colorectal cancer risk in a large, prospective cohort. Am J Gastroenterol 106(7):1340–1350. doi:10.1038/ajg.2011.38

33. Sandler RS, Halabi S, Baron JA, Budinger S, Paskett E, Keresztes R, Petrelli N, Pipas JM, Karp DD, Loprinzi CL, Steinbach G, Schilsky R (2003) A randomized trial of aspirin to prevent colorectal adenomas in patients with previous colorectal cancer. N Engl J Med 348 (10):883–890. doi:10.1056/NEJMoa021633

34. Alexander D, Chatla C, Funkhouser E, Meleth S, Grizzle WE, Manne U (2004) Postsurgical disparity in survival between African Americans and Caucasians with colonic adenocarcinoma. Cancer 101(1):66–76. doi:10.1002/cncr.20337

35. Cheng X, Chen VW, Steele B, Ruiz B, Fulton J, Liu L, Carozza SE, Greenlee R (2001) Subsite-specific incidence rate and stage of disease in colorectal cancer by race, gender, and age group in the United States, 1992-1997. Cancer 92(10):2547–2554

36. Rim SH, Seeff L, Ahmed F, King JB, Coughlin SS (2009) Colorectal cancer incidence in the United States, 1999–2004: an updated analysis of data from the National Program of Cancer Registries and the Surveillance, Epidemiology, and End Results Program. Cancer 115 (9):1967–1976. doi:10.1002/cncr.24216

37. Wallace K, Hill EG, Lewin DN, Williamson G, Oppenheimer S, Ford ME, Wargovich MJ, Berger FG, Bolick SW, Thomas MB, Alberg AJ (2013) Racial disparities in advanced-stage colorectal cancer survival. Cancer Causes Control 24(3):463–471. doi:10.1007/s10552-012-0133-5

38. Wassira LN, Pinheiro PS, Symanowski J, Hansen A (2013) Racial-ethnic colorectal cancer survival disparities in the mountain west region: the case of Blacks compared to Whites. Ethn Dis 23(1):103–109

39. Yothers G, Sargent DJ, Wolmark N, Goldberg RM, O'Connell MJ, Benedetti JK, Saltz LB, Dignam JJ, Blackstock AW (2011) Outcomes among black patients with stage II and III colon cancer receiving chemotherapy: an analysis of ACCENT adjuvant trials. J Natl Cancer Inst 103(20):1498–1506. doi:10.1093/jnci/djr310

40. Pereira MA, Kartashov AI, Ebbeling CB, Van Horn L, Slattery ML, Jacobs DR Jr, Ludwig DS (2005) Fast-food habits, weight gain, and insulin resistance (the CARDIA study): 15-year prospective analysis. Lancet 365(9453):36–42. doi:10.1016/S0140-6736(04)17663-0

41. Flegal KM, Carroll MD, Kit BK, Ogden CL (2012) Prevalence of obesity and trends in the distribution of body mass index among US adults, 1999-2010. JAMA 307(5):491–497. doi:10.1001/jama.2012.39

42. Flegal KM, Carroll MD, Ogden CL, Johnson CL (2002) Prevalence and trends in obesity among US adults, 1999-2000. JAMA 288(14):1723–1727

43. Aleksandrova K, Nimptsch K, Pischon T (2013) Influence of obesity and related metabolic alterations on colorectal cancer risk. Curr Nutr Rep 2(1):1–9. doi:10.1007/s13668-012-0036-9

44. Alberti KG, Zimmet P, Shaw J (2006) Metabolic syndrome—a new world-wide definition. A consensus statement from the International Diabetes Federation. Diabet Med 23(5):469–480. doi:10.1111/j.1464-5491.2006.01858.x

45. Admiraal WM, Celik F, Gerdes VE, Dallal RM, Hoekstra JB, Holleman F (2012) Ethnic differences in weight loss and diabetes remission after bariatric surgery: a meta-analysis. Diabetes Care 35(9):1951–1958. doi:10.2337/dc12-0260

46. Bayham BE, Bellanger DE, Hargroder AG, Johnson WD, Greenway FL (2012) Racial differences in weight loss, payment method, and complications following Roux-en-Y gastric bypass and sleeve gastrectomy. Adv Ther 29(11):970–978. doi:10.1007/s12325-012-0062-4

47. Buffington CK, Marema RT (2006) Ethnic differences in obesity and surgical weight loss between African-American and Caucasian females. Obes Surg 16(2):159–165. doi:10.1381/096089206775565258

48. Burke GL, Bild DE, Hilner JE, Folsom AR, Wagenknecht LE, Sidney S (1996) Differences in weight gain in relation to race, gender, age and education in young adults: the CARDIA

Study. Coronary Artery Risk Development in Young Adults. Ethn Health 1(4):327–335. doi:10.1080/13557858.1996.9961802

49. Byrne NM, Weinsier RL, Hunter GR, Desmond R, Patterson MA, Darnell BE, Zuckerman PA (2003) Influence of distribution of lean body mass on resting metabolic rate after weight loss and weight regain: comparison of responses in white and black women. Am J Clin Nutr 77(6):1368–1373

50. Kumanyika SK, Obarzanek E, Stevens VJ, Hebert PR, Whelton PK (1991) Weight-loss experience of black and white participants in NHLBI-sponsored clinical trials. Am J Clin Nutr 53(6 Suppl):1631S–1638S

51. Wing RR, Anglin K (1996) Effectiveness of a behavioral weight control program for blacks and whites with NIDDM. Diabetes Care 19(5):409–413

52. Call R, Grimsley M, Cadwallader L, Cialone L, Hill M, Hreish V, King ST, Riche DM (2010) Insulin—carcinogen or mitogen? Preclinical and clinical evidence from prostate, breast, pancreatic, and colorectal cancer research. Postgrad Med 122(3):158–165. doi:10.3810/pgm.2010.05.2153

53. Durai R, Yang W, Gupta S, Seifalian AM, Winslet MC (2005) The role of the insulin-like growth factor system in colorectal cancer: review of current knowledge. Int J Colorectal Dis 20(3):203–220. doi:10.1007/s00384-004-0675-4

54. Nagel JM, Staffa J, Renner-Muller I, Horst D, Vogeser M, Langkamp M, Hoeflich A, Goke B, Kolligs FT, Mantzoros CS (2010) Insulin glargine and NPH insulin increase to a similar degree epithelial cell proliferation and aberrant crypt foci formation in colons of diabetic mice. Horm Cancer 1(6):320–330. doi:10.1007/s12672-010-0020-z

55. Tran TT, Naigamwalla D, Oprescu AI, Lam L, McKeown-Eyssen G, Bruce WR, Giacca A (2006) Hyperinsulinemia, but not other factors associated with insulin resistance, acutely enhances colorectal epithelial proliferation in vivo. Endocrinology 147(4):1830–1837. doi:10.1210/en.2005-1012

56. Weinstein D, Simon M, Yehezkel E, Laron Z, Werner H (2009) Insulin analogues display IGF-I-like mitogenic and anti-apoptotic activities in cultured cancer cells. Diabetes Metab Res Rev 25(1):41–49. doi:10.1002/dmrr.912

57. Brancati FL, Kao WH, Folsom AR, Watson RL, Szklo M (2000) Incident type 2 diabetes mellitus in African American and white adults: the Atherosclerosis Risk in Communities Study. JAMA 283(17):2253–2259

58. Cavicchia PP, Adams SA, Steck SE, Hussey JR, Liu J, Daguise VG, Hebert JR (2013) Racial disparities in colorectal cancer incidence by type 2 diabetes mellitus status. Cancer Causes Control 24(2):277–285. doi:10.1007/s10552-012-0095-7

59. Cohen DH, LeRoith D (2012) Obesity, type 2 diabetes, and cancer: the insulin and IGF connection. Endocr Relat Cancer 19(5):F27–F45. doi:10.1530/ERC-11-0374

60. Wong HL, Koh WP, Probst-Hensch NM, Van den Berg D, Yu MC, Ingles SA (2008) Insulin-like growth factor-1 promoter polymorphisms and colorectal cancer: a functional genomics approach. Gut 57(8):1090–1096. doi:10.1136/gut.2007.140855

61. DeLellis K, Rinaldi S, Kaaks RJ, Kolonel LN, Henderson B, Le Marchand L (2004) Dietary and lifestyle correlates of plasma insulin-like growth factor-I (IGF-I) and IGF binding protein-3 (IGFBP-3): the multiethnic cohort. Cancer Epidemiol Biomarkers Prev 13 (9):1444–1451

62. Jernstrom H, Chu W, Vesprini D, Tao Y, Majeed N, Deal C, Pollak M, Narod SA (2001) Genetic factors related to racial variation in plasma levels of insulin-like growth factor-1: implications for premenopausal breast cancer risk. Mol Genet Metab 72(2):144–154. doi:10.1006/mgme.2000.3130

63. Albu JB, Kovera AJ, Allen L, Wainwright M, Berk E, Raja-Khan N, Janumala I, Burkey B, Heshka S, Gallagher D (2005) Independent association of insulin resistance with larger amounts of intermuscular adipose tissue and a greater acute insulin response to glucose in African American than in white nondiabetic women. Am J Clin Nutr 82(6):1210–1217

64. Arslanian SA (2002) Metabolic differences between Caucasian and African-American children and the relationship to type 2 diabetes mellitus. J Pediatr Endocrinol Metab 15 Suppl 1:509–517
65. Bacha F, Saad R, Gungor N, Janosky J, Arslanian SA (2003) Obesity, regional fat distribution, and syndrome X in obese black versus white adolescents: race differential in diabetogenic and atherogenic risk factors. J Clin Endocrinol Metab 88(6):2534–2540
66. Goran MI, Nagy TR, Treuth MS, Trowbridge C, Dezenberg C, McGloin A, Gower BA (1997) Visceral fat in white and African American prepubertal children. Am J Clin Nutr 65(6):1703–1708
67. Haffner SM, D'Agostino R, Saad MF, Rewers M, Mykkanen L, Selby J, Howard G, Savage PJ, Hamman RF, Wagenknecht LE et al (1996) Increased insulin resistance and insulin secretion in nondiabetic African-Americans and Hispanics compared with non-Hispanic whites. The Insulin Resistance Atherosclerosis Study. Diabetes 45(6):742–748
68. Hyatt TC, Phadke RP, Hunter GR, Bush NC, Munoz AJ, Gower BA (2009) Insulin sensitivity in African-American and white women: association with inflammation. Obesity (Silver Spring) 17(2):276–282. doi:10.1038/oby.2008.549
69. Liska D, Dufour S, Zern TL, Taksali S, Cali AM, Dziura J, Shulman GI, Pierpont BM, Caprio S (2007) Interethnic differences in muscle, liver and abdominal fat partitioning in obese adolescents. PLoS One 2(6):e569. doi:10.1371/journal.pone.0000569
70. Harris MI, Cowie CC, Gu K, Francis ME, Flegal K, Eberhardt MS (2002) Higher fasting insulin but lower fasting C-peptide levels in African Americans in the US population. Diabetes Metab Res Rev 18(2):149–155. doi:10.1002/dmrr.273
71. Walker SE, Gurka MJ, Oliver MN, Johns DW, DeBoer MD (2012) Racial/ethnic discrepancies in the metabolic syndrome begin in childhood and persist after adjustment for environmental factors. Nutr Metab Cardiovasc Dis 22(2):141–148. doi:10.1016/j.numecd.2010.05.006
72. Wang D, Chen J, Chen H, Duan Z, Xu Q, Wei M, Wang L, Zhong M (2012) Leptin regulates proliferation and apoptosis of colorectal carcinoma through PI3K/Akt/mTOR signalling pathway. J Biosci 37(1):91–101
73. Hoda MR, Keely SJ, Bertelsen LS, Junger WG, Dharmasena D, Barrett KE (2007) Leptin acts as a mitogenic and antiapoptotic factor for colonic cancer cells. Br J Surg 94(3):346–354. doi:10.1002/bjs.5530
74. Drew JE (2012) Molecular mechanisms linking adipokines to obesity-related colon cancer: focus on leptin. Proc Nutr Soc 71(1):175–180. doi:10.1017/S0029665111003259
75. Endo H, Hosono K, Uchiyama T, Sakai E, Sugiyama M, Takahashi H, Nakajima N, Wada K, Takeda K, Nakagama H, Nakajima A (2011) Leptin acts as a growth factor for colorectal tumours at stages subsequent to tumour initiation in murine colon carcinogenesis. Gut 60(10):1363–1371. doi:10.1136/gut.2010.235754
76. Stattin P, Lukanova A, Biessy C, Soderberg S, Palmqvist R, Kaaks R, Olsson T, Jellum E (2004) Obesity and colon cancer: does leptin provide a link? Int J Cancer 109(1):149–152. doi:10.1002/ijc.11668
77. Tamakoshi K, Toyoshima H, Wakai K, Kojima M, Suzuki K, Watanabe Y, Hayakawa N, Yatsuya H, Kondo T, Tokudome S, Hashimoto S, Suzuki S, Kawado M, Ozasa K, Ito Y, Tamakoshi A (2005) Leptin is associated with an increased female colorectal cancer risk: a nested case-control study in Japan. Oncology 68(4–6):454–461. doi:10.1159/000086988
78. Tutino V, Notarnicola M, Guerra V, Lorusso D, Caruso MG (2011) Increased soluble leptin receptor levels are associated with advanced tumor stage in colorectal cancer patients. Anticancer Res 31(10):3381–3383
79. Kim HH, Kim YS, Kang YK, Moon JS (2012) Leptin and peroxisome proliferator-activated receptor gamma expression in colorectal adenoma. World J Gastroenterol 18(6):557–562. doi:10.3748/wjg.v18.i6.557
80. Healy LA, Howard JM, Ryan AM, Beddy P, Mehigan B, Stephens R, Reynolds JV (2012) Metabolic syndrome and leptin are associated with adverse pathological features in male

colorectal cancer patients. Colorectal Dis 14(2):157–165. doi:10.1111/j.1463-1318.2011. 02562.x

81. Petty KH, Li K, Dong Y, Fortenberry J, Stallmann-Jorgensen I, Guo D, Zhu H (2010) Sex dimorphisms in inflammatory markers and adiposity in African-American youth. Int J Pediatr Obes 5(4):327–333. doi:10.3109/17477160903497019

82. Zuo H, Shi Z, Yuan B, Dai Y, Wu G, Hussain A (2013) Association between serum leptin concentrations and insulin resistance: a population-based study from China. PLoS One 8(1): e54615. doi:10.1371/journal.pone.0054615

83. Rasmussen-Torvik LJ, Wassel CL, Ding J, Carr J, Cushman M, Jenny N, Allison MA (2012) Associations of body mass index and insulin resistance with leptin, adiponectin, and the leptin-to-adiponectin ratio across ethnic groups: the Multi-Ethnic Study of Atherosclerosis (MESA). Ann Epidemiol 22(10):705–709. doi:10.1016/j.annepidem.2012.07.011

84. Khan UI, Wang D, Sowers MR, Mancuso P, Everson-Rose SA, Scherer PE, Wildman RP (2012) Race-ethnic differences in adipokine levels: the Study of Women's Health Across the Nation (SWAN). Metabolism 61(9):1261–1269. doi:10.1016/j.metabol.2012.02.005

85. Ruhl CE, Everhart JE (2001) Leptin concentrations in the United States: relations with demographic and anthropometric measures. Am J Clin Nutr 74(3):295–301

86. Azrad M, Gower BA, Hunter GR, Nagy TR (2013) Racial differences in adiponectin and leptin in healthy premenopausal women. Endocrine 43(3):586–592. doi:10.1007/s12020-012-9797-6

87. Ziemke F, Mantzoros CS (2010) Adiponectin in insulin resistance: lessons from translational research. Am J Clin Nutr 91(1):258S–261S. doi:10.3945/ajcn.2009.28449C

88. Nakajima TE, Yamada Y, Hamano T, Furuta K, Matsuda T, Fujita S, Kato K, Hamaguchi T, Shimada Y (2010) Adipocytokines as new promising markers of colorectal tumors: adiponectin for colorectal adenoma, and resistin and visfatin for colorectal cancer. Cancer Sci 101(5):1286–1291. doi:10.1111/j.1349-7006.2010.01518.x

89. Touvier M, Fezeu L, Ahluwalia N, Julia C, Charnaux N, Sutton A, Mejean C, Latino-Martel P, Hercberg S, Galan P, Czernichow S (2012) Pre-diagnostic levels of adiponectin and soluble vascular cell adhesion molecule-1 are associated with colorectal cancer risk. World J Gastroenterol 18(22):2805–2812. doi:10.3748/wjg.v18.i22.2805

90. Ukkola O, Santaniemi M (2002) Adiponectin: a link between excess adiposity and associated comorbidities? J Mol Med (Berl) 80(11):696–702. doi:10.1007/s00109-002-0378-7

91. Xu XT, Xu Q, Tong JL, Zhu MM, Huang ML, Ran ZH, Xiao SD (2011) Meta-analysis: circulating adiponectin levels and risk of colorectal cancer and adenoma. J Dig Dis 12 (4):234–244. doi:10.1111/j.1751-2980.2011.00504.x

92. Sugiyama M, Takahashi H, Hosono K, Endo H, Kato S, Yoneda K, Nozaki Y, Fujita K, Yoneda M, Wada K, Nakagama H, Nakajima A (2009) Adiponectin inhibits colorectal cancer cell growth through the AMPK/mTOR pathway. Int J Oncol 34(2):339–344

93. Saxena A, Chumanevich A, Fletcher E, Larsen B, Lattwein K, Kaur K, Fayad R (2012) Adiponectin deficiency: role in chronic inflammation induced colon cancer. Biochim Biophys Acta 1822(4):527–536. doi:10.1016/j.bbadis.2011.12.006

94. Mutoh M, Teraoka N, Takasu S, Takahashi M, Onuma K, Yamamoto M, Kubota N, Iseki T, Kadowaki T, Sugimura T, Wakabayashi K (2011) Loss of adiponectin promotes intestinal carcinogenesis in Min and wild-type mice. Gastroenterology 140(7):2000–2008, 2008 e2001-2002. doi:10.1053/j.gastro.2011.02.019

95. Moon HS, Liu X, Nagel JM, Chamberland JP, Diakopoulos KN, Brinkoetter MT, Hatziapostolou M, Wu Y, Robson SC, Iliopoulos D, Mantzoros CS (2013) Salutary effects of adiponectin on colon cancer: in vivo and in vitro studies in mice. Gut 62(4):561–570. doi:10.1136/gutjnl-2012-302092

96. Bacha F, Saad R, Gungor N, Arslanian SA (2005) Does adiponectin explain the lower insulin sensitivity and hyperinsulinemia of African-American children? Pediatr Diabetes 6(2):100–102. doi:10.1111/j.1399-543X.2005.00108.x

97. Degawa-Yamauchi M, Dilts JR, Bovenkerk JE, Saha C, Pratt JH, Considine RV (2003) Lower serum adiponectin levels in African-American boys. Obes Res 11(11):1384–1390. doi:10.1038/oby.2003.187
98. Schutte AE, Huisman HW, Schutte R, Malan L, van Rooyen JM, Malan NT, Schwarz PE (2007) Differences and similarities regarding adiponectin investigated in African and Caucasian women. Eur J Endocrinol 157(2):181–188. doi:10.1530/EJE-07-0044
99. Balkwill F (2002) Tumor necrosis factor or tumor promoting factor? Cytokine Growth Factor Rev 13(2):135–141
100. Kaiser GC, Polk DB (1997) Tumor necrosis factor alpha regulates proliferation in a mouse intestinal cell line. Gastroenterology 112(4):1231–1240
101. Grimm M, Lazariotou M, Kircher S, Hofelmayr A, Germer CT, von Rahden BH, Waaga-Gasser AM, Gasser M (2011) Tumor necrosis factor-alpha is associated with positive lymph node status in patients with recurrence of colorectal cancer-indications for anti-TNF-alpha agents in cancer treatment. Cell Oncol (Dordr) 34(4):315–326. doi:10.1007/s13402-011-0027-7
102. Kim S, Keku TO, Martin C, Galanko J, Woosley JT, Schroeder JC, Satia JA, Halabi S, Sandler RS (2008) Circulating levels of inflammatory cytokines and risk of colorectal adenomas. Cancer Res 68(1):323–328. doi:10.1158/0008-5472.CAN-07-2924
103. Caballero AE, Bousquet-Santos K, Robles-Osorio L, Montagnani V, Soodini G, Porramatikul S, Hamdy O, Nobrega AC, Horton ES (2008) Overweight Latino children and adolescents have marked endothelial dysfunction and subclinical vascular inflammation in association with excess body fat and insulin resistance. Diabetes Care 31(3):576–582. doi:10.2337/dc07-1540
104. Gonzalez F, Minium J, Rote NS, Kirwan JP (2006) Altered tumor necrosis factor alpha release from mononuclear cells of obese reproductive-age women during hyperglycemia. Metabolism 55(2):271–276. doi:10.1016/j.metabol.2005.08.022
105. Hirashita T, Ohta M, Endo Y, Masuda T, Iwashita Y, Kitano S (2012) Effects of visceral fat resection and gastric banding in an obese diabetic rat model. Surgery 151(1):6–12. doi:10.1016/j.surg.2011.06.025
106. Janowska J, Zahorska-Markiewicz B, Olszanecka-Glinianowicz M (2006) Relationship between serum resistin concentration and proinflammatory cytokines in obese women with impaired and normal glucose tolerance. Metabolism 55(11):1495–1499. doi:10.1016/j.metabol.2006.06.020
107. Reinehr T, Stoffel-Wagner B, Roth CL, Andler W (2005) High-sensitive C-reactive protein, tumor necrosis factor alpha, and cardiovascular risk factors before and after weight loss in obese children. Metabolism 54(9):1155–1161. doi:10.1016/j.metabol.2005.03.022
108. Ruan H, Lodish HF (2003) Insulin resistance in adipose tissue: direct and indirect effects of tumor necrosis factor-alpha. Cytokine Growth Factor Rev 14(5):447–455
109. Sheu WH, Chang TM, Lee WJ, Ou HC, Wu CM, Tseng LN, Lang HF, Wu CS, Wan CJ, Lee IT (2008) Effect of weight loss on proinflammatory state of mononuclear cells in obese women. Obesity (Silver Spring) 16(5):1033–1038. doi:10.1038/oby.2008.37
110. Hotamisligil GS, Arner P, Caro JF, Atkinson RL, Spiegelman BM (1995) Increased adipose tissue expression of tumor necrosis factor-alpha in human obesity and insulin resistance. J Clin Invest 95(5):2409–2415. doi:10.1172/JCI117936
111. Flores MB, Rocha GZ, Damas-Souza DM, Osorio-Costa F, Dias MM, Ropelle ER, Camargo JA, de Carvalho RB, Carvalho HF, Saad MJ, Carvalheira JB (2012) Obesity-induced increase in tumor necrosis factor-alpha leads to development of colon cancer in mice. Gastroenterology 143(3):741–753 e741-744. doi:10.1053/j.gastro.2012.05.045
112. Pendyala S, Neff LM, Suarez-Farinas M, Holt PR (2011) Diet-induced weight loss reduces colorectal inflammation: implications for colorectal carcinogenesis. Am J Clin Nutr 93 (2):234–242. doi:10.3945/ajcn.110.002683

113. Fisher G, Hyatt TC, Hunter GR, Oster RA, Desmond RA, Gower BA (2012) Markers of inflammation and fat distribution following weight loss in African-American and white women. Obesity (Silver Spring) 20(4):715–720. doi:10.1038/oby.2011.85

114. Olson NC, Callas PW, Hanley AJ, Festa A, Haffner SM, Wagenknecht LE, Tracy RP (2012) Circulating levels of TNF-alpha are associated with impaired glucose tolerance, increased insulin resistance, and ethnicity: the Insulin Resistance Atherosclerosis Study. J Clin Endocrinol Metab 97(3):1032–1040. doi:10.1210/jc.2011-2155

115. Hopkins MH, Flanders WD, Bostick RM (2012) Associations of circulating inflammatory biomarkers with risk factors for colorectal cancer in colorectal adenoma patients. Biomark Insights 7:143–150. doi:10.4137/BMI.S10092

116. Ho GY, Wang T, Gunter MJ, Strickler HD, Cushman M, Kaplan RC, Wassertheil-Smoller S, Xue X, Rajpathak SN, Chlebowski RT, Vitolins MZ, Scherer PE, Rohan TE (2012) Adipokines linking obesity with colorectal cancer risk in postmenopausal women. Cancer Res 72(12):3029–3037. doi:10.1158/0008-5472.CAN-11-2771

117. Kantola T, Klintrup K, Vayrynen JP, Vornanen J, Bloigu R, Karhu T, Herzig KH, Napankangas J, Makela J, Karttunen TJ, Tuomisto A, Makinen MJ (2012) Stage-dependent alterations of the serum cytokine pattern in colorectal carcinoma. Br J Cancer 107(10):1729–1736. doi:10.1038/bjc.2012.456

118. Yeh KY, Li YY, Hsieh LL, Lu CH, Chou WC, Liaw CC, Tang RP, Liao SK (2010) Analysis of the effect of serum interleukin-6 (IL-6) and soluble IL-6 receptor levels on survival of patients with colorectal cancer. Jpn J Clin Oncol 40(6):580–587. doi:10.1093/jjco/hyq010

119. Waldner MJ, Foersch S, Neurath MF (2012) Interleukin-6—a key regulator of colorectal cancer development. Int J Biol Sci 8(9):1248–1253. doi:10.7150/ijbs.4614

120. Knupfer H, Preiss R (2010) Serum interleukin-6 levels in colorectal cancer patients—a summary of published results. Int J Colorectal Dis 25(2):135–140. doi:10.1007/s00384-009-0818-8

121. Takasu C, Shimada M, Kurita N, Iwata T, Nishioka M, Morimoto S, Yoshikawa K, Miyatani T, Kashihara H (2013) Impact of C-reactive protein on prognosis of patients with colorectal carcinoma. Hepatogastroenterology 60(123):507–511. doi:10.5754/hge11425

122. Ishizuka M, Nagata H, Takagi K, Iwasaki Y, Kubota K (2013) Inflammation-based prognostic system predicts survival after surgery for stage IV colorectal cancer. Am J Surg 205 (1):22–28. doi:10.1016/j.amjsurg.2012.04.012

123. Wong VK, Malik HZ, Hamady ZZ, Al-Mukhtar A, Gomez D, Prasad KR, Toogood GJ, Lodge JP (2007) C-reactive protein as a predictor of prognosis following curative resection for colorectal liver metastases. Br J Cancer 96(2):222–225. doi:10.1038/sj.bjc.6603558

124. van de Poll MC, Klaver YL, Lemmens VE, Leenders BJ, Nienhuijs SW, de Hingh IH (2011) C-reactive protein concentration is associated with prognosis in patients suffering from peritoneal carcinomatosis of colorectal origin. Int J Colorectal Dis 26(8):1067–1073. doi:10.1007/s00384-011-1187-7

125. Hopkins MH, Owen J, Ahearn T, Fedirko V, Flanders WD, Jones DP, Bostick RM (2011) Effects of supplemental vitamin D and calcium on biomarkers of inflammation in colorectal adenoma patients: a randomized, controlled clinical trial. Cancer Prev Res (Phila) 4 (10):1645–1654. doi:10.1158/1940-6207.CAPR-11-0105

126. Colangelo LA, Chiu B, Kopp P, Liu K, Gapstur SM (2009) Serum IGF-I and C-reactive protein in healthy black and white young men: the CARDIA male hormone study. Growth Horm IGF Res 19(5):420–425. doi:10.1016/j.ghir.2008.12.002

127. Kraus VB, Stabler TV, Luta G, Renner JB, Dragomir AD, Jordan JM (2007) Interpretation of serum C-reactive protein (CRP) levels for cardiovascular disease risk is complicated by race, pulmonary disease, body mass index, gender, and osteoarthritis. Osteoarthritis Cartilage 15 (8):966–971. doi:10.1016/j.joca.2007.02.014

128. Paalani M, Lee JW, Haddad E, Tonstad S (2011) Determinants of inflammatory markers in a bi-ethnic population. Ethn Dis 21(2):142–149

129. Morita I (2002) Distinct functions of COX-1 and COX-2. Prostaglandins Other Lipid Mediat 68–69:165–175

130. Brand L, Munding J, Pox CP, Ziebarth W, Reiser M, Huppe D, Schmiegel W, Reinacher-Schick A, Tannapfel A (2013) ss-catenin, Cox-2 and p53 immunostaining in colorectal adenomas to predict recurrence after endoscopic polypectomy. Int J Colorectal Dis 28 (8):1091–1098. doi:10.1007/s00384-013-1667-z

131. Al-Maghrabi J, Buhmeida A, Emam E, Syrjanen K, Sibiany A, Al-Qahtani M, Al-Ahwal M (2012) Cyclooxygenase-2 expression as a predictor of outcome in colorectal carcinoma. World J Gastroenterol 18(15):1793–1799. doi:10.3748/wjg.v18.i15.1793

132. Xiong B, Sun TJ, Hu WD, Cheng FL, Mao M, Zhou YF (2005) Expression of cyclooxygenase-2 in colorectal cancer and its clinical significance. World J Gastroenterol 11(8):1105–1108

133. Eberhart CE, Coffey RJ, Radhika A, Giardiello FM, Ferrenbach S, DuBois RN (1994) Up-regulation of cyclooxygenase 2 gene expression in human colorectal adenomas and adenocarcinomas. Gastroenterology 107(4):1183–1188

134. Min BS, Choi YJ, Pyo HR, Kim H, Seong J, Chung HC, Rha SY, Kim NK (2008) Cyclooxygenase-2 expression in pretreatment biopsy as a predictor of tumor responses after preoperative chemoradiation in rectal cancer. Arch Surg 143(11):1091–1097. doi:10.1001/archsurg.143.11.1091, discussion 1097

135. Ogino S, Kirkner GJ, Nosho K, Irahara N, Kure S, Shima K, Hazra A, Chan AT, Dehari R, Giovannucci EL, Fuchs CS (2008) Cyclooxygenase-2 expression is an independent predictor of poor prognosis in colon cancer. Clin Cancer Res 14(24):8221–8227. doi:10.1158/1078-0432.CCR-08-1841

136. Soumaoro LT, Uetake H, Higuchi T, Takagi Y, Enomoto M, Sugihara K (2004) Cyclooxygenase-2 expression: a significant prognostic indicator for patients with colorectal cancer. Clin Cancer Res 10(24):8465–8471. doi:10.1158/1078-0432.CCR-04-0653

137. Risch NJ (2000) Searching for genetic determinants in the new millennium. Nature 405 (6788):847–856. doi:10.1038/35015718

138. Risch N (2001) The genetic epidemiology of cancer: interpreting family and twin studies and their implications for molecular genetic approaches. Cancer Epidemiol Biomarkers Prev 10 (7):733–741

139. Abuli A, Bessa X, Gonzalez JR, Ruiz-Ponte C, Caceres A, Munoz J, Gonzalo V, Balaguer F, Fernandez-Rozadilla C, Gonzalez D, de Castro L, Clofent J, Bujanda L, Cubiella J, Rene JM, Morillas JD, Lanas A, Rigau J, Garcia AM, Latorre M, Salo J, Fernandez Banares F, Arguello L, Pena E, Vilella A, Riestra S, Carreno R, Paya A, Alenda C, Xicola RM, Doyle BJ, Jover R, Llor X, Carracedo A, Castells A, Castellvi-Bel S, Andreu M (2010) Susceptibility genetic variants associated with colorectal cancer risk correlate with cancer phenotype. Gastroenterology 139(3):788–796, 796 e781-786. doi:10.1053/j.gastro.2010.05.072

140. Goel A, Boland CR (2010) Recent insights into the pathogenesis of colorectal cancer. Curr Opin Gastroenterol 26(1):47–52. doi:10.1097/MOG.0b013e328332b850

141. Haerian MS, Baum L, Haerian BS (2011) Association of 8q24.21 loci with the risk of colorectal cancer: a systematic review and meta-analysis. J Gastroenterol Hepatol 26 (10):1475–1484. doi:10.1111/j.1440-1746.2011.06831.x

142. Hoskins JM, Ong PS, Keku TO, Galanko JA, Martin CF, Coleman CA, Wolfe M, Sandler RS, McLeod HL (2012) Association of eleven common, low-penetrance colorectal cancer susceptibility genetic variants at six risk loci with clinical outcome. PLoS One 7(7):e41954. doi:10.1371/journal.pone.0041954

143. Houlston RS, Webb E, Broderick P, Pittman AM, Di Bernardo MC, Lubbe S, Chandler I, Vijayakrishnan J, Sullivan K, Penegar S, Carvajal-Carmona L, Howarth K, Jaeger E, Spain SL, Walther A, Barclay E, Martin L, Gorman M, Domingo E, Teixeira AS, Kerr D, Cazier JB, Niittymaki I, Tuupanen S, Karhu A, Aaltonen LA, Tomlinson IP, Farrington SM, Tenesa A, Prendergast JG, Barnetson RA, Cetnarskyj R, Porteous ME, Pharoah PD, Koessler T, Hampe J, Buch S, Schafmayer C, Tepel J, Schreiber S, Volzke H, Chang-Claude J,

Hoffmeister M, Brenner H, Zanke BW, Montpetit A, Hudson TJ, Gallinger S, Campbell H, Dunlop MG (2008) Meta-analysis of genome-wide association data identifies four new susceptibility loci for colorectal cancer. Nat Genet 40(12):1426–1435. doi:10.1038/ng.262

144. Tenesa A, Farrington SM, Prendergast JG, Porteous ME, Walker M, Haq N, Barnetson RA, Theodoratou E, Cetnarskyj R, Cartwright N, Semple C, Clark AJ, Reid FJ, Smith LA, Kavoussanakis K, Koessler T, Pharoah PD, Buch S, Schafmayer C, Tepel J, Schreiber S, Volzke H, Schmidt CO, Hampe J, Chang-Claude J, Hoffmeister M, Brenner H, Wilkening S, Canzian F, Capella G, Moreno V, Deary IJ, Starr JM, Tomlinson IP, Kemp Z, Howarth K, Carvajal-Carmona L, Webb E, Broderick P, Vijayakrishnan J, Houlston RS, Rennert G, Ballinger D, Rozek L, Gruber SB, Matsuda K, Kidokoro T, Nakamura Y, Zanke BW, Greenwood CM, Rangrej J, Kustra R, Montpetit A, Hudson TJ, Gallinger S, Campbell H, Dunlop MG (2008) Genome-wide association scan identifies a colorectal cancer susceptibility locus on 11q23 and replicates risk loci at 8q24 and 18q21. Nat Genet 40(5):631–637. doi:10.1038/ng.133

145. Tomlinson IP, Webb E, Carvajal-Carmona L, Broderick P, Howarth K, Pittman AM, Spain S, Lubbe S, Walther A, Sullivan K, Jaeger E, Fielding S, Rowan A, Vijayakrishnan J, Domingo E, Chandler I, Kemp Z, Qureshi M, Farrington SM, Tenesa A, Prendergast JG, Barnetson RA, Penegar S, Barclay E, Wood W, Martin L, Gorman M, Thomas H, Peto J, Bishop DT, Gray R, Maher ER, Lucassen A, Kerr D, Evans DG, Schafmayer C, Buch S, Volzke H, Hampe J, Schreiber S, John U, Koessler T, Pharoah P, van Wezel T, Morreau H, Wijnen JT, Hopper JL, Southey MC, Giles GG, Severi G, Castellvi-Bel S, Ruiz-Ponte C, Carracedo A, Castells A, Forsti A, Hemminki K, Vodicka P, Naccarati A, Lipton L, Ho JW, Cheng KK, Sham PC, Luk J, Agundez JA, Ladero JM, de la Hoya M, Caldes T, Niittymaki I, Tuupanen S, Karhu A, Aaltonen L, Cazier JB, Campbell H, Dunlop MG, Houlston RS (2008) A genome-wide association study identifies colorectal cancer susceptibility loci on chromosomes 10p14 and 8q23.3. Nat Genet 40(5):623–630. doi:10.1038/ng.111

146. Takatsuno Y, Mimori K, Yamamoto K, Sato T, Niida A, Inoue H, Imoto S, Kawano S, Yamaguchi R, Toh H, Iinuma H, Ishimaru S, Ishii H, Suzuki S, Tokudome S, Watanabe M, Tanaka J, Kudo SE, Mochizuki H, Kusunoki M, Yamada K, Shimada Y, Moriya Y, Miyano S, Sugihara K, Mori M (2013) The rs6983267 SNP is associated with MYC transcription efficiency, which promotes progression and worsens prognosis of colorectal cancer. Ann Surg Oncol 20(4):1395–1402. doi:10.1245/s10434-012-2657-z

147. Dai J, Gu J, Huang M, Eng C, Kopetz ES, Ellis LM, Hawk E, Wu X (2012) GWAS-identified colorectal cancer susceptibility loci associated with clinical outcomes. Carcinogenesis 33 (7):1327–1331. doi:10.1093/carcin/bgs147

148. He J, Wilkens LR, Stram DO, Kolonel LN, Henderson BE, Wu AH, Le Marchand L, Haiman CA (2011) Generalizability and epidemiologic characterization of eleven colorectal cancer GWAS hits in multiple populations. Cancer Epidemiol Biomarkers Prev 20(1):70–81. doi:10.1158/1055-9965.EPI-10-0892

149. Kupfer SS, Torres JB, Hooker S, Anderson JR, Skol AD, Ellis NA, Kittles RA (2009) Novel single nucleotide polymorphism associations with colorectal cancer on chromosome 8q24 in African and European Americans. Carcinogenesis 30(8):1353–1357. doi:10.1093/carcin/bgp123

150. Frayling TM, Timpson NJ, Weedon MN, Zeggini E, Freathy RM, Lindgren CM, Perry JR, Elliott KS, Lango H, Rayner NW, Shields B, Harries LW, Barrett JC, Ellard S, Groves CJ, Knight B, Patch AM, Ness AR, Ebrahim S, Lawlor DA, Ring SM, Ben-Shlomo Y, Jarvelin MR, Sovio U, Bennett AJ, Melzer D, Ferrucci L, Loos RJ, Barroso I, Wareham NJ, Karpe F, Owen KR, Cardon LR, Walker M, Hitman GA, Palmer CN, Doney AS, Morris AD, Smith GD, Hattersley AT, McCarthy MI (2007) A common variant in the FTO gene is associated with body mass index and predisposes to childhood and adult obesity. Science 316 (5826):889–894. doi:10.1126/science.1141634

151. Scuteri A, Sanna S, Chen WM, Uda M, Albai G, Strait J, Najjar S, Nagaraja R, Orru M, Usala G, Dei M, Lai S, Maschio A, Busonero F, Mulas A, Ehret GB, Fink AA, Weder AB,

Cooper RS, Galan P, Chakravarti A, Schlessinger D, Cao A, Lakatta E, Abecasis GR (2007) Genome-wide association scan shows genetic variants in the FTO gene are associated with obesity-related traits. PLoS Genet 3(7):e115. doi:10.1371/journal.pgen.0030115

152. Hertel JK, Johansson S, Raeder H, Midthjell K, Lyssenko V, Groop L, Molven A, Njolstad PR (2008) Genetic analysis of recently identified type 2 diabetes loci in 1,638 unselected patients with type 2 diabetes and 1,858 control participants from a Norwegian population-based cohort (the HUNT study). Diabetologia 51(6):971–977. doi:10.1007/s00125-008-0982-3

153. Bressler J, Kao WH, Pankow JS, Boerwinkle E (2010) Risk of type 2 diabetes and obesity is differentially associated with variation in FTO in whites and African-Americans in the ARIC study. PLoS One 5(5):e10521. doi:10.1371/journal.pone.0010521

154. Song Y, You NC, Hsu YH, Howard BV, Langer RD, Manson JE, Nathan L, Niu T, F Tinker L, Liu S (2008) FTO polymorphisms are associated with obesity but not diabetes risk in postmenopausal women. Obesity (Silver Spring) 16(11):2472–2480. doi:10.1038/oby.2008.408

155. Dorajoo R, Blakemore AI, Sim X, Ong RT, Ng DP, Seielstad M, Wong TY, Saw SM, Froguel P, Liu J, Tai ES (2012) Replication of 13 obesity loci among Singaporean Chinese, Malay and Asian-Indian populations. Int J Obes (Lond) 36(1):159–163. doi:10.1038/ijo.2011.86

156. Ng MC, Park KS, Oh B, Tam CH, Cho YM, Shin HD, Lam VK, Ma RC, So WY, Cho YS, Kim HL, Lee HK, Chan JC, Cho NH (2008) Implication of genetic variants near TCF7L2, SLC30A8, HHEX, CDKAL1, CDKN2A/B, IGF2BP2, and FTO in type 2 diabetes and obesity in 6,719 Asians. Diabetes 57(8):2226–2233. doi:10.2337/db07-1583

157. Yajnik CS, Janipalli CS, Bhaskar S, Kulkarni SR, Freathy RM, Prakash S, Mani KR, Weedon MN, Kale SD, Deshpande J, Krishnaveni GV, Veena SR, Fall CH, McCarthy MI, Frayling TM, Hattersley AT, Chandak GR (2009) FTO gene variants are strongly associated with type 2 diabetes in South Asian Indians. Diabetologia 52(2):247–252. doi:10.1007/s00125-008-1186-6

158. Li H, Kilpelainen TO, Liu C, Zhu J, Liu Y, Hu C, Yang Z, Zhang W, Bao W, Cha S, Wu Y, Yang T, Sekine A, Choi BY, Yajnik CS, Zhou D, Takeuchi F, Yamamoto K, Chan JC, Mani KR, Been LF, Imamura M, Nakashima E, Lee N, Fujisawa T, Karasawa S, Wen W, Joglekar CV, Lu W, Chang Y, Xiang Y, Gao Y, Liu S, Song Y, Kwak SH, Shin HD, Park KS, Fall CH, Kim JY, Sham PC, Lam KS, Zheng W, Shu X, Deng H, Ikegami H, Krishnaveni GV, Sanghera DK, Chuang L, Liu L, Hu R, Kim Y, Daimon M, Hotta K, Jia W, Kooner JS, Chambers JC, Chandak GR, Ma RC, Maeda S, Dorajoo R, Yokota M, Takayanagi R, Kato N, Lin X, Loos RJ (2012) Association of genetic variation in FTO with risk of obesity and type 2 diabetes with data from 96,551 East and South Asians. Diabetologia 55(4):981–995. doi:10.1007/s00125-011-2370-7

159. Adeyemo A, Chen G, Zhou J, Shriner D, Doumatey A, Huang H, Rotimi C (2010) FTO genetic variation and association with obesity in West Africans and African Americans. Diabetes 59(6):1549–1554. doi:10.2337/db09-1252

160. Hester JM, Wing MR, Li J, Palmer ND, Xu J, Hicks PJ, Roh BH, Norris JM, Wagenknecht LE, Langefeld CD, Freedman BI, Bowden DW, Ng MC (2012) Implication of European-derived adiposity loci in African Americans. Int J Obes (Lond) 36(3):465–473. doi:10.1038/ijo.2011.131

161. Nock NL, Plummer SJ, Thompson CL, Casey G, Li L (2011) FTO polymorphisms are associated with adult body mass index (BMI) and colorectal adenomas in African-Americans. Carcinogenesis 32(5):748–756. doi:10.1093/carcin/bgr026

162. Cheng I, Caberto CP, Lum-Jones A, Seifried A, Wilkens LR, Schumacher FR, Monroe KR, Lim U, Tiirikainen M, Kolonel LN, Henderson BE, Stram DO, Haiman CA, Le Marchand L (2011) Type 2 diabetes risk variants and colorectal cancer risk: the Multiethnic Cohort and PAGE studies. Gut 60(12):1703–1711. doi:10.1136/gut.2011.237727

163. Diorio C, Brisson J, Berube S, Pollak M (2008) Genetic polymorphisms involved in insulin-like growth factor (IGF) pathway in relation to mammographic breast density and IGF levels. Cancer Epidemiol Biomarkers Prev 17(4):880–888. doi:10.1158/1055-9965.EPI-07-2500

164. Verheus M, McKay JD, Kaaks R, Canzian F, Biessy C, Johansson M, Grobbee DE, Peeters PH, van Gils CH (2008) Common genetic variation in the IGF-1 gene, serum IGF-I levels and breast density. Breast Cancer Res Treat 112(1):109–122. doi:10.1007/s10549-007-9827-x

165. Gu F, Schumacher FR, Canzian F, Allen NE, Albanes D, Berg CD, Berndt SI, Boeing H, Bueno-de-Mesquita HB, Buring JE, Chabbert-Buffet N, Chanock SJ, Clavel-Chapelon F, Dumeaux V, Gaziano JM, Giovannucci EL, Haiman CA, Hankinson SE, Hayes RB, Henderson BE, Hunter DJ, Hoover RN, Johansson M, Key TJ, Khaw KT, Kolonel LN, Lagiou P, Lee IM, LeMarchand L, Lund E, Ma J, Onland-Moret NC, Overvad K, Rodriguez L, Sacerdote C, Sanchez MJ, Stampfer MJ, Stattin P, Stram DO, Thomas G, Thun MJ, Tjonneland A, Trichopoulos D, Tumino R, Virtamo J, Weinstein SJ, Willett WC, Yeager M, Zhang SM, Kaaks R, Riboli E, Ziegler RG, Kraft P (2010) Eighteen insulin-like growth factor pathway genes, circulating levels of IGF-I and its binding protein, and risk of prostate and breast cancer. Cancer Epidemiol Biomarkers Prev 19(11):2877–2887. doi:10.1158/1055-9965.EPI-10-0507

166. Terry KL, Tworoger SS, Gates MA, Cramer DW, Hankinson SE (2009) Common genetic variation in IGF1, IGFBP1 and IGFBP3 and ovarian cancer risk. Carcinogenesis 30 (12):2042–2046. doi:10.1093/carcin/bgp257

167. D'Aloisio AA, Schroeder JC, North KE, Poole C, West SL, Travlos GS, Baird DD (2009) IGF-I and IGFBP-3 polymorphisms in relation to circulating levels among African American and Caucasian women. Cancer Epidemiol Biomarkers Prev 18(3):954–966. doi:10.1158/1055-9965.EPI-08-0856

168. Keku TO, Vidal A, Oliver S, Hoyo C, Hall IJ, Omofoye O, McDoom M, Worley K, Galanko J, Sandler RS, Millikan R (2012) Genetic variants in IGF-I, IGF-II, IGFBP-3, and adiponectin genes and colon cancer risk in African Americans and Whites. Cancer Causes Control 23(7):1127–1138. doi:10.1007/s10552-012-9981-2

169. Su X, Colditz GA, Willett WC, Collins LC, Schnitt SJ, Connolly JL, Pollak MN, Rosner B, Tamimi RM (2010) Genetic variation and circulating levels of IGF-I and IGFBP-3 in relation to risk of proliferative benign breast disease. Int J Cancer 126(1):180–190. doi:10.1002/ijc.24674

170. Wong HL, Delellis K, Probst-Hensch N, Koh WP, Van Den Berg D, Lee HP, Yu MC, Ingles SA (2005) A new single nucleotide polymorphism in the insulin-like growth factor I regulatory region associates with colorectal cancer risk in Singapore Chinese. Cancer Epidemiol Biomarkers Prev 14(1):144–151

171. Morimoto LM, Newcomb PA, White E, Bigler J, Potter JD (2005) Insulin-like growth factor polymorphisms and colorectal cancer risk. Cancer Epidemiol Biomarkers Prev 14(5):1204–1211. doi:10.1158/1055-9965.EPI-04-0695

172. Pechlivanis S, Wagner K, Chang-Claude J, Hoffmeister M, Brenner H, Forsti A (2007) Polymorphisms in the insulin like growth factor 1 and IGF binding protein 3 genes and risk of colorectal cancer. Cancer Detect Prev 31(5):408–416. doi:10.1016/j.cdp.2007.10.001

173. Hager J, Clement K, Francke S, Dina C, Raison J, Lahlou N, Rich N, Pelloux V, Basdevant A, Guy-Grand B, North M, Froguel P (1998) A polymorphism in the 5' untranslated region of the human ob gene is associated with low leptin levels. Int J Obes Relat Metab Disord 22(3):200–205

174. Jiang Y, Wilk JB, Borecki I, Williamson S, DeStefano AL, Xu G, Liu J, Ellison RC, Province M, Myers RH (2004) Common variants in the 5' region of the leptin gene are associated with body mass index in men from the National Heart, Lung, and Blood Institute Family Heart Study. Am J Hum Genet 75(2):220–230. doi:10.1086/422699

175. Le Stunff C, Le Bihan C, Schork NJ, Bougneres P (2000) A common promoter variant of the leptin gene is associated with changes in the relationship between serum leptin and fat mass in obese girls. Diabetes 49(12):2196–2200

176. Li WD, Reed DR, Lee JH, Xu W, Kilker RL, Sodam BR, Price RA (1999) Sequence variants in the 5′ flanking region of the leptin gene are associated with obesity in women. Ann Hum Genet 63(Pt 3):227–234
177. Mammes O, Betoulle D, Aubert R, Herbeth B, Siest G, Fumeron F (2000) Association of the G-2548A polymorphism in the 5′ region of the LEP gene with overweight. Ann Hum Genet 64(Pt 5):391–394
178. Quinton ND, Lee AJ, Ross RJ, Eastell R, Blakemore AI (2001) A single nucleotide polymorphism (SNP) in the leptin receptor is associated with BMI, fat mass and leptin levels in postmenopausal Caucasian women. Hum Genet 108(3):233–236
179. Gallicchio L, Chang HH, Christo DK, Thuita L, Huang HY, Strickland P, Ruczinski I, Clipp S, Helzlsouer KJ (2009) Single nucleotide polymorphisms in obesity-related genes and all-cause and cause-specific mortality: a prospective cohort study. BMC Med Genet 10:103. doi:10.1186/1471-2350-10-103
180. Friedlander Y, Li G, Fornage M, Williams OD, Lewis CE, Schreiner P, Pletcher MJ, Enquobahrie D, Williams M, Siscovick DS (2010) Candidate molecular pathway genes related to appetite regulatory neural network, adipocyte homeostasis and obesity: results from the CARDIA Study. Ann Hum Genet 74(5):387–398. doi:10.1111/j.1469-1809.2010.00596.x
181. Slattery ML, Wolff RK, Herrick J, Caan BJ, Potter JD (2008) Leptin and leptin receptor genotypes and colon cancer: gene-gene and gene-lifestyle interactions. Int J Cancer 122 (7):1611–1617. doi:10.1002/ijc.23135
182. Vasku A, Vokurka J, Bienertova-Vasku J (2009) Obesity-related genes variability in Czech patients with sporadic colorectal cancer: preliminary results. Int J Colorectal Dis 24(3):289–294. doi:10.1007/s00384-008-0553-6
183. Liu L, Zhong R, Wei S, Xiang H, Chen J, Xie D, Yin J, Zou L, Sun J, Chen W, Miao X, Nie S (2013) The leptin gene family and colorectal cancer: interaction with smoking behavior and family history of cancer. PLoS One 8(4):e60777. doi:10.1371/journal.pone.0060777
184. Chung HK, Chae JS, Hyun YJ, Paik JK, Kim JY, Jang Y, Kwon HM, Song YD, Lee HC, Lee JH (2009) Influence of adiponectin gene polymorphisms on adiponectin level and insulin resistance index in response to dietary intervention in overweight-obese patients with impaired fasting glucose or newly diagnosed type 2 diabetes. Diabetes Care 32(4):552–558. doi:10.2337/dc08-1605
185. Filippi E, Sentinelli F, Trischitta V, Romeo S, Arca M, Leonetti F, Di Mario U, Baroni MG (2004) Association of the human adiponectin gene and insulin resistance. Eur J Hum Genet 12(3):199–205. doi:10.1038/sj.ejhg.5201120
186. Gonzalez-Sanchez JL, Zabena CA, Martinez-Larrad MT, Fernandez-Perez C, Perez-Barba M, Laakso M, Serrano-Rios M (2005) An SNP in the adiponectin gene is associated with decreased serum adiponectin levels and risk for impaired glucose tolerance. Obes Res 13 (5):807–812. doi:10.1038/oby.2005.91
187. Menzaghi C, Ercolino T, Di Paola R, Berg AH, Warram JH, Scherer PE, Trischitta V, Doria A (2002) A haplotype at the adiponectin locus is associated with obesity and other features of the insulin resistance syndrome. Diabetes 51(7):2306–2312
188. Nakatani K, Noma K, Nishioka J, Kasai Y, Morioka K, Katsuki A, Hori Y, Yano Y, Sumida Y, Wada H, Nobori T (2005) Adiponectin gene variation associates with the increasing risk of type 2 diabetes in non-diabetic Japanese subjects. Int J Mol Med 15 (1):173–177
189. Panagopoulou P, Stamna E, Tsolkas G, Galli-Tsinopoulou A, Pavlitou-Tsiontsi E, Nousia-Arvanitakis S, Vavatsi-Christaki N (2009) Adiponectin gene polymorphisms in obese Greek youth. J Pediatr Endocrinol Metab 22(10):955–959
190. Yang WS, Yang YC, Chen CL, Wu IL, Lu JY, Lu FH, Tai TY, Chang CJ (2007) Adiponectin SNP276 is associated with obesity, the metabolic syndrome, and diabetes in the elderly. Am J Clin Nutr 86(2):509–513

191. Gunter MJ, Leitzmann MF (2006) Obesity and colorectal cancer: epidemiology, mechanisms and candidate genes. J Nutr Biochem 17(3):145–156. doi:10.1016/j.jnutbio.2005.06.011

192. Bik W, Ostrowski J, Baranowska-Bik A, Wolinska-Witort E, Bialkowska M, Martynska L, Baranowska B (2010) Adipokines and genetic factors in overweight or obese but metabolically healthy Polish women. Neuro Endocrinol Lett 31(4):497–506

193. Bouatia-Naji N, Meyre D, Lobbens S, Seron K, Fumeron F, Balkau B, Heude B, Jouret B, Scherer PE, Dina C, Weill J, Froguel P (2006) ACDC/adiponectin polymorphisms are associated with severe childhood and adult obesity. Diabetes 55(2):545–550

194. Petrone A, Zavarella S, Caiazzo A, Leto G, Spoletini M, Potenziani S, Osborn J, Vania A, Buzzetti R (2006) The promoter region of the adiponectin gene is a determinant in modulating insulin sensitivity in childhood obesity. Obesity (Silver Spring) 14(9):1498–1504. doi:10.1038/oby.2006.172

195. Schwarz PE, Govindarajalu S, Towers W, Schwanebeck U, Fischer S, Vasseur F, Bornstein SR, Schulze J (2006) Haplotypes in the promoter region of the ADIPOQ gene are associated with increased diabetes risk in a German Caucasian population. Horm Metab Res 38(7):447–451. doi:10.1055/s-2006-947842

196. Warodomwichit D, Shen J, Arnett DK, Tsai MY, Kabagambe EK, Peacock JM, Hixson JE, Straka RJ, Province MA, An P, Lai CQ, Parnell LD, Borecki IB, Ordovas JM (2009) ADIPOQ polymorphisms, monounsaturated fatty acids, and obesity risk: the GOLDN study. Obesity (Silver Spring) 17(3):510–517. doi:10.1038/oby.2008.583

197. Woo JG, Dolan LM, Deka R, Kaushal RD, Shen Y, Pal P, Daniels SR, Martin LJ (2006) Interactions between noncontiguous haplotypes in the adiponectin gene ACDC are associated with plasma adiponectin. Diabetes 55(2):523–529

198. Liu L, Zhong R, Wei S, Yin JY, Xiang H, Zou L, Chen W, Chen JG, Zheng XW, Huang LJ, Zhu BB, Chen Q, Duan SY, Rui R, Yang BF, Sun JW, Xie DS, Xu YH, Miao XP, Nie SF (2011) Interactions between genetic variants in the adiponectin, adiponectin receptor 1 and environmental factors on the risk of colorectal cancer. PLoS One 6(11):e27301. doi:10.1371/journal.pone.0027301

199. Hajeer AH, Hutchinson IV (2000) TNF-alpha gene polymorphism: clinical and biological implications. Microsc Res Tech 50(3):216–228. doi:10.1002/1097-0029(20000801) 50:3<216::AID-JEMT5>3.0.CO;2-Q

200. Dalziel B, Gosby AK, Richman RM, Bryson JM, Caterson ID (2002) Association of the TNF-alpha -308 G/A promoter polymorphism with insulin resistance in obesity. Obes Res 10 (5):401–407. doi:10.1038/oby.2002.55

201. Sookoian SC, Gonzalez C, Pirola CJ (2005) Meta-analysis on the G-308A tumor necrosis factor alpha gene variant and phenotypes associated with the metabolic syndrome. Obes Res 13(12):2122–2131. doi:10.1038/oby.2005.263

202. Fontaine-Bisson B, Wolever TM, Chiasson JL, Rabasa-Lhoret R, Maheux P, Josse RG, Leiter LA, Rodger NW, Ryan EA, El-Sohemy A (2007) Tumor necrosis factor alpha -238G > A genotype alters postprandial plasma levels of free fatty acids in obese individuals with type 2 diabetes mellitus. Metabolism 56(5):649–655. doi:10.1016/j.metabol.2006.12.013

203. de Luis DA, Aller R, Izaola O, Sagrado MG, Conde R (2009) Influence of G308A promoter variant of tumor necrosis factor-alpha gene on insulin resistance and weight loss secondary to two hypocaloric diets: a randomized clinical trial. Arch Med Res 40(1):36–41. doi:10.1016/j.arcmed.2008.10.012

204. Rasmussen SK, Urhammer SA, Jensen JN, Hansen T, Borch-Johnsen K, Pedersen O (2000) The -238 and -308G → A polymorphisms of the tumor necrosis factor alpha gene promoter are not associated with features of the insulin resistance syndrome or altered birth weight in Danish Caucasians. J Clin Endocrinol Metab 85(4):1731–1734

205. Joffe YT, van der Merwe L, Evans J, Collins M, Lambert EV, September A, Goedecke JH (2012) The tumor necrosis factor-alpha gene -238G > A polymorphism, dietary fat intake, obesity risk and serum lipid concentrations in black and white South African women. Eur J Clin Nutr 66(12):1295–1302. doi:10.1038/ejcn.2012.156

206. Fan W, Maoqing W, Wangyang C, Fulan H, Dandan L, Jiaojiao R, Xinshu D, Binbin C, Yashuang Z (2011) Relationship between the polymorphism of tumor necrosis factor-alpha-308G > A and susceptibility to inflammatory bowel diseases and colorectal cancer: a meta-analysis. Eur J Hum Genet 19(4):432–437. doi:10.1038/ejhg.2010.159

207. Jang WH, Yang YI, Yea SS, Lee YJ, Chun JH, Kim HI, Kim MS, Paik KH (2001) The -238 tumor necrosis factor-alpha promoter polymorphism is associated with decreased susceptibility to cancers. Cancer Lett 166(1):41–46

208. Landi S, Moreno V, Gioia-Patricola L, Guino E, Navarro M, de Oca J, Capella G, Canzian F (2003) Association of common polymorphisms in inflammatory genes interleukin (IL)6, IL8, tumor necrosis factor alpha, NFKB1, and peroxisome proliferator-activated receptor gamma with colorectal cancer. Cancer Res 63(13):3560–3566

209. Macarthur M, Sharp L, Hold GL, Little J, El-Omar EM (2005) The role of cytokine gene polymorphisms in colorectal cancer and their interaction with aspirin use in the northeast of Scotland. Cancer Epidemiol Biomarkers Prev 14(7):1613–1618. doi:10.1158/1055-9965. EPI-04-0878

210. Theodoropoulos G, Papaconstantinou I, Felekouras E, Nikiteas N, Karakitsos P, Panoussopoulos D, Lazaris A, Patsouris E, Bramis J, Gazouli M (2006) Relation between common polymorphisms in genes related to inflammatory response and colorectal cancer. World J Gastroenterol 12(31):5037–5043

211. Wang J, Cao C, Luo H, Xiong S, Xu Y, Xiong W (2011) Tumour necrosis factor alpha -308G/A polymorphism and risk of the four most frequent cancers: a meta-analysis. Int J Immunogenet 38(4):311–320. doi:10.1111/j.1744-313X.2011.01014.x

212. Berthier MT, Paradis AM, Tchernof A, Bergeron J, Prud'homme D, Despres JP, Vohl MC (2003) The interleukin 6-174G/C polymorphism is associated with indices of obesity in men. J Hum Genet 48(1):14–19. doi:10.1007/s100380300002

213. Goyenechea E, Parra D, Martinez JA (2007) Impact of interleukin 6 −174G > C polymorphism on obesity-related metabolic disorders in people with excess in body weight. Metabolism 56(12):1643–1648. doi:10.1016/j.metabol.2007.07.005

214. Razquin C, Martinez JA, Martinez-Gonzalez MA, Fernandez-Crehuet J, Santos JM, Marti A (2010) A Mediterranean diet rich in virgin olive oil may reverse the effects of the -174G/C IL6 gene variant on 3-year body weight change. Mol Nutr Food Res 54(Suppl 1):S75–S82. doi:10.1002/mnfr.200900257

215. Riikola A, Sipila K, Kahonen M, Jula A, Nieminen MS, Moilanen L, Kesaniemi YA, Lehtimaki T, Hulkkonen J (2009) Interleukin-6 promoter polymorphism and cardiovascular risk factors: the Health 2000 Survey. Atherosclerosis 207(2):466–470. doi:10.1016/j.athero sclerosis.2009.06.004

216. Stephens JW, Hurel SJ, Cooper JA, Acharya J, Miller GJ, Humphries SE (2004) A common functional variant in the interleukin-6 gene is associated with increased body mass index in subjects with type 2 diabetes mellitus. Mol Genet Metab 82(2):180–186. doi:10.1016/j. ymgme.2004.04.001

217. Wernstedt I, Eriksson AL, Berndtsson A, Hoffstedt J, Skrtic S, Hedner T, Hulten LM, Wiklund O, Ohlsson C, Jansson JO (2004) A common polymorphism in the interleukin-6 gene promoter is associated with overweight. Int J Obes Relat Metab Disord 28(10):1272–1279. doi:10.1038/sj.ijo.0802763

218. McKenzie JA, Weiss EP, Ghiu IA, Kulaputana O, Phares DA, Ferrell RE, Hagberg JM (2004) Influence of the interleukin-6 −174 G/C gene polymorphism on exercise training-induced changes in glucose tolerance indexes. J Appl Physiol 97(4):1338–1342. doi:10.1152/ japplphysiol.00199.2004

219. Nakagoe T, Tsuji T, Sawai T, Tanaka K, Hidaka S, Shibasaki S, Nanashima A, Ohbatake M, Yamaguchi H, Yasutake T, Sugawara K, Inokuchi N, Kamihira S (2003) Increased serum levels of interleukin-6 in malnourished patients with colorectal cancer. Cancer Lett 202 (1):109–115

220. Esfandi F, Mohammadzadeh Ghobadloo S, Basati G (2006) Interleukin-6 level in patients with colorectal cancer. Cancer Lett 244(1):76–78. doi:10.1016/j.canlet.2005.12.003

221. Tsilidis KK, Helzlsouer KJ, Smith MW, Grinberg V, Hoffman-Bolton J, Clipp SL, Visvanathan K, Platz EA (2009) Association of common polymorphisms in IL10, and in other genes related to inflammatory response and obesity with colorectal cancer. Cancer Causes Control 20(9):1739–1751. doi:10.1007/s10552-009-9427-7

222. Belluco C, Olivieri F, Bonafe M, Giovagnetti S, Mammano E, Scalerta R, Ambrosi A, Franceschi C, Nitti D, Lise M (2003) 174 G > C polymorphism of interleukin 6 gene promoter affects interleukin 6 serum level in patients with colorectal cancer. Clin Cancer Res 9(6):2173–2176

223. Slattery ML, Wolff RK, Herrick JS, Caan BJ, Potter JD (2007) IL6 genotypes and colon and rectal cancer. Cancer Causes Control 18(10):1095–1105. doi:10.1007/s10552-007-9049-x

224. Wilkening S, Tavelin B, Canzian F, Enquist K, Palmqvist R, Altieri A, Hallmans G, Hemminki K, Lenner P, Forsti A (2008) Interleukin promoter polymorphisms and prognosis in colorectal cancer. Carcinogenesis 29(6):1202–1206. doi:10.1093/carcin/bgn101

225. Yu Y, Wang W, Zhai S, Dang S, Sun M (2012) IL6 gene polymorphisms and susceptibility to colorectal cancer: a meta-analysis and review. Mol Biol Rep 39(8):8457–8463. doi:10.1007/s11033-012-1699-4

226. Hoffmann SC, Stanley EM, Cox ED, DiMercurio BS, Koziol DE, Harlan DM, Kirk AD, Blair PJ (2002) Ethnicity greatly influences cytokine gene polymorphism distribution. Am J Transplant 2(6):560–567

227. Huang HY, Thuita L, Strickland P, Hoffman SC, Comstock GW, Helzlsouer KJ (2007) Frequencies of single nucleotide polymorphisms in genes regulating inflammatory responses in a community-based population. BMC Genet 8:7. doi:10.1186/1471-2156-8-7

228. Almeida OP, Norman PE, Allcock R, van Bockxmeer F, Hankey GJ, Jamrozik K, Flicker L (2009) Polymorphisms of the CRP gene inhibit inflammatory response and increase susceptibility to depression: the Health in Men Study. Int J Epidemiol 38(4):1049–1059. doi:10.1093/ije/dyp199

229. Ben Assayag E, Shenhar-Tsarfaty S, Bova I, Berliner S, Usher S, Peretz H, Shapira I, Bornstein NM (2009) Association of the -757 T > C polymorphism in the CRP gene with circulating C-reactive protein levels and carotid atherosclerosis. Thromb Res 124(4):458–462. doi:10.1016/j.thromres.2009.04.008

230. Elliott P, Chambers JC, Zhang W, Clarke R, Hopewell JC, Peden JF, Erdmann J, Braund P, Engert JC, Bennett D, Coin L, Ashby D, Tzoulaki I, Brown IJ, Mt-Isa S, McCarthy MI, Peltonen L, Freimer NB, Farrall M, Ruokonen A, Hamsten A, Lim N, Froguel P, Waterworth DM, Vollenweider P, Waeber G, Jarvelin MR, Mooser V, Scott J, Hall AS, Schunkert H, Anand SS, Collins R, Samani NJ, Watkins H, Kooner JS (2009) Genetic Loci associated with C-reactive protein levels and risk of coronary heart disease. JAMA 302(1):37–48. doi:10.1001/jama.2009.954

231. Faucher G, Guenard F, Bouchard L, Garneau V, Turcot V, Houde A, Tchernof A, Bergeron J, Deshaies Y, Hould FS, Lebel S, Marceau P, Vohl MC (2012) Genetic contribution to C-reactive protein levels in severe obesity. Mol Genet Metab 105(3):494–501. doi:10.1016/j.ymgme.2011.11.198

232. Ferrari SL, Ahn-Luong L, Garnero P, Humphries SE, Greenspan SL (2003) Two promoter polymorphisms regulating interleukin-6 gene expression are associated with circulating levels of C-reactive protein and markers of bone resorption in postmenopausal women. J Clin Endocrinol Metab 88(1):255–259

233. Kilpelainen TO, Laaksonen DE, Lakka TA, Herder C, Koenig W, Lindstrom J, Eriksson JG, Uusitupa M, Kolb H, Laakso M, Tuomilehto J (2010) The rs1800629 polymorphism in the TNF gene interacts with physical activity on the changes in C-reactive protein levels in the Finnish Diabetes Prevention Study. Exp Clin Endocrinol Diabetes 118(10):757–759. doi:10.1055/s-0030-1249686

234. Pai JK, Mukamal KJ, Rexrode KM, Rimm EB (2008) C-reactive protein (CRP) gene polymorphisms, CRP levels, and risk of incident coronary heart disease in two nested case-control studies. PLoS One 3(1):e1395. doi:10.1371/journal.pone.0001395

235. Perry TE, Muehlschlegel JD, Liu KY, Fox AA, Collard CD, Body SC, Shernan SK (2009) C-Reactive protein gene variants are associated with postoperative C-reactive protein levels after coronary artery bypass surgery. BMC Med Genet 10:38. doi:10.1186/1471-2350-10-38

236. Wang L, Lu X, Li Y, Li H, Chen S, Gu D (2009) Functional analysis of the C-reactive protein (CRP) gene -717A > G polymorphism associated with coronary heart disease. BMC Med Genet 10:73. doi:10.1186/1471-2350-10-73

237. Lange LA, Carlson CS, Hindorff LA, Lange EM, Walston J, Durda JP, Cushman M, Bis JC, Zeng D, Lin D, Kuller LH, Nickerson DA, Psaty BM, Tracy RP, Reiner AP (2006) Association of polymorphisms in the CRP gene with circulating C-reactive protein levels and cardiovascular events. JAMA 296(22):2703–2711. doi:10.1001/jama.296.22.2703

238. Crawford DC, Sanders CL, Qin X, Smith JD, Shephard C, Wong M, Witrak L, Rieder MJ, Nickerson DA (2006) Genetic variation is associated with C-reactive protein levels in the Third National Health and Nutrition Examination Survey. Circulation 114(23):2458–2465. doi:10.1161/CIRCULATIONAHA.106.615740

239. Eiriksdottir G, Smith AV, Aspelund T, Hafsteinsdottir SH, Olafsdottir E, Launer LJ, Harris TB, Gudnason V (2009) The interaction of adiposity with the CRP gene affects CRP levels: age, gene/environment susceptibility-Reykjavik study. Int J Obes (Lond) 33(2):267–272. doi:10.1038/ijo.2008.274

240. Teng MS, Hsu LA, Wu S, Chang HH, Chou HH, Ko YL (2009) Association between C-reactive protein gene haplotypes and C-reactive protein levels in Taiwanese: interaction with obesity. Atherosclerosis 204(2):e64–e69. doi:10.1016/j.atherosclerosis.2008.10.034

241. Martinez-Calleja A, Quiroz-Vargas I, Parra-Rojas I, Munoz-Valle JF, Leyva-Vazquez MA, Fernandez-Tilapa G, Vences-Velazquez A, Cruz M, Salazar-Martinez E, Flores-Alfaro E (2012) Haplotypes in the CRP gene associated with increased BMI and levels of CRP in subjects with type 2 diabetes or obesity from Southwestern Mexico. Exp Diabetes Res 2012:982683. doi:10.1155/2012/982683

242. Yang SH, Huang CJ, Chang SC, Lin JK (2011) Association of C-reactive protein gene polymorphisms and colorectal cancer. Ann Surg Oncol 18(7):1907–1915. doi:10.1245/s10434-011-1575-9

243. Lee SG, Yum JS, Moon HM, Kim HJ, Yang YJ, Kim HL, Yoon Y, Lee S, Song K (2005) Analysis of mannose-binding lectin 2 (MBL2) genotype and the serum protein levels in the Korean population. Mol Immunol 42(8):969–977. doi:10.1016/j.molimm.2004.09.036

244. Madsen HO, Garred P, Thiel S, Kurtzhals JA, Lamm LU, Ryder LP, Svejgaard A (1995) Interplay between promoter and structural gene variants control basal serum level of mannan-binding protein. J Immunol 155(6):3013–3020

245. Madsen HO, Satz ML, Hogh B, Svejgaard A, Garred P (1998) Different molecular events result in low protein levels of mannan-binding lectin in populations from southeast Africa and South America. J Immunol 161(6):3169–3175

246. Zanetti KA, Haznadar M, Welsh JA, Robles AI, Ryan BM, McClary AC, Bowman ED, Goodman JE, Bernig T, Chanock SJ, Harris CC (2012) 3′-UTR and functional secretor haplotypes in mannose-binding lectin 2 are associated with increased colon cancer risk in African Americans. Cancer Res 72(6):1467–1477. doi:10.1158/0008-5472.CAN-11-3073

247. Nakagawa T, Kawasaki N, Ma Y, Uemura K, Kawasaki T (2003) Antitumor activity of mannan-binding protein. Methods Enzymol 363:26–33. doi:10.1016/S0076-6879(03)01041-3

248. Nakagawa T, Ma BY, Uemura K, Oka S, Kawasaki N, Kawasaki T (2003) Role of mannan-binding protein, MBP, in innate immunity. Anticancer Res 23(6a):4467–4471

249. Hoff JH, te Morsche RH, Roelofs HM, van der Logt EM, Nagengast FM, Peters WH (2009) COX-2 polymorphisms -765G → C and -1195A → G and colorectal cancer risk. World J Gastroenterol 15(36):4561–4565

250. Siemes C, Visser LE, Coebergh JW, Hofman A, Uitterlinden AG, Stricker BH (2008) Protective effect of NSAIDs on cancer and influence of COX-2 C(-765G) genotype. Curr Cancer Drug Targets 8(8):753–764

251. Lurje G, Nagashima F, Zhang W, Yang D, Chang HM, Gordon MA, El-Khoueiry A, Husain H, Wilson PM, Ladner RD, Mauro DJ, Langer C, Rowinsky EK, Lenz HJ (2008) Polymorphisms in cyclooxygenase-2 and epidermal growth factor receptor are associated with progression-free survival independent of K-ras in metastatic colorectal cancer patients treated with single-agent cetuximab. Clin Cancer Res 14(23):7884–7895. doi:10.1158/1078-0432.CCR-07-5165

252. Zhu W, Wei BB, Shan X, Liu P (2010) 765G > C and 8473 T > C polymorphisms of COX-2 and cancer risk: a meta-analysis based on 33 case-control studies. Mol Biol Rep 37(1):277–288. doi:10.1007/s11033-009-9685-1

253. Xing LL, Wang ZN, Jiang L, Zhang Y, Xu YY, Li J, Luo Y, Zhang X (2008) Cyclooxygenase 2 polymorphism and colorectal cancer: -765G > C variant modifies risk associated with smoking and body mass index. World J Gastroenterol 14(11):1785–1789

254. Tan W, Wu J, Zhang X, Guo Y, Liu J, Sun T, Zhang B, Zhao D, Yang M, Yu D, Lin D (2007) Associations of functional polymorphisms in cyclooxygenase-2 and platelet 12-lipoxygenase with risk of occurrence and advanced disease status of colorectal cancer. Carcinogenesis 28 (6):1197–1201. doi:10.1093/carcin/bgl242

255. Goodman JE, Bowman ED, Chanock SJ, Alberg AJ, Harris CC (2004) Arachidonate lipoxygenase (ALOX) and cyclooxygenase (COX) polymorphisms and colon cancer risk. Carcinogenesis 25(12):2467–2472. doi:10.1093/carcin/bgh260

256. Sansbury LB, Millikan RC, Schroeder JC, North KE, Moorman PG, Keku TO, de Cotret AR, Player J, Sandler RS (2006) COX-2 polymorphism, use of nonsteroidal anti-inflammatory drugs, and risk of colon cancer in African Americans (United States). Cancer Causes Control 17(3):257–266. doi:10.1007/s10552-005-0417-0

257. Lin HJ, Lakkides KM, Keku TO, Reddy ST, Louie AD, Kau IH, Zhou H, Gim JS, Ma HL, Matthies CF, Dai A, Huang HF, Materi AM, Lin JH, Frankl HD, Lee ER, Hardy SI, Herschman HR, Henderson BE, Kolonel LN, Le Marchand L, Garavito RM, Sandler RS, Haile RW, Smith WL (2002) Prostaglandin H synthase 2 variant (Val511Ala) in African Americans may reduce the risk for colorectal neoplasia. Cancer Epidemiol Biomarkers Prev 11(11):1305–1315

258. Ashktorab H, Tsang S, Luke B, Sun Z, Adam-Campbell L, Kwagyan J, Poirier R, Akter S, Akhgar A, Smoot D, Munroe DJ, Ali IU (2008) Protective effect of Cox-2 allelic variants on risk of colorectal adenoma development in African Americans. Anticancer Res 28(5B):3119–3123

259. Kwagyan J, Apprey V, Ashktorab H (2012) Linkage disequilibrium and haplotype analysis of COX-2 and risk of colorectal adenoma development. Clin Transl Sci 5(1):60–64. doi:10.1111/j.1752-8062.2011.00373.x

260. Erichsen HC, Chanock SJ (2004) SNPs in cancer research and treatment. Br J Cancer 90 (4):747–751. doi:10.1038/sj.bjc.6601574

261. Frazer KA, Murray SS, Schork NJ, Topol EJ (2009) Human genetic variation and its contribution to complex traits. Nat Rev Genet 10(4):241–251. doi:10.1038/nrg2554

Chapter 8
Ethnic Differences in Insulin Resistance as a Mediator of Cancer Disparities

Rebecca E. Hasson and Michael I. Goran

Abstract Ethnic differences in the incidence and prevalence of certain obesity-related cancers are well established. African Americans have increased risk of prostate, breast (premenopausal), and colorectal cancer and myeloma, compared to Caucasians with the lowest rates in Latinos, Asians, and Native Americans. Prior work in this area suggests that there are distinct ethnic differences in obesity-related metabolic risk factors for cancer, insulin resistance in particular, that are evident early in life, and may help explain ethnic differences in the incidence and prevalence of obesity-related cancers. The focus of this chapter is to review and discuss ethnic differences in insulin resistance and its link with other cancer-related metabolic risk factors including hyperinsulinemia, insulin-like growth factors, body fat distribution, adipose tissue biology, low-grade inflammation, non-esterified fatty acids, and oxidative stress. This chapter places a particular emphasis on ethnic differences between African Americans and Latinos for two reasons: (1) African Americans and Latinos are the two largest ethnic minority groups in the USA, and (2) these populations share a similar propensity for obesity and insulin resistance but markedly different profiles for obesity-related cancers, creating an informative comparative contrast. Although the literature is limited by an inconsistency in the terminology used for various ethnicities, in most cases we refer to Caucasian for any study using the terms Caucasian, White, or non-Hispanic White; Latino to describe people of Hispanic, Latino, or Mexican-American descent; African American to describe people of African, African American, or Black-Caribbean descent; Asian to describe people of Asian, South Asian, East Asian, and Southeast

R.E. Hasson (✉)
University of Michigan, Schools of Kinesiology and Public Health, 1402 Washington Heights, 2110 Observatory Lodge, Ann Arbor, MI 48109, USA
e-mail: hassonr@umich.edu

M.I. Goran
University of Southern California, Keck School of Medicine, Department of Preventive Medicine, 2250 Alcazar Street CSC 212, Los Angeles, CA 90089, USA
e-mail: goran@usc.edu

D.J. Bowen et al. (eds.), *Impact of Energy Balance on Cancer Disparities*, Energy Balance and Cancer 9, DOI 10.1007/978-3-319-06103-0_8, © Springer International Publishing Switzerland 2014

Asian descent or any other specific Asian ethnicity; and Native American to describe people of American Indian, Pima Indian, Aboriginal, First Nation, or Alaska Native ethnicity. We also recognize that there may be variation within these subgroups; however, comprehensive review of this literature is beyond the scope of this chapter.

Keywords Insulin • Insulin resistance • Hyperinsulinemia • Insulin-like growth factor • Non-esterified fatty acids • Oxidative stress • Psychological stress • Cortisol-induced obesity • Body fat distribution • Intramyocellular lipid • Hepatic fat • Pancreatic fat • Ectopic fat • Adipose tissue biology

The Scope of the Problem: Obesity and Cancer Disparities

According to the 2010 US Current Population Survey, there are 53 million people of Latino origin and 41 million African Americans in the USA, comprising 17 % and 13 % of the total population, respectively. Latinos are the fastest growing ethnic group in this country adding almost 13 million people to the population and increasing in size by 41 % in the last decade. Obesity is a significant problem in both African Americans and Latinos with the most recent National Health and Nutrition Examination Survey (NHANES) estimates from 2009 to 2010 suggesting higher rates of overweight and obesity in African American and Latino adults compared to Caucasians [1]. In adults, 20 years of age and older, African Americans had the highest age-adjusted rates of obesity (49.5 %), followed by Mexican Americans (40.4 %), all Latinos (39.1 %), and Caucasians (34.3 %). Of note, the prevalence of grade 2 [body mass index (BMI) of at least 35 kg/m^2] and grade 3 obesity (BMI greater than or equal to 40 kg/m^2) were highest among African Americans (26 % for grade 2, and 13.1 % for grade 3), compared to Caucasians (14.4 % for grade 2, and 5.7 % for grade 3) and Latinos (14.9 % for grade 2, and 5.4 % for grade 3). Although American Indians comprise a smaller proportion of the total US population (1.2 %), obesity is also a significant problem in this ethnic group with 39.4 % of American Indian men and women categorized as obese [2]. Among Asians, this ethnic group is 60 % less likely to be obese compared to Caucasians; however, there is substantial variation in the prevalence of overweight and obesity within this ethnic group [3]. Filipino Americans are 70 % more likely to be obese as compared to the overall Asian population. Interestingly, Southeast Asians have one of the highest prevalences of type 2 diabetes in the USA, yet the prevalence of obesity in this group is 6 % with 30–35 % of Southeast Asians classified as overweight [4]. In contrast, Chinese, Korean, and Vietnamese Americans have the lowest rates of overweight (BMI, 25 to <30 kg/m^2) and one in ten Korean and Vietnamese Americans are classified as underweight [3].

In 2010, pediatric obesity rates in the USA also showed a well-defined disparity by ethnicity, where 42 % of Latinos, 41 % of African Americans, and 30 % of Caucasians between the ages of 12 and 19 years were classified as overweight or obese [5]. Of note, Native American adolescents had the highest prevalence of obesity than those in all other ethnic groups combined [6]. As a result, obesity-related complications such as prediabetes and type 2 diabetes are more common in ethnic minority children and adults compared to Caucasians [7–12]. Specifically, the risk of diagnosed diabetes is 1.8 times higher among African Americans, and 1.7 times higher among Hispanics compared to Caucasians [13]. Moreover, 16.1 % of the total adult American Indian population has diagnosed diabetes [13]. A similar trend is noted in children, with African American, Latino, and Native American children reporting the highest rates of type 2 diabetes compared to other ethnicities [11, 12, 14]. The higher risk and prevalence of type 2 diabetes among these ethnic minority groups have been attributed to more severe insulin resistance and hyperinsulinemia (relative to Caucasians [8, 15–18]).

There is convincing evidence that overweight and obesity are also associated with cancers of the kidney, breast, colon, esophagus, endometrium, prostate, and colorectum, whereas studies on the relation between obesity and other forms of cancers are less consistent [19–23].

Despite a similar predisposition towards obesity, insulin resistance, and type 2 diabetes among African Americans, Latinos and Native Americans, there are marked differences in cancer incidence across different ethnic groups [24]. African Americans have increased risk of certain forms of obesity-related cancers, whereas for these same outcomes, Latinos and Native Americans appear to be somewhat "protected." In support of this hypothesis, data from the Surveillance Epidemiology and End Results (SEER) Database suggest that African American men have the highest incidence of cancer (all cancers combined) followed by Caucasians, with lower cancer rates among Latino, Native American and Asian men [25]. More specifically, African American men in the USA have the highest rates of prostate cancer worldwide. The prevalence rate is almost two times higher compared to Caucasians and Latinos and almost three times higher compared to Native Americans and Asians [25]. Breast cancer—the most common cancer among women—is highest among African Americans and Caucasian women compared to Latinas, Native Americans, and Asians. Interestingly, African American women have the highest rates of breast cancer before age 40 whereas Caucasians have the highest rates at older ages [26]. For both men and women, rates of colorectal cancer and myeloma are highest among African Americans followed by Caucasians with the lowest rates among Latinos, Native Americans, and Asians [25, 27]. Similar trends are observed for most other types of cancer, with rates among African Americans or Caucasians higher than those for other ethnic minority groups including Latinos [25]. Taken together, distinct differences in obesity-related cancer outcomes persist between African Americans, Latinos, and Native Americans despite all three groups having an increased propensity for obesity and similar risk for type 2 diabetes. This chapter reviews ethnic differences in cancer-related metabolic risk factors, insulin resistance, and hyperinsulinemia in particular and their potential contributions to ethnic differences in obesity-related cancer outcomes.

Obesity and Cancer Risk: Potential Mechanisms

Insulin Resistance

Obesity is the strongest contributing factor to insulin resistance and hyperinsulinemia, and this is evident early in life [8, 15, 28–30]. Many studies have shown that body fatness is positively associated with circulating fasting insulin levels in both animals and humans [31]. Insulin is a critical hormone for regulating metabolism, and its concentration in circulation is carefully coordinated, varying acutely in response to glucose and meal consumption. Insulin resistance is a condition in which muscle, fat, and liver cells are less sensitive to the metabolic effect of insulin. As a result, physiologic actions of insulin are inhibited but can be compensated for by increased insulin levels in circulation (i.e., hyperinsulinemia) to clear glucose from circulation [32, 33]. In addition, elevated insulin may stimulate cellular proliferation in pancreatic beta cells and fat cells, ensuring additional insulin production and fat storage, respectively [34]. This mechanism may have substantial advantages because it provides fat cells that can hold on to ingested fat and prevent its ectopic distribution elsewhere in the body [35, 36]. Thus, obesity results in continuous exposure of body tissues to elevated background and glucose-stimulated levels of insulin.

One of the leading hypotheses explaining why "fat is bad" relates to the role of insulin resistance and hyperinsulinemia as the mediating link between obesity and cancer risk. As mentioned above, besides its metabolic effects, insulin has promitotic and anti-apoptotic effects that may be tumorigenic [23, 37, 38]. Moreover, increased insulin resistance and hyperinsulinemia have been associated with increased risk of breast, endometrial, and colon cancer [20, 39–45]. Hence, detailed studies comparing ethnic differences in insulin resistance and hyperinsulinemia have been helpful in understanding why certain subgroups of the population are at increased cancer risk.

Research has consistently demonstrated that African Americans are more insulin resistant compared to Caucasians, which is only partially explained by greater overall adiposity in this ethnic group [8, 18, 46–56]. The Insulin Resistance Atherosclerosis Study (IRAS), a large-scale multicenter epidemiological study, was the first to provide compelling evidence in support of a metabolic predisposition towards insulin resistance in African American adults [57]. Compared to

Caucasians, African Americans had significantly higher fasting and 2-h postprandial insulin concentrations, higher acute insulin responses to glucose, and greater insulin resistance [57]. These ethnic differences persisted after adjusting for differences in age, obesity, body fat distribution, self-reported physical activity, and percent calories from fat and fiber. Data from the NHANES III subsequently confirmed ethnic differences in mean fasting insulin concentrations between African American and Caucasian men and women at each BMI category [55].

Similar to African Americans, large-scale studies of obesity, insulin resistance, insulin secretion, and beta-cell response in Latino and Native American populations have consistently reported an increased insulin response to glucose [8, 50, 58–61]. Glucose-tolerant Native Americans and Latinos were found to have greater insulin resistance and fasting hyperinsulinemia compared to Caucasians [62–65]. In addition, both groups were found to have exaggerated early insulin secretory responses to both intravenous and oral glucose challenges [50, 58, 59, 66, 67]. Others have confirmed that Latino adults have greater fasting and post-challenge insulin and greater insulin resistance than Caucasians [8, 62].

Studies in children are of increased significance because they allow examination of potentially underlying biological differences across subgroups of the population to be performed in the absence of potential confounding factors such as smoking, alcohol, aging, and menopausal status. Data from the Bogalusa Heart Study were the first to report increased insulin resistance in African American compared to Caucasian children based on measures of fasting insulin [68]. Subsequently, other studies have demonstrated greater insulin resistance and greater acute insulin response to glucose in African American compared to Caucasian children [30, 69]; these differences were independent of body fat, visceral fat, dietary factors, and physical activity. A recent study, using a hyperglycemic clamp technique, supported these observations where overweight African American compared to Caucasian youth had up to a 75 % higher insulin secretion relative to their insulin sensitivity [15], an indicator of increased or up-regulated pancreatic beta-cell responsiveness.

Ethnic differences in insulin resistance have been well documented in Latino, Asian, and Native American youth, where, independent of overall adiposity, these ethnic minority groups exhibit more severe insulin resistance but an enhanced insulin secretory response when compared to Caucasian children [8, 11]. Studies comparing multiple ethnic groups confirmed greater insulin resistance during an intravenous glucose tolerance test in Native Americans compared to African Americans and Caucasians [70]. Another study reported equally greater insulin resistance assessed via hyperglycemic clamp among African Americans, Latinos, and Asians than in Caucasians [62]. In addition, Asians were the most insulin resistant followed by Latinos, African Americans, and Caucasians [62]. In prepubertal children, African American and Latino children were found to be equally more insulin resistant than Caucasian children [8]. However, in peripubertal adolescents, obese African Americans were more insulin resistant than Latinos, independent of body composition and fat distribution [60]. Pancreatic beta-cell function and the acute insulin response to a glucose challenge were also higher in African American than in Latino adolescents, suggesting that ethnic differences in pubertal induced insulin resistance may be an important contributor to ethnic differences in insulin resistance [71]. Of interest, the compensatory responses to insulin resistance were different in African American compared to Latino children and adolescents [8]. African American children tend to compensate with a higher acute insulin response to glucose, and this effect was in part due to a reduction in hepatic insulin extraction [8]. Following the ingestion of oral glucose, lower extraction rates have also been reported in African American adults [54]. In contrast, Latino children and adolescents compensate to the same degree of insulin resistance with greater second-phase insulin secretion [8]. Both beta-cell secretion and/or insulin clearance by the liver determine peripheral insulin levels and help to maintain normal glucose levels in circulation [72]. The mechanisms by which Native American and Asian populations compensate for insulin resistance is understudied; nevertheless, increased insulin resistance and secretion as well as hyperinsulinemia are present among ethnic minority children, adolescents, and adults compared to Caucasians, and these findings have been confirmed using a variety of methodologies.

The well-documented ethnic differences in insulin resistance and secretion in children and adults have been explained in part by genetic, behavioral, and/or environmental factors. Previous research has reported a positive association between African genetic admixture and insulin resistance [73]. In contrast, recent work has demonstrated that socio-behavioral factors including physical activity and self-reported racial discrimination, but not African genetic admixture, were associated with increased cardiometabolic risk (i.e., blood pressure) among African Americans [74]. Moreover, research in the area of molecular epigenetic mechanisms of gene expression has also suggested that the genome is subject to environmental regulation [75], suggesting that ethnic differences in insulin resistance may have a gene-environmental origin. Consequently, in addition to nutrition and physical activity (which is further discussed in the next chapter), research has begun to investigate the role of the social environment, particularly psychosocial stress, and its implications for obesity and insulin resistance. The physiological stress response originates from the hypothalamic-pituitary-adrenal axis and undergoes a cascade of reactions including the release of corticotrophin-releasing hormone from the hypothalamus, causing the release of adrenocorticotrophic hormone by the adrenal pituitary, and ultimately the release of cortisol by the adrenal cortex into circulation [76]. Cortisol levels increase in response to both stressors in the laboratory [77] and naturalistic social environments [78]. Designed to increase energy availability in the short term, cortisol acutely impairs insulin secretion and increases hepatic glucose output [79]. An environment of prolonged glucocorticoid exposure (i.e., chronic stress) exerts diabetogenic effects by interfering with insulin action on several different levels [80–82], including a direct inhibition of insulin secretion from pancreatic beta cells [83], impaired insulin-mediated glucose uptake [84], and disruption of the insulin signaling cascade in skeletal muscle [85]. Under chronic conditions, healthy lean individuals appear able to compensate for glucocorticoid-induced insulin resistance with increased beta-cell function or increased insulin release [86–88]. However, in the obese or the insulin-resistant state, those compensatory mechanisms fail to counteract glucocorticoid-induced insulin resistance, resulting in hyperglycemia [87, 88]. Hence, prolonged glucocorticoid exposure may further compromise the already lower insulin sensitivity in obese African Americans by exacerbating the progression towards insulin resistance in these populations. Previous research has demonstrated the negative association between cortisol and obesity in adults [89, 90], and a recent study showed that cortisol contributes to the reduction in insulin sensitivity over a 1-year period in overweight Latino children and adolescents [91], underlining the relevance of reducing glucocorticoid-induced insulin resistance in ethnic minority populations.

Prolonged glucocorticoid exposure also leads to weight gain and visceral fat accumulation [92–94], not only through behavioral pathways such as increased food consumption [92, 95, 96] and sedentariness [97–100] but also directly via the release of neuropeptide Y [93, 96]. Several longitudinal studies have reported a positive association between psychological stress and BMI in adults [101, 102]. Another study reported that higher levels of psychological stress over a 10-year period predicted significantly greater increases in BMI over time compared to lower levels of stress, and this relationship was significantly stronger for African American compared to Caucasian girls [103]. In Latino youth, a significant association between cortisol, total fat mass, and visceral fat accumulation has not found [91], suggesting that the mechanisms by which cortisol induced obesity and insulin resistance may differ by ethnicity.

In addition to responding to stressful events, the HPA axis also follows a strong circadian rhythm [78, 104]. Typically, cortisol levels are high upon waking; reach a peak about 30–40 min after waking; and then decline throughout the remainder of the day, reaching a nadir around midnight [104, 105]. The scientific literature examining ethnic differences in cortisol is not extensive but demonstrates divergent diurnal cortisol patterns for African Americans compared to Caucasians [106–110]. African Americans tend to have flatter diurnal cortisol slopes, with lower morning levels and higher evening levels, than Caucasians [106–110]. These findings have been replicated across studies of adolescents [107], pregnant women [110], adults [108, 111], and elderly populations [109]. Two studies examining ethnic differences in cortisol diurnal patterns in normal-weight African American, Latino, and Caucasian children and adolescents also reported flatter morning-to-evening cortisol slopes among African Americans and lower evening cortisol levels for Latinos relative to Caucasians [107, 112]. Deviations from the typical diurnal patterns have important implications for insulin resistance [113]. Specifically, flattened diurnal patterns previously reported in chronically stressed individuals are associated with insulin resistance and cancer-related metabolic risk factors (i.e., inflammation) [113]. Hence, greater exposure to psychosocial and environmental stressors (e.g., socioeconomic burden and racial discrimination) in African American populations may contribute to the increased obesity and insulin resistance, hyperinsulinemia, and subsequent cancer risk in this population.

Hyperinsulinemia and the IGF-1 Pathway

The direct effects of insulin resistance on cancer risk are unclear and likely do not solely explain the increased cancer risk among African Americans compared to Latinos and Native Americans since all three ethnic minority groups appear to be similar in degree of insulin resistance. Accordingly, the effect of insulin resistance is postulated to be mediated by the effects of chronic hyperinsulinemia on insulin-like growth factor (IGF)-1 bioactivity [23]. IGF-1 is a growth factor that is regulated by growth hormone levels [114, 115], present in circulation, and has insulin-like properties and functions [116]. The bioactivity of IGF-1 is determined by the circulating IGF-1 and IGF-binding protein (BPs) produced by the liver as well as paracrine effects of IGF-1, IGFBPs, and IGFBP proteases [23]. Insulin can also affect IGF-1 bioactivity via increasing IGF-1 secretion, IGF-1/IGFBP-3, IGFBP-3 proteolysis, and secretion of IGFBP-1 and IGFBP-2 and increased responsiveness of cells to IGF-1 and other growth factors. Numerous studies suggest that high level of IGF-1 is a risk factor for several cancers including breast, prostate, colon, and lung cancer [117–122].

IGF-1 bioactivity has been implicated in carcinogenesis as a function of its ability to stimulate the differentiation and proliferation of myoblasts as well as inhibit apoptosis [38]. Moreover, increasing evidence suggests that chronic hyperinsulinemia increases the risk of colon and endometrial cancer [20]. Thus, chronic exposure to high levels of insulin and IGF-1 is hypothesized to mediate many cancer risk factors [23], and as a result the IGF/insulin system has been suggested as a potential target for cancer therapy [37].

While obesity status is known to correlate with serum IGF-1 levels [123, 124], studies have reported an independent effect of ethnicity on IGF-1 bioactivity in children and adults, potentially explaining ethnic specific differences in cancer risk. Previous research has reported higher levels of IGF-1 and IGFBP-3 in African Americans compared to Caucasian and Latino adults, independent of adiposity [125]. Another study reported race by gender differences where African American females had higher IGF-1 levels compared to Caucasians with similar IGF-1 levels in males in both ethnic groups [126]. The lower IGF-1 levels in Latinos relative to African American have also been shown in prepubertal females [127].

It is important to note that previous studies have been inconsistent with respect to the relationships between obesity and circulating levels of IGF-1 [128]. Studies among healthy adults have reported a null association [129–131], a positive association [132], an inverse association [128, 133–135], and a nonlinear association [136, 137] between BMI and IGF-1 levels. However, data from studies examining ethnic differences in the relationship between obesity and circulating IGF-1 have shown more consistent trends and may help to explain the abovementioned inconsistencies in obesity–IGF relationships. In a multiethnic cohort study of 200,000 adults in Los Angeles and Hawaii, researchers reported a decline in plasma IGF-1 levels with increasing BMI in Latinos and Asians; this decline was attenuated in Caucasians and absent in African Americans [138]. After adjustment for age and

BMI, African Americans had the highest IGF-1 bioactivity compared to other ethnic groups. Taken together, there appears to be a progressive increase in IGF-1 levels with increasing obesity status in African Americans compared to a decline in IGF-1 with increasing obesity in other ethnic minority groups, particularly Latinos.

Ethnic differences in IGF-1 bioactivity among children are generally similar to those observed in adults. It has been shown that African American prepubertal females have higher IGF-1 levels compared to Caucasian and Latino females [125, 127]. An inverse relationship between IGF-1 and IGFBP-3 with total fat mass and body fat distribution has been reported in overweight Latino children, whereas others have demonstrated a positive association between total body fat and IGF-1 levels in both African American and Caucasian children [139, 140]. These findings were not explained by diet, physical activity, socioeconomic status, or adiposity but were related to the degree of African admixture [141], suggesting a potential genetic basis for this difference. Taken together, these results demonstrate that African American children and adults have the highest levels of IGF-1 and exhibit a positive relationship between IGF-1 and obesity, likely contributing to the increased risk of obesity-related cancers in this population.

A possible biological mechanism mediating the association between obesity and IGF-1 may be through the effect of growth hormone. Typically, obesity results in lower circulating IGFBP-1 and IGFBP-2 levels, leading to an increased negative feedback by free IGF-1 on pituitary growth hormone secretion and a decreased IGF-1 synthesis [142]. Given the positive association between obesity and IGF-1 levels in African Americans, it is possible that the growth hormone–IGF axis may be regulated differently in this population compared to other ethnic groups. Another possible mechanism may be through the effects of cortisol on IGF-1 and growth hormone levels. IGF-1 is mainly derived from the liver, which also is the sole site of splanchnic cortisol production, which suggests a close interaction between cortisol and IGF-1 [143]. Previous research has reported a negative association between cortisol and IGF-1 in obese Latino children and adolescents [80]. Hence, high cortisol and low IGF-1 may act in concert to reduce cancer risk in Latino children and adolescents. A final mechanism centers on the relationship between IGF-1, IGFBP-1, and body fat distribution. A recent study identified a modifying effect of ethnicity on the relationship between IGF-1 and subcutaneous fat as well as IGFBP-1 and hepatic fat in overweight African American and Latino adolescents, respectively [144]. IGF-1 and IGFBP-1 were inversely correlated with BMI, total fat mass, visceral fat, and hepatic fat, while IGFBP-1 was inversely correlated with subcutaneous fat. These relationships did not differ by ethnicity; however, the relationship between IGF-1 and subcutaneous fat, as well as IGFBP-1 and hepatic fat, was stronger in African Americans compared to Latinos [144]. These results suggest that the relationship between IGF-1, IGFBP-1, and body fat distribution differs among African American and Latino adolescents, which may contribute to the higher IGF-1 levels and subsequent cancer risk in African Americans. Hence, a more in-depth discussion regarding the role of body fat distribution and its association with cancer risk is given in the section below.

Body Fat Distribution

Visceral Fat

The location of body fat is important, especially with regard to how it might affect insulin resistance. Visceral fat (adipose tissue inside the abdominal cavity) in particular has been hypothesized to be one of the major factors linking increased obesity to increased insulin resistance and subsequent cancer risk mainly due to the effects of free fatty acids released from visceral fat into the hepatic portal vein with direct exposure to the liver [145]. In addition, several studies have found that insulin sensitivity is negatively associated with adipose stores in the abdominal region [146–151], particularly visceral fat, and this is consistent across age and ethnicity [152, 153], with one notable exception [154]. Increases in visceral adipose tissue in Native American adults do not explain the greater insulin resistance and hyperinsulinemia in this population when compared to equally obese Caucasians [154].

Emerging evidence however suggests that there are ethnic differences in the relationships between BMI, waist circumference, percent body fat, and visceral fat. Much research has focused on comparisons between Caucasians and Asians, with greater visceral fat in Southeast Asian women compared with their Caucasian counterparts even at the same BMI [155–158]. In addition, Latino children and adults also have greater visceral fat compared to similarly obese Caucasians [146, 159]. In contrast, several studies have reported lower amounts of visceral fat for a given waist circumference, BMI, or waist-to-hip ratio in African American compared to Caucasian women [152, 160–163]. One study confirmed similar BMIs and waist circumference measurements in middle-aged and older African American men and women compared with Caucasians and Latinos but lower visceral fat (total visceral fat and measured at the L4L5 spinal level) in African Americans. Other studies confirmed these findings and consistently reported ethnic differences in fat distribution between African Americans and Caucasians even after significant weight gain [279] and weight loss [164, 165]. Moreover, these differences are evident before puberty, both cross-sectionally and longitudinally, with a lower growth-related increase in visceral adipose tissue in African Americans compared to Caucasians [166, 167]. Taken together, these data suggest that visceral fat is associated with insulin resistance; however, the lower volumes of visceral fat previously reported in African Americans do not appear to explain the greater insulin resistance and subsequent cancer risk in this population. On the other hand, African Americans tend to have more subcutaneous fat, which may provide a better explanation for ethnic differences in cancer-related outcomes.

Subcutaneous Fat

Although some studies suggest that visceral fat plays a larger role in the development of insulin resistance [146, 147], other studies in adults suggest that subcutaneous fat has a significant impact on metabolic disease risk given its larger volume and functional characteristics, making it more susceptible to inflammation and subsequent deposition of ectopic fat [149, 168]. More specifically, subcutaneous fat has two distinct compartments, the deep and superficial depots, which differ in their contribution to metabolic disease risk [169, 170]. For example, a study in lean and obese adults found that deep subcutaneous fat and visceral fat, but not superficial subcutaneous fat, were inversely correlated with insulin sensitivity as measured by euglycemic clamp [169]. At the same time, recent studies have identified ethnic differences in the distribution of deep and superficial subcutaneous fat with Asians reporting the lowest BMI, but the largest accumulation of visceral fat and deep subcutaneous fat when compared to Caucasian, African American, and Latino adults [171–174]. In another study, higher amounts of deep subcutaneous fat were reported in Native American and Asian adults compared to Caucasians [172]. With respect to African Americans, higher levels of subcutaneous fat have been consistently reported across populations of African descent including residents in the USA, the Caribbean, South America, or Europe [175]. Taken together, these findings suggest that ethnic differences in deep and superficial subcutaneous fat could partially explain ethnic differences in insulin sensitivity and secretion. More importantly, the greater volumes of subcutaneous fat and the previously reported stronger relationship between this fat depot and IGF-1 in African Americans offer another potential explanation for the greater insulin resistance and cancer risk previously reported in this ethnic group.

Intramyocellular Lipid

More recently evidence suggests that fat deposition outside of adipose tissue (e.g., in muscle, liver, or pancreas) contributes to increased insulin resistance [176–183]. Intramyocellular lipid, for example, has been shown to be a major determinant of insulin resistance in adults [179], obese individuals [176, 178], and obese adolescents [183]. Several studies have also reported an inverse relationship between intramyocellular lipid and insulin sensitivity in inactive individuals, independent of total body fat in both animal [184] and human models [185]. Reductions in intramyocellular lipid content have also been implicated in the improvements of insulin sensitivity in response to a short-term hypocaloric diet in both normoglycemic and type 2 diabetic patients [186]. Similar improvements in insulin sensitivity have also been observed in parallel with intramyocellular lipid depletion in morbidly obese subjects after surgical treatment of obesity [187]. These findings highlight the importance of intramyocellular lipid as a metabolically active fat depot that influences insulin resistance independent of total body fat.

Few studies have examined ethnic differences in intramyocellular lipid in adults. One study in Asian and Caucasian men reported higher intramyocellular lipid content in Asians compared to age- and BMI-matched Caucasians [178]. Interestingly, intramyocellular lipid in Asians was not related to insulin sensitivity or adiposity; this relationship was present in Caucasians [178]. Similar differences by ethnicity were reported between African Americans and Caucasians, with intramyocellular lipid content related to insulin sensitivity and adiposity in Caucasians, but not African Americans [188]. Another study in Native Americans also noted that intramyocellular lipid did not predict a reduction in peripheral or hepatic insulin sensitivity [189]. Hence, intramyocellular lipid content does not appear to explain or contribute to the increased insulin resistance in ethnic minority adults. To date, the relationship between intramyocellular lipid content and insulin sensitivity in Latino adults has not been studied.

Many more ethnic comparison studies of intramyocellular lipid content have been conducted in overweight and obese youth. One recent report demonstrated that African Americans and Latinos have more intramyocellular lipid than Caucasians, even after controlling for BMI and visceral fat [181]. Another study in African American, Latino, and Caucasian children observed an inverse relationship between intramyocellular lipid and markers of inflammation; however, the majority of these relationships were eliminated after controlling for BMI and subcutaneous and visceral fat [181], suggesting that other fat depots may be more strongly associated with low-grade inflammation and insulin resistance in ethnic minority groups. To our knowledge there are no studies examining intramyocellular lipid in Native American or Asian children. Taken together, these studies suggest that increases in intramyocellular lipid may contribute to insulin resistance in an ethnic specific manner; however, the documented correlation between intramyocellular lipid, subcutaneous, visceral, and hepatic fat makes it difficult to tease apart the exact influence of each fat depot [177, 181, 190, 191]. Hence, additional studies comparing the contribution of intramyocellular, subcutaneous, and visceral fat are warranted to better understand the relationship between body fat distribution and observed ethnic differences in insulin resistance and subsequent cancer risk in ethnic minority populations.

Hepatic Fat

Numerous studies have documented inverse associations between hepatic fat, insulin sensitivity, and pancreatic beta-cell function [171, 192–197]. In a previous study of normal-weight, overweight, and obese Caucasian adolescents, those with hepatic steatosis had lower insulin sensitivity and a twofold greater prevalence of metabolic syndrome compared to those without hepatic steatosis [196]. In another study in both Canadian Caucasian and Native American adolescents, those with type 2 diabetes had higher hepatic fat compared to those without type 2 diabetes; moreover, hepatic fat was negatively associated with insulin sensitivity [197]. A US study that included Caucasian, African American, and Asian adolescents found that obese adolescents with nonalcoholic fatty liver disease (NAFLD) had a lower pancreatic beta-cell function compared to those who were obese and without NAFLD [193]. Others have confirmed these relationships in obese Latino adolescents where those with elevated hepatic fat (>5.5 %) had a significantly lower insulin sensitivity and higher acute insulin response to intravenous glucose compared to those with lower hepatic fat [192]. These results suggest that hepatic fat is associated with metabolic abnormalities including insulin resistance and the deleterious effects of hepatic fat on insulin resistance appear consistent across different ethnic groups [171, 194, 195, 198, 199].

When making ethnic comparisons of hepatic fat content, similar to visceral fat, both African American adolescents and adults have lower amounts of hepatic fat compared to Latinos and Caucasians [200–202]. Nevertheless, the relationship between hepatic fat and insulin resistance appears to be stronger in this ethnic group. In one study, hepatic fat, not visceral fat, was inversely associated with insulin sensitivity and the effect of high hepatic fat (>5.5 %) compared to low hepatic fat was more pronounced in African American compared to Latino children [192]. In Latinos, high hepatic fat was associated with a 24 % lower insulin sensitivity, whereas in African Americans, high hepatic fat was associated with a 49 % lower insulin sensitivity [195]. These results suggest a stronger relationship between hepatic fat and insulin resistance in African Americans. Similar studies have not been performed in children belonging to other ethnic groups. Taken together, these findings suggest that for African Americans who have greater volumes of hepatic fat, this depot may contribute to increased insulin resistance. However, for the majority of African Americans who tend to have extremely low volumes of hepatic fat, this depot is not likely to be a major contributor to the increased insulin resistance and subsequent cancer risk in this population.

Pancreatic Fat

Accumulation of fat in the pancreas has also been associated with insulin resistance and hyperinsulinemia in both normal-weight and obese/type 2 diabetic individuals; this relationship appears to be independent of total body fat [195, 199, 203]. Moreover, pancreatic fat has been used as a marker of pancreatic beta-cell dysfunction, especially in Latinos [199]. A recent study examining ethnic differences in pancreatic fat determined that when comparing Caucasian, African American, and Latino adults at similar levels of adiposity, Latinos had a twofold greater volume of pancreatic fat compared to African Americans; Latinos and Caucasians had similar levels of pancreatic fat [199].

Studies in children and adolescents are limited, and no studies to date have been conducted in Asians or Native Americans. In African American and Latino overweight and obese adolescents and young adults [195, 198], one study reported greater hepatic and pancreatic fat volumes in those with prediabetes compared to those with normal glucose tolerance [195]. However, pancreatic fat predicted prediabetes in African Americans whereas hepatic fat predicted prediabetes in Latinos [195]. These results suggest that ethnic differences in the relationship between ectopic fat depots and metabolic disease risk are present with pancreatic fat playing a larger role in the metabolic abnormalities previously reported in African Americans. Of note, visceral, hepatic, and pancreatic fat are highly correlated; hence, future studies should aim to examine fat depots in an effort to elucidate the exact contributions of each fat depot, particularly pancreatic fat, to the increases in insulin resistance and subsequent cancer risk in African American populations.

Adipose Tissue Biology

There is increasing evidence to suggest that differences in body fat accumulation and patterning may result from fundamental differences in adipose tissue biology [145, 204]. The increase in body fat content with obesity can occur by either an increase in adipocyte cell size or number or the spillover of triglycerides to ectopic tissues [145, 204]. When adipocyte cell size increases with progressing obesity, it is an indication of the inability of adipocytes to expand in number to accommodate the extra triglyceride accumulation [204]. Increased adipocyte cell size is also related to greater insulin resistance independent of total body fat [67]. Larger adipocytes have also been shown to be associated with more lipid deposition in visceral and hepatic fat depots (but not muscle), and this may also contribute to insulin resistance [205]. Furthermore, it is now evident that adipose tissue is infiltrated with macrophages [206]. One animal study has shown that accumulation of excess body fat in response to excess caloric intake leads to increasing fat cell size and then to adipocyte death, with the excess fat deposited in the liver [207].

Despite the important role that adipose tissue biology appears to play in the link between obesity, insulin resistance, and related cancer risk, there are no studies to date examining potential ethnic differences in the metabolic risk factor. Some studies have compared adipocyte cell size in African Americans and Caucasians but have not shown any difference in subcutaneous abdominal or gluteal adipocytes from obese women [208]. There are no data in the literature comparing ethnic differences in adipose tissue biology in Latinos and the potential relationship between adipocyte cell size and spillover of triglycerides to other ectopic storage depots like liver and pancreas. It is plausible that Latinos may have larger fat cells than African Americans that are more likely to die due to greater macrophage infiltration, thus leading to the greater likelihood of ectopic fat accumulation in Latinos. On the other hand, the higher circulating IGF-1 present in African Americans may contribute to a greater likelihood for adipocyte proliferation during obesity [209], leading to less likelihood for spillover of fat into ectopic depots; the opposite scenario is present in Latinos (lower obesity-related IGF-1 profile).

Thus, differences in the obesity–IGF pathway and adipocyte differentiation/growth factor pathways may also elucidate mechanisms explaining ethnic differences in body fat accumulation, body fat patterning, and subsequent cancer risk; additional research is warranted.

Adipose Tissue Inflammation

In conjunction with the accumulation and distribution of fat throughout the body, another potential explanation for ethnic differences in insulin resistance and subsequent cancer risk involves inflammation. Studies have shown that obesity is associated with a state of chronic low-grade inflammation, which is correlated with increased insulin resistance, and impaired glucose metabolism [210–213]. Although it was once believed that adipose tissue was only involved in the storage of free fatty acids as triglycerides, researchers now recognized that this tissue also acts as a dynamic endocrine organ, contributing to the chronic low-grade inflammation seen during obesity. For instance, during excess weight gain there is a marked increase in adipose tissue inflammation, which has been shown to be associated with insulin resistance seen during obesity [214]. Obesity is characterized by elevated circulating levels of acute-phase proteins, for example leptin, tumor necrosis factor (TNF)-alpha, interleukin (IL)-6, and decreased adiponectin [215]. Although the cause and effect nature of these proteins on insulin action is not clear, it has been suggested that these inflammatory markers affect disease processes in part by causing or exacerbating insulin resistance. Epidemiologic studies have demonstrated a positive association between acute-phase proteins and insulin resistance [216]. For example, leptin serves as part of an "adipostat" mechanism, whereby increased fat mass sets in motion responses that will eventually reduce adiposity. Hence, the reduced responsiveness to leptin that accompanies obesity may play a role in causing obesity and also contribute to insulin resistance [217, 218]. Another example is TNF-alpha, which has been shown to impair insulin signaling by activating serine/threonine kinases in skeletal muscle and downregulate glucose transporter type 4 (GLUT 4) in adipose tissue [216]. Circulating levels of IL-6 increase hepatic glucose production and stimulate the release of free fatty acids; however IL-6 also appears to have anti-inflammatory actions since it decreases TNF-alpha [219]. Adiponectin is exclusively produced in adipose tissue, and in humans its production is slightly higher in subcutaneous fat than visceral fat [220]. Adiponectin levels are negatively correlated with BMI and body fat, and this protein has been shown to play a role in hepatic insulin sensitivity and whole-body metabolism [221]. Both experiments in humans [222] and in animals [223] have demonstrated that low-grade inflammation predicts the development of insulin resistance.

Recent studies have also examined low-grade inflammation from adipose tissue biopsies in young adults. Specifically, subcutaneous adipose tissue biopsies performed in Caucasian, African American, Latino, and Native American adults have shown that in addition to elevations in plasma markers of inflammation, increases in pro-inflammatory immune cells in adipose tissue are associated with systemic and local inflammation [224–227]. In another study, subcutaneous adipose tissue inflammation was assessed by the presence of crown-like structures in obese African American and Latino young adults. Individuals with subcutaneous adipose tissue inflammation had greater levels of visceral fat, hepatic fat, TNF-alpha, and fasting insulin and glucose and a lower beta-cell function compared to those without subcutaneous inflammation [226].

Although there are no studies in children involving adipose tissue biopsies, one study in obese youth observed macrophages and lymphocytes in perivascular positions in the adipose tissue [228] while another study in children found macrophages in the subcutaneous adipose tissue of normal-weight, overweight, and obese children as young as 5 years of age [229]. Studies using plasma markers of inflammation have also found strong associations with insulin resistance in overweight and obese youth from various ethnic groups. For example, a study in boys found that those who were overweight had higher serum levels of IL-6, IL-8, interferon-γ, monocyte chemoattractant protein (MCP)-1, and C-reactive protein (CRP) compared to those of normal weight [230]. Compared to normal-weight Latino children, higher levels of CRP and IL-1beta were reported in obese Latino children [210]. Another study in African American and Latino peripubertal females demonstrated that CRP was positively related to BMI, percent body fat, fasting insulin, and acute insulin response to glucose as well as negatively correlated with insulin sensitivity [211]. One of the few recent studies including Asian children found that, after controlling for adiposity, Asians had higher levels of CRP, A1C, and insulin levels compared to Caucasian and African American children [213]. To our knowledge, there is only one study examining inflammation in Native American children. This study found elevated levels of CRP that were associated with increased adiposity, insulin resistance, worsening lipid profile, and decreased adiponectin levels [231]. Findings from these studies in children suggest that obesity is accompanied by chronic levels of low-grade inflammation starting at an early age into adulthood, possibly contributing to increased insulin resistance in these populations.

There are only sparse data on inflammatory profiles in multiethnic cohorts in the USA. These studies suggested that inflammation may be higher in African Americans [232–234], although not all studies showed this trend [235]. Specifically, CRP concentrations were higher in African Americans than in Caucasians in several large studies [232, 234, 236]. The Women's Health Study reported higher levels of CRP in African Americans than in Caucasians [232]. In contrast, NHANES data did not show this trend and instead observed higher CRP in Latina women compared with Caucasians [237]. In another study that measured visceral fat, the negative association between visceral adipose tissue and adiponectin was stronger in African Americans [237]. However, overall body fatness may still have played a role in inflammation because subcutaneous fat also had significant independent association with CRP in this ethnic group. Of note, African American women consistently exhibited greater markers of inflammation even after controlling for both L4L5 visceral and subcutaneous fat [159]. More importantly, the greater inflammation among these African American women was present despite similar or lower self-reported rates of smoking and similar or higher self-reported rates of taking lipid-lowering medications and nonsteroidal anti-inflammatory drugs [159]. The mechanisms contributing to greater low-grade inflammation in African Americans are unclear, but possibilities include higher intrinsic activity of cytokine pathways and/or different behavioral influences (i.e., high-fat diet and physical inactivity) on inflammation.

Aside from intrinsic cytokine production pathways, lifestyle factors such as diet or exercise may play a role in the altered visceral fat/body fat–inflammatory biomarker relationship. An observational study found that diets high in glycemic load were associated with increased concentrations of inflammation and that the dose–response gradient between glycemic load and inflammation was more exaggerated in overweight women [238]. Other dietary factors that have been shown to increase low-grade inflammation include sucrose, artificial sweeteners, fats, and processed meats [239]. In contrast, fiber, fruits, and vegetables have been associated with reduced inflammation [240]. Previous research has reported eating patterns reflecting higher consumption of fat and calories and lower consumption of fruits and vegetables in African Americans [241], which may contribute to the greater inflammation in this ethnic group. Moreover, African American women in particular have been shown to have lower rates of physical activity participation compared to Caucasians [242–245], which may independently contribute to inflammation. Hence, studies examining whether ethnic differences in exercise or dietary patterns account for the altered visceral fat–inflammation relationships among African Americans are warranted to better understand the increased cancer risk in this population.

Non-esterified Fatty Acids

Studies in obese adults have documented a relationship between adipose tissue insulin resistance and non-esterified fatty acids (NEFA) [246]. Given that increased hepatic fat, intramyocellular lipid [247, 248], and inflamed adipose tissue [249] are associated with increased whole-body insulin resistance, it is possible that NEFA play a mediating role in the link between ethnic differences in ectopic fat, inflammation, and insulin resistance. However, most of the research in this area has been conducted in children. Studies in overweight and obese youth have observed elevations in fasting NEFA and NEFA levels after an oral glucose or intravenous lipid challenge. Longitudinal data has confirmed an inverse relationship between fasting NEFA and insulin secretion following a 30-min oral glucose challenge in children with normal glucose tolerance [250]. Other researchers have shown that when compared to normoglycemic Latino children, those with prediabetes had higher fasting NEFA that were also inversely related to insulin secretion [195].

The earliest work in this field with regard to ethnicity first showed that after an intravenous lipid infusion, elevations in NEFA were associated with increased insulin resistance in African American and Caucasian adolescents [251]. Of note, ethnicity did not modify the relationship between NEFA and insulin resistance despite lower insulin sensitivity in African Americans compared to Caucasians [251]. Another study reported ethnic differences in NEFA during an intravenous glucose tolerance test [181, 252]. Independent of insulin secretion, African American women and girls had lower NEFA than Caucasian women and girls [181, 252]. To our knowledge, there are no studies examining these relationships in Asian or NA children, warranting their inclusion in future studies. Hence, NEFA contributes to insulin resistance and ethnicity does not appear to modify this relationship. However, African Americans tend to have lower NEFA suggesting that this mechanism does not explain the increased insulin resistance and subsequent cancer risk in this population.

Oxidative Stress

The potential role of oxidative stress in carcinogenesis is rapidly evolving, which may also link obesity and insulin resistance to increased cancer risk. Oxidative stress occurs when there is excessive production of reactive oxygen species (ROS) or insufficient in vivo antioxidant defense mechanisms [253]. This results in damage to DNA as well as lipid peroxidation, protein modification, membrane disruption, and mitochondrial damage [218, 254]. Data support the notion that increased formation of ROS may play an important role in carcinogenesis as well as atherosclerosis, diabetes, and neurodegenerative diseases [255]. Although ROS-induced lipid peroxides are usually described as harmful to cellular systems, they are also critical mediators of apoptosis [256] and have been shown to inhibit cancer growth in a number of experimental studies [257]. More specifically, factors that increase lipid peroxidation could also increase cancer and other degenerative diseases in people with innate or acquired high levels of ROS. However, factors that increase lipid peroxidation can increase apoptosis of precancerous and cancerous cells and thus protect against cancer, particularly in people with a low innate baseline level of ROS [256]. Thus, antioxidants may protect against certain cancers if background levels of ROS are higher in "at-risk" populations, but not if background ROS levels are lower because this may place a greater importance on the suppression of oxidation-induced apoptosis [256].

The relationship between obesity, insulin resistance, and oxidative stress has not been widely explored, but some supporting evidence suggests a link. Obese adults have elevated levels of lipid peroxidation that is reversible with weight reduction [255]. Metabolic conditions associated with insulin resistance are associated with elevated lipid peroxidation, including hypertension [255], impaired glucose tolerance [258, 259], and type 2 diabetes [258, 260–268]. In addition, increased oxidized low-density lipoprotein or susceptibility to oxidation has been reported in patients with type 2 diabetes [261, 262, 265, 268, 269]. Small dense low-density lipoprotein particles, which are also a component of the metabolic syndrome, are more susceptible than larger ones to oxidative modification [270, 271]. Finally, lipid peroxidation and oxidative stress, induced by elevations in glucose and possibly free fatty acid levels, may play a key role in causing insulin resistance by their ability to activate stress-sensitive signaling pathways [272].

Relatively few studies have compared lipid peroxidation and oxidative stress in different ethnic groups. In adults with type 2 diabetes, increased levels of lipid peroxidation were found in African Caribbeans compared to Caucasians [273]. Previous work showed greater lipid peroxidation in Latinos compared to Caucasians with [274] and without type 2 diabetes [275]. In another study, lipid peroxidation was higher in African Americans than in Caucasians during hyperlipidemia induced by lipid infusion [276]. Of note, recent data from the multiethnic IRAS cohort reported lower urinary F2-isoprostane levels, a marker of lipid peroxidation, among African American compared with Caucasians and Latinos [277, 278]. When stratified by BMI, ethnic differences in F2-isoprostane levels were not observed among participants with normal BMI but appeared among overweight participants and increased among obese participants [278]. Hence, additional studies comparing the markers of oxidative stress are warranted to better understand its potential contributions to ethnic differences in cancer risk.

Summary and Conclusions

Obesity is a predisposing risk factor for certain forms of cancer, and the link between obesity and cancer appears to be particularly complex. Obesity is associated with increased insulin resistance, and hyperinsulinemia may play a critical role in influencing cancer risk. It is notable that obesity-related cancer risk differs dramatically by ethnicity. African Americans appear particularly prone to obesity-related cancers including prostate, breast, and colorectal and myeloma, whereas Latinos appear relatively protected. Based on previous literature, it is plausible that ethnic differences in the insulin response to obesity may contribute to ethnic differences in obesity-related cancer profiles. Obese Latinos seem more prone to an ectopic fat pattern (increased visceral, hepatic, and pancreatic fat), and this might be driven by greater fat cell size, greater likelihood of adipocyte macrophage infiltration and cell death, and decreased capacity for fat cells to differentiate, possibly due to a lower obesity-related IGF-1 profile. On the other hand, obese African Americans seem more prone to some forms of obesity (subcutaneous fat pattern) and insulin-related cancers compared to Latinos and have less likelihood of ectopic fat. These differences could be driven by the much higher obesity-related hyperinsulinemia (especially in response to glucose) and IGF-1 profile in African Americans. This is important because it suggests that reducing levels of insulin in obesity in this population as a strategy to prevent obesity-related cancers may have the unwanted side effect of reducing fat cell proliferation and promotion of hepatic fat, and other ectopic fat deposition, unless it is combined with behavioral interventions to influence energy balance (reduce energy intake and increase physical activity) and subsequent weight status. Additional factors that contribute to increased insulin resistance and cancer risk in African Americans include chronic glucocorticoid exposure, chronic inflammation, and possibly greater oxidative stress. Hence, additional therapies that reduce multiple cancer-related metabolic risk factors in African American children and adults are warranted.

In summary, the causes and consequences of obesity and insulin resistance differ by ethnicity of people and much more work is needed to establish the specific mechanisms linking obesity and insulin to various cancer outcomes. These mechanistic issues are fundamental to understanding the basic pathophysiology of why increased body fat and hyperinsulinemia are related to cancer outcomes in some ethnic groups but not others and will ultimately have widespread implications for the application of more individualized prevention and treatment approaches to reduce the disparity in obesity-related cancers.

References

1. Flegal KM, Carroll MD, Kit BK, Ogden CL (2012) Prevalence of obesity and trends in the distribution of body mass index among US adults, 1999-2010. JAMA 307:491–497
2. Barnes PM, Adams PF, Powell-Griner E (2010) Health characteristics of the American Indian or Alaska Native adult population: United States, 2004–2008 National health statistics reports, no. 20. National Center for Health Statistics, Hyattsville, MD
3. Barnes PM, Adams PF, Powell-Griner E (2008) Health characteristics of the Asian adult population: United States, 2004–2006. Advance data from vital and health statistics; no 394. National Center for Health Statistics, Hyattsville, MD
4. Narayan KM, Aviles-Santa L, Oza-Frank R, Pandey M, Curb JD, McNeely M, Araneta MR, Palaniappan L, Rajpathak S, Barrett-Connor E, Cardiovascular Disease in A, Pacific Islander Populations NWG (2010) Report of a National Heart, Lung, And Blood Institute Workshop: heterogeneity in cardiometabolic risk in Asian Americans in the U.S. opportunities for research. J Am Coll Cardiol 55:966–973
5. Ogden CL, Carroll MD, Kit BK, Flegal KM (2012) Prevalence of obesity and trends in body mass index among US children and adolescents, 1999-2010. JAMA 307:483–490
6. Broussard BA, Johnson A, Himes JH, Story M, Fichtner R, Hauck F, Bachman-Carter K, Hayes J, Frohlich K, Gray N et al (1991) Prevalence of obesity in American Indians and Alaska Natives. Am J Clin Nutr 53:1535S–1542S
7. Cowie CC, Rust KF, Ford ES, Eberhardt MS, Byrd-Holt DD, Li C, Williams DE, Gregg EW, Bainbridge KE, Saydah SH, Geiss LS (2009) Full accounting of diabetes and pre-diabetes in the U.S. population in 1988-1994 and 2005-2006. Diabetes Care 32:287–294
8. Goran MI, Bergman RN, Cruz ML, Watanabe R (2002) Insulin resistance and associated compensatory responses in African-American and Hispanic children. Diabetes Care 25:2184–2190
9. Moore K (2010) Youth-onset type 2 diabetes among American Indians and Alaska Natives. J Public Health Manag Pract 16:388–393
10. Narayan KM, Boyle JP, Thompson TJ, Sorensen SW, Williamson DF (2003) Lifetime risk for diabetes mellitus in the United States. JAMA 290:1884–1890

11. Nsiah-Kumi PA, Lasley S, Whiting M, Brushbreaker C, Erickson JM, Qiu F, Yu F, Larsen JL (2013) Diabetes, pre-diabetes and insulin resistance screening in Native American children and youth. Int J Obes (Lond) 37:540–545

12. Writing Group for the SfDiYSG, Dabelea D, Bell RA, D'Agostino RB Jr, Imperatore G, Johansen JM, Linder B, Liu LL, Loots B, Marcovina S, Mayer-Davis EJ, Pettitt DJ, Waitzfelder B (2007) Incidence of diabetes in youth in the United States. JAMA 297:2716–2724

13. Schiller JS, Lucas JW, Ward BW, Peregoy JA (2012) Summary health statistics for U.S. adults: National Health Interview Survey, 2010. National Center for Health Statistics. Vital Health Stat 10(252)

14. American Diabetes Association (2000) Type 2 diabetes in children and adolescents. Pediatrics 105:671–680

15. Arslanian SA, Saad R, Lewy V, Danadian K, Janosky J (2002) Hyperinsulinemia in African-American children: decreased insulin clearance and increased insulin secretion and its relationship to insulin sensitivity. Diabetes 51:3014–3019

16. Hannon TS, Bacha F, Lin Y, Arslanian SA (2008) Hyperinsulinemia in African-American adolescents compared with their American white peers despite similar insulin sensitivity: a reflection of upregulated beta-cell function? Diabetes Care 31:1445–1447

17. Ku CY, Gower BA, Hunter GR, Goran MI (2000) Racial differences in insulin secretion and sensitivity in prepubertal children: role of physical fitness and physical activity. Obes Res 8:506–515

18. Lindquist CH, Gower BA, Goran MI (2000) Role of dietary factors in ethnic differences in early risk of cardiovascular disease and type 2 diabetes. Am J Clin Nutr 71:725–732

19. IARC Working Group (2005) IARC working group on the evaluation of cancer-preventive strategies. Breast cancer screening. IARC Press, Lyon

20. Calle EE, Kaaks R (2004) Overweight, obesity and cancer: epidemiological evidence and proposed mechanisms. Nat Rev Cancer 4:579–591

21. Prentice RL, Willett WC, Greenwald P, Alberts D, Bernstein L, Boyd NF, Byers T, Clinton SK, Fraser G, Freedman L, Hunter D, Kipnis V, Kolonel LN, Kristal BS, Kristal A, Lampe JW, McTiernan A, Milner J, Patterson RE, Potter JD, Riboli E, Schatzkin A, Yates A, Yetley E (2004) Nutrition and physical activity and chronic disease prevention: research strategies and recommendations. J Natl Cancer Inst 96:1276–1287

22. Fuemmeler BF, Pendzich MK, Tercyak KP (2009) Weight, dietary behavior, and physical activity in childhood and adolescence: implications for adult cancer risk. Obes Facts 2:179–186

23. Giovannucci E (2003) Nutrition, insulin, insulin-like growth factors and cancer. Horm Metab Res 35:694–704

24. Miller B, Kolonel LN, Bernstein L, Young JJ, Swanson G (1996) Racial/ethnic patterns of cancer in the United States 1988-1992. National Cancer Institute, Bethesda, MD

25. Howlader N, Noone AM, Krapcho M, Garshell J, Miller D, Altekruse SF, Kosary CL, Yu M, Ruhl J, Tatalovich Z, Mariotto A, Lewis DR, Chen HS, Feuer EJ, Cronin KA (eds) (2014) SEER cancer statistics review, 1975–2011. National Cancer Institute, Bethesda, MD, http://seer.cancer.gov/csr/1975_2011/, based on November 2013 SEER data submission, posted to the SEER web site, April 2014

26. Boyle P (2012) Triple-negative breast cancer: epidemiological considerations and recommendations. Ann Oncol 23 Suppl 6:vi7–vi12

27. Kolonel LN, Altshuler D, Henderson BE (2004) The multiethnic cohort study: exploring genes, lifestyle and cancer risk. Nat Rev Cancer 4:519–527

28. Caprio S, Hyman LD, Limb C, McCarthy S, Lange R, Sherwin RS, Shulman G, Tamborlane WV (1995) Central adiposity and its metabolic correlates in obese adolescent girls. Am J Physiol 269:E118–E126

29. Goran MI, Bergman RN, Gower BA (2001) Influence of total vs. visceral fat on insulin action and secretion in African American and white children. Obes Res 9:423–431

30. Gower BA, Nagy TR, Goran MI (1999) Visceral fat, insulin sensitivity, and lipids in prepubertal children. Diabetes 48:1515–1521
31. Goran MI, Ball GD, Cruz ML (2003) Obesity and risk of type 2 diabetes and cardiovascular disease in children and adolescents. J Clin Endocrinol Metab 88:1417–1427
32. Bogardus C, Lillioja S, Howard BV, Reaven G, Mott D (1984) Relationships between insulin secretion, insulin action, and fasting plasma glucose concentration in nondiabetic and noninsulin-dependent diabetic subjects. J Clin Invest 74:1238–1246
33. Cutfield WS, Bergman RN, Menon RK, Sperling MA (1990) The modified minimal model: application to measurement of insulin sensitivity in children. J Clin Endocrinol Metab 70:1644–1650
34. Hershko DD (2008) Oncogenic properties and prognostic implications of the ubiquitin ligase Skp2 in cancer. Cancer 112:1415–1424
35. Auld CA, Caccia CD, Morrison RF (2007) Hormonal induction of adipogenesis induces Skp2 expression through PI3K and MAPK pathways. J Cell Biochem 100:204–216
36. Cooke PS, Holsberger DR, Cimafranca MA, Meling DD, Beals CM, Nakayama K, Nakayama KI, Kiyokawa H (2007) The F box protein S phase kinase-associated protein 2 regulates adipose mass and adipocyte number in vivo. Obesity (Silver Spring) 15:1400–1408
37. Gray SG, Stenfeldt Mathiasen I, De Meyts P (2003) The insulin-like growth factors and insulin-signalling systems: an appealing target for breast cancer therapy? Horm Metab Res 35:857–871
38. Pollak MN, Schernhammer ES, Hankinson SE (2004) Insulin-like growth factors and neoplasia. Nat Rev Cancer 4:505–518
39. Cheney KE, Liu RK, Smith GS, Leung RE, Mickey MR, Walford RL (1980) Survival and disease patterns in C57BL/6J mice subjected to undernutrition. Exp Gerontol 15:237–258
40. Grasl-Kraupp B, Bursch W, Ruttkay-Nedecky B, Wagner A, Lauer B, Schulte-Hermann R (1994) Food restriction eliminates preneoplastic cells through apoptosis and antagonizes carcinogenesis in rat liver. Proc Natl Acad Sci U S A 91:9995–9999
41. Ross MH, Bras G (1965) Tumor incidence patterns and nutrition in the rat. J Nutr 87:245–260
42. Spindler SR (2005) Rapid and reversible induction of the longevity, anticancer and genomic effects of caloric restriction. Mech Ageing Dev 126:960–966
43. Tannenbaum A, Silverstone H (1953) The genesis and growth of tumors. VI. Effects of varying the level of minerals in the diet. Cancer Res 13:460–463
44. Volk MJ, Pugh TD, Kim M, Frith CH, Daynes RA, Ershler WB, Weindruch R (1994) Dietary restriction from middle age attenuates age-associated lymphoma development and interleukin 6 dysregulation in C57BL/6 mice. Cancer Res 54:3054–3061
45. Weindruch R, Walford RL (1982) Dietary restriction in mice beginning at 1 year of age: effect on life-span and spontaneous cancer incidence. Science 215:1415–1418
46. Albu JB, Kovera AJ, Allen L, Wainwright M, Berk E, Raja-Khan N, Janumala I, Burkey B, Heshka S, Gallagher D (2005) Independent association of insulin resistance with larger amounts of intermuscular adipose tissue and a greater acute insulin response to glucose in African American than in white nondiabetic women. Am J Clin Nutr 82:1210–1217
47. Cossrow N, Falkner B (2004) Race/ethnic issues in obesity and obesity-related comorbidities. J Clin Endocrinol Metab 89:2590–2594
48. Donahue RP, Bean JA, Donahue RA, Goldberg RB, Prineas RJ (1997) Insulin response in a triethnic population: effects of sex, ethnic origin, and body fat. Miami Community Health Study. Diabetes Care 20:1670–1676
49. Haffner SM, D'Agostino R Jr, Goff D, Howard B, Festa A, Saad MF, Mykkanen L (1999) LDL size in African Americans, Hispanics, and non-Hispanic whites: the insulin resistance atherosclerosis study. Arterioscler Thromb Vasc Biol 19:2234–2240
50. Haffner SM, D'Agostino R, Saad MF, Rewers M, Mykkanen L, Selby J, Howard G, Savage PJ, Hamman RF, Wagenknecht LE et al (1996) Increased insulin resistance and insulin

secretion in nondiabetic African-Americans and Hispanics compared with non-Hispanic whites. The Insulin Resistance Atherosclerosis Study. Diabetes 45:742–748

51. Howard BV, Mayer-Davis EJ, Goff D, Zaccaro DJ, Laws A, Robbins DC, Saad MF, Selby J, Hamman RF, Krauss RM, Haffner SM (1998) Relationships between insulin resistance and lipoproteins in nondiabetic African Americans, Hispanics, and non-Hispanic whites: the Insulin Resistance Atherosclerosis Study. Metabolism 47:1174–1179

52. Kasim-Karakas SE (2000) Ethnic differences in the insulin resistance syndrome. Am J Clin Nutr 71:670–671

53. Osei K, Cottrell DA, Harris B (1992) Differences in basal and poststimulation glucose homeostasis in nondiabetic first degree relatives of black and white patients with type 2 diabetes mellitus. J Clin Endocrinol Metab 75:82–86

54. Osei K, Schuster DP (1994) Ethnic differences in secretion, sensitivity, and hepatic extraction of insulin in black and white Americans. Diabet Med 11:755–762

55. Palaniappan LP, Carnethon MR, Fortmann SP (2002) Heterogeneity in the relationship between ethnicity, BMI, and fasting insulin. Diabetes Care 25:1351–1357

56. Sirikul B, Gower BA, Hunter GR, Larson-Meyer DE, Newcomer BR (2006) Relationship between insulin sensitivity and in vivo mitochondrial function in skeletal muscle. Am J Physiol Endocrinol Metab 291:E724–E728

57. Bogardus C, Thuillez P, Ravussin E, Vasquez B, Narimiga M, Azhar S (1983) Effect of muscle glycogen depletion on in vivo insulin action in man. J Clin Invest 72:1605–1610

58. Haffner SM, Miettinen H, Gaskill SP, Stern MP (1995) Decreased insulin secretion and increased insulin resistance are independently related to the 7-year risk of NIDDM in Mexican-Americans. Diabetes 44:1386–1391

59. Haffner SM, Stern MP, Mitchell BD, Hazuda HP, Patterson JK (1990) Incidence of type II diabetes in Mexican Americans predicted by fasting insulin and glucose levels, obesity, and body-fat distribution. Diabetes 39:283–288

60. Hasson RE, Adam TC, Davis JN, Weigensberg MJ, Ventura EE, Lane CJ, Roberts CK, Goran MI (2010) Ethnic differences in insulin action in obese African-American and Latino adolescents. J Clin Endocrinol Metab 95:4048–4051

61. Stefan N, Stumvoll M, Weyer C, Bogardus C, Tataranni PA, Pratley RE (2004) Exaggerated insulin secretion in Pima Indians and African-Americans but higher insulin resistance in Pima Indians compared to African-Americans and Caucasians. Diabet Med 21:1090–1095

62. Chiu KC, Cohan P, Lee NP, Chuang LM (2000) Insulin sensitivity differs among ethnic groups with a compensatory response in beta-cell function. Diabetes Care 23:1353–1358

63. Lillioja S, Mott DM, Spraul M, Ferraro R, Foley JE, Ravussin E, Knowler WC, Bennett PH, Bogardus C (1993) Insulin resistance and insulin secretory dysfunction as precursors of non-insulin-dependent diabetes mellitus. Prospective studies of Pima Indians. N Engl J Med 329:1988–1992

64. Lillioja S, Mott DM, Zawadzki JK, Young AA, Abbott WG, Knowler WC, Bennett PH, Moll P, Bogardus C (1987) In vivo insulin action is familial characteristic in nondiabetic Pima Indians. Diabetes 36:1329–1335

65. Pratley RE, Weyer C, Bogardus C (1999) Metabolic abnormalities in the development of type 2 diabetes mellitus. In: LeRoith D, Taylor SI, Olefsky JM (eds) Diabetes mellitus: a fundamental and clinical text. Lippincott, Williams & Wilkins, Philadelphia, pp 548–557

66. Lillioja S (1996) Impaired glucose tolerance in Pima Indians. Diabet Med 13:S127–S132

67. Weyer C, Bogardus C, Mott DM, Pratley RE (1999) The natural history of insulin secretory dysfunction and insulin resistance in the pathogenesis of type 2 diabetes mellitus. J Clin Invest 104:787–794

68. Freedman DS, Srinivasan SR, Burke GL, Shear CL, Smoak CG, Harsha DW, Webber LS, Berenson GS (1987) Relation of body fat distribution to hyperinsulinemia in children and adolescents: the Bogalusa Heart Study. Am J Clin Nutr 46:403–410

69. Arslanian S, Suprasongsin C, Janosky JE (1997) Insulin secretion and sensitivity in black versus white prepubertal healthy children. J Clin Endocrinol Metab 82:1923–1927

70. Moore E, Copeland KC, Parker D, Burgin C, Blackett PR (2006) Ethnic differences in fasting glucose, insulin resistance and lipid profiles in obese adolescents. J Okla State Med Assoc 99:439–443

71. Goran MI, Gower BA (2001) Longitudinal study on pubertal insulin resistance. Diabetes 50:2444–2450

72. Cobelli C, Toffolo GM, Dalla Man C, Campioni M, Denti P, Caumo A, Butler P, Rizza R (2007) Assessment of beta-cell function in humans, simultaneously with insulin sensitivity and hepatic extraction, from intravenous and oral glucose tests. Am J Physiol Endocrinol Metab 293:E1–E15

73. Gower BA, Fernandez JR, Beasley TM, Shriver MD, Goran MI (2003) Using genetic admixture to explain racial differences in insulin-related phenotypes. Diabetes 52:1047–1051

74. Klimentidis YC, Dulin-Keita A, Casazza K, Willig AL, Allison DB, Fernandez JR (2012) Genetic admixture, social-behavioural factors and body composition are associated with blood pressure differently by racial-ethnic group among children. J Hum Hypertens 26:98–107

75. Francis DD (2009) Conceptualizing child health disparities: a role for developmental neurogenomics. Pediatrics 124 Suppl 3:S196–S202

76. Pervanidou P, Chrousos GP (2012) Metabolic consequences of stress during childhood and adolescence. Metabolism 61:611–619

77. Dickerson SS, Kemeny ME (2004) Acute stressors and cortisol responses: a theoretical integration and synthesis of laboratory research. Psychol Bull 130:355–391

78. Adam EK (2006) Transactions among adolescent trait and state emotion and diurnal and momentary cortisol activity in naturalistic settings. Psychoneuroendocrinology 31:664–679

79. De Vriendt T, Moreno LA, De Henauw S (2009) Chronic stress and obesity in adolescents: scientific evidence and methodological issues for epidemiological research. Nutr Metab Cardiovasc Dis 19:511–519

80. Adam TC, Epel ES (2007) Stress, eating and the reward system. Physiol Behav 91:449–458

81. Dallman MF, Strack AM, Akana SF, Bradbury MJ, Hanson ES, Scribner KA, Smith M (1993) Feast and famine: critical role of glucocorticoids with insulin in daily energy flow. Front Neuroendocrinol 14:303–347

82. Rosmond R (2003) Stress induced disturbances of the HPA axis: a pathway to Type 2 diabetes? Med Sci Monit 9:RA35–RA39

83. Lambillotte C, Gilon P, Henquin JC (1997) Direct glucocorticoid inhibition of insulin secretion. An in vitro study of dexamethasone effects in mouse islets. J Clin Invest 99:414–423

84. Coderre L, Vallega GA, Pilch PF, Chipkin SR (1996) In vivo effects of dexamethasone and sucrose on glucose transport (GLUT-4) protein tissue distribution. Am J Physiol 271:E643–E648

85. van Raalte DH, Ouwens DM, Diamant M (2009) Novel insights into glucocorticoid-mediated diabetogenic effects: towards expansion of therapeutic options? Eur J Clin Invest 39:81–93

86. Beard JC, Halter JB, Best JD, Pfeifer MA, Porte D Jr (1984) Dexamethasone-induced insulin resistance enhances B cell responsiveness to glucose level in normal men. Am J Physiol 247:E592–E596

87. Grill V, Pigon J, Hartling SG, Binder C, Efendic S (1990) Effects of dexamethasone on glucose-induced insulin and proinsulin release in low and high insulin responders. Metabolism 39:251–258

88. Larsson H, Ahren B (1999) Insulin resistant subjects lack islet adaptation to short-term dexamethasone-induced reduction in insulin sensitivity. Diabetologia 42:936–943

89. Bjorntorp P (1997) Neuroendocrine factors in obesity. J Endocrinol 155:193–195

90. Bjorntorp P, Rosmond R (2000) Obesity and cortisol. Nutrition 16:924–936

91. Adam TC, Hasson RE, Ventura EE, Toledo-Corral C, Le KA, Mahurkar S, Lane CJ, Weigensberg MJ, Goran MI (2010) Cortisol is negatively associated with insulin sensitivity in overweight Latino youth. J Clin Endocrinol Metab 95:4729–4735

92. Dallman MF (2010) Stress-induced obesity and the emotional nervous system. Trends Endocrinol Metab 21:159–165
93. Kuo LE, Kitlinska JB, Tilan JU, Li L, Baker SB, Johnson MD, Lee EW, Burnett MS, Fricke ST, Kvetnansky R, Herzog H, Zukowska Z (2007) Neuropeptide Y acts directly in the periphery on fat tissue and mediates stress-induced obesity and metabolic syndrome. Nat Med 13:803–811
94. Weigensberg MJ, Toledo-Corral CM, Goran MI (2008) Association between the metabolic syndrome and serum cortisol in overweight Latino youth. J Clin Endocrinol Metab 93:1372–1378
95. Epel E, Lapidus R, McEwen B, Brownell K (2001) Stress may add bite to appetite in women: a laboratory study of stress-induced cortisol and eating behavior. Psychoneuroendocrinology 26:37–49
96. Newman E, O'Connor DB, Conner M (2007) Daily hassles and eating behaviour: the role of cortisol reactivity status. Psychoneuroendocrinology 32:125–132
97. Boutelle KN, Murray DM, Jeffery RW, Hennrikus DJ, Lando HA (2000) Associations between exercise and health behaviors in a community sample of working adults. Prev Med 30:217–224
98. Ng DM, Jeffery RW (2003) Relationships between perceived stress and health behaviors in a sample of working adults. Health Psychol 22:638–642
99. Steptoe A, Wardle J, Pollard TM, Canaan L, Davies GJ (1996) Stress, social support and health-related behavior: a study of smoking, alcohol consumption and physical exercise. J Psychosom Res 41:171–180
100. Stetson BA, Rahn JM, Dubbert PM, Wilner BI, Mercury MG (1997) Prospective evaluation of the effects of stress on exercise adherence in community-residing women. Health Psychol 16:515–520
101. Block JP, He Y, Zaslavsky AM, Ding L, Ayanian JZ (2009) Psychosocial stress and change in weight among US adults. Am J Epidemiol 170:181–192
102. Brunner EJ, Chandola T, Marmot MG (2007) Prospective effect of job strain on general and central obesity in the Whitehall II Study. Am J Epidemiol 165:828–837
103. Tomiyama AJ, Puterman E, Epel ES, Rehkopf DH, Laraia BA (2013) Chronic psychological stress and racial disparities in body mass index change between Black and White girls aged 10-19. Ann Behav Med 45:3–12
104. Kirschbaum C, Hellhammer DH (1989) Salivary cortisol in psychobiological research: an overview. Neuropsychobiology 22:150–169
105. Pruessner JC, Wolf OT, Hellhammer DH, Buske-Kirschbaum A, von Auer K, Jobst S, Kaspers F, Kirschbaum C (1997) Free cortisol levels after awakening: a reliable biological marker for the assessment of adrenocortical activity. Life Sci 61:2539–2549
106. Cohen S, Doyle WJ, Baum A (2006) Socioeconomic status is associated with stress hormones. Psychosom Med 68:414–420
107. DeSantis AS, Adam EK, Doane LD, Mineka S, Zinbarg RE, Craske MG (2007) Racial/ethnic differences in cortisol diurnal rhythms in a community sample of adolescents. J Adolesc Health 41:3–13
108. Hajat A, Diez-Roux A, Franklin TG, Seeman T, Shrager S, Ranjit N, Castro C, Watson K, Sanchez B, Kirschbaum C (2010) Socioeconomic and race/ethnic differences in daily salivary cortisol profiles: the multi-ethnic study of atherosclerosis. Psychoneuroendocrinology 35:932–943
109. McCallum TJ, Sorocco KH, Fritsch T (2006) Mental health and diurnal salivary cortisol patterns among African American and European American female dementia family caregivers. Am J Geriatr Psychiatry 14:684–693
110. Suglia SF, Staudenmayer J, Cohen S, Enlow MB, Rich-Edwards JW, Wright RJ (2010) Cumulative stress and cortisol disruption among Black and Hispanic pregnant women in an Urban Cohort. Psychol Trauma 2:326–334

111. Cohen S, Schwartz JE, Epel E, Kirschbaum C, Sidney S, Seeman T (2006) Socioeconomic status, race, and diurnal cortisol decline in the Coronary Artery Risk Development in Young Adults (CARDIA) Study. Psychosom Med 68:41–50

112. Martin CG, Bruce J, Fisher PA (2012) Racial and ethnic differences in diurnal cortisol rhythms in preadolescents: the role of parental psychosocial risk and monitoring. Horm Behav 61:661–668

113. Rosmond R, Bjorntorp P (2000) The hypothalamic-pituitary-adrenal axis activity as a predictor of cardiovascular disease, type 2 diabetes and stroke. J Intern Med 247:188–197

114. LeRoith D, Roberts CT Jr (2003) The insulin-like growth factor system and cancer. Cancer Lett 195:127–137

115. Vottero A, Guzzetti C, Loche S (2013) New aspects of the physiology of the GH-IGF-1 axis. Endocr Dev 24:96–105

116. Clemmons DR (2012) Metabolic actions of insulin-like growth factor-I in normal physiology and diabetes. Endocrinol Metab Clin North Am 41:425–443, vii-viii

117. Cohen P (1998) Serum insulin-like growth factor-I levels and prostate cancer risk–interpreting the evidence. J Natl Cancer Inst 90:876–879

118. Hankinson SE, Willett WC, Colditz GA, Hunter DJ, Michaud DS, Deroo B, Rosner B, Speizer FE, Pollak M (1998) Circulating concentrations of insulin-like growth factor-I and risk of breast cancer. Lancet 351:1393–1396

119. Ma J, Pollak MN, Giovannucci E, Chan JM, Tao Y, Hennekens CH, Stampfer MJ (1999) Prospective study of colorectal cancer risk in men and plasma levels of insulin-like growth factor (IGF)-I and IGF-binding protein-3. J Natl Cancer Inst 91:620–625

120. Vadgama JV, Wu Y, Datta G, Khan H, Chillar R (1999) Plasma insulin-like growth factor-I and serum IGF-binding protein 3 can be associated with the progression of breast cancer, and predict the risk of recurrence and the probability of survival in African-American and Hispanic women. Oncology 57:330–340

121. Wolk A, Mantzoros CS, Andersson SO, Bergstrom R, Signorello LB, Lagiou P, Adami HO, Trichopoulos D (1998) Insulin-like growth factor 1 and prostate cancer risk: a population-based, case-control study. J Natl Cancer Inst 90:911–915

122. Yu H, Spitz MR, Mistry J, Gu J, Hong WK, Wu X (1999) Plasma levels of insulin-like growth factor-I and lung cancer risk: a case-control analysis. J Natl Cancer Inst 91:151–156

123. Frystyk J, Skjaerbaek C, Vestbo E, Fisker S, Orskov H (1999) Circulating levels of free insulin-like growth factors in obese subjects: the impact of type 2 diabetes. Diabetes Metab Res Rev 15:314–322

124. Nam SY, Lee EJ, Kim KR, Cha BS, Song YD, Lim SK, Lee HC, Huh KB (1997) Effect of obesity on total and free insulin-like growth factor (IGF)-1, and their relationship to IGF-binding protein (BP)-1, IGFBP-2, IGFBP-3, insulin, and growth hormone. Int J Obes Relat Metab Disord 21:355–359

125. Yanovski JA, Sovik KN, Nguyen TT, Sebring NG (2000) Insulin-like growth factors and bone mineral density in African American and White girls. J Pediatr 137:826–832

126. DeLellis K, Rinaldi S, Kaaks RJ, Kolonel LN, Henderson B, Le Marchand L (2004) Dietary and lifestyle correlates of plasma insulin-like growth factor-I (IGF-I) and IGF binding protein-3 (IGFBP-3): the multiethnic cohort. Cancer Epidemiol Biomarkers Prev 13:1444–1451

127. Girgis R, Abrams SA, Castracane VD, Gunn SK, Ellis KJ, Copeland KC (2000) Ethnic differences in androgens, IGF-I and body fat in healthy prepubertal girls. J Pediatr Endocrinol Metab 13:497–503

128. Copeland KC, Colletti RB, Devlin JT, McAuliffe TL (1990) The relationship between insulin-like growth factor-I, adiposity, and aging. Metabolism 39:584–587

129. Jernstrom H, Deal C, Wilkin F, Chu W, Tao Y, Majeed N, Hudson T, Narod SA, Pollak M (2001) Genetic and nongenetic factors associated with variation of plasma levels of insulin-like growth factor-I and insulin-like growth factor-binding protein-3 in healthy premenopausal women. Cancer Epidemiol Biomarkers Prev 10:377–384

130. Lukanova A, Toniolo P, Akhmedkhanov A, Hunt K, Rinaldi S, Zeleniuch-Jacquotte A, Haley NJ, Riboli E, Stattin P, Lundin E, Kaaks R (2001) A cross-sectional study of IGF-I determinants in women. Eur J Cancer Prev 10:443–452

131. Schoen RE, Schragin J, Weissfeld JL, Thaete FL, Evans RW, Rosen CJ, Kuller LH (2002) Lack of association between adipose tissue distribution and IGF-1 and IGFBP-3 in men and women. Cancer Epidemiol Biomarkers Prev 11:581–586

132. Teramukai S, Rohan T, Eguchi H, Oda T, Shinchi K, Kono S (2002) Anthropometric and behavioral correlates of insulin-like growth factor I and insulin-like growth factor binding protein 3 in middle-aged Japanese men. Am J Epidemiol 156:344–348

133. Chang S, Wu X, Yu H, Spitz MR (2002) Plasma concentrations of insulin-like growth factors among healthy adult men and postmenopausal women: associations with body composition, lifestyle, and reproductive factors. Cancer Epidemiol Biomarkers Prev 11:758–766

134. Landin-Wilhelmsen K, Wilhelmsen L, Lappas G, Rosen T, Lindstedt G, Lundberg PA, Bengtsson BA (1994) Serum insulin-like growth factor I in a random population sample of men and women: relation to age, sex, smoking habits, coffee consumption and physical activity, blood pressure and concentrations of plasma lipids, fibrinogen, parathyroid hormone and osteocalcin. Clin Endocrinol (Oxf) 41:351–357

135. Morimoto LM, Newcomb PA, White E, Bigler J, Potter JD (2005) Variation in plasma insulin-like growth factor-1 and insulin-like growth factor binding protein-3: personal and lifestyle factors (United States). Cancer Causes Control 16:917–927

136. Lukanova A, Lundin E, Zeleniuch-Jacquotte A, Muti P, Mure A, Rinaldi S, Dossus L, Micheli A, Arslan A, Lenner P, Shore RE, Krogh V, Koenig KL, Riboli E, Berrino F, Hallmans G, Stattin P, Toniolo P, Kaaks R (2004) Body mass index, circulating levels of sex-steroid hormones, IGF-I and IGF-binding protein-3: a cross-sectional study in healthy women. Eur J Endocrinol 150:161–171

137. Lukanova A, Soderberg S, Stattin P, Palmqvist R, Lundin E, Biessy C, Rinaldi S, Riboli E, Hallmans G, Kaaks R (2002) Nonlinear relationship of insulin-like growth factor (IGF)-I and IGF-I/IGF-binding protein-3 ratio with indices of adiposity and plasma insulin concentrations (Sweden). Cancer Causes Control 13:509–516

138. Henderson KD, Goran MI, Kolonel LN, Henderson BE, Le Marchand L (2006) Ethnic disparity in the relationship between obesity and plasma insulin-like growth factors: the multiethnic cohort. Cancer Epidemiol Biomarkers Prev 15:2298–2302

139. Garnett SP, Hogler W, Blades B, Baur LA, Peat J, Lee J, Cowell CT (2004) Relation between hormones and body composition, including bone, in prepubertal children. Am J Clin Nutr 80:966–972

140. Ong K, Kratzsch J, Kiess W, Dunger D, Team AS (2002) Circulating IGF-I levels in childhood are related to both current body composition and early postnatal growth rate. J Clin Endocrinol Metab 87:1041–1044

141. Higgins PB, Fernandez JR, Goran MI, Gower BA (2005) Early ethnic difference in insulin-like growth factor-1 is associated with African genetic admixture. Pediatr Res 58:850–854

142. Rosmond R, Dallman MF, Bjorntorp P (1998) Stress-related cortisol secretion in men: relationships with abdominal obesity and endocrine, metabolic and hemodynamic abnormalities. J Clin Endocrinol Metab 83:1853–1859

143. Misra M, Bredella MA, Tsai P, Mendes N, Miller KK, Klibanski A (2008) Lower growth hormone and higher cortisol are associated with greater visceral adiposity, intramyocellular lipids, and insulin resistance in overweight girls. Am J Physiol Endocrinol Metab 295:E385–E392

144. Alderete TL, Byrd-Williams CE, Toledo-Corral CM, Conti DV, Weigensberg MJ, Goran MI (2011) Relationships between IGF-1 and IGFBP-1 and adiposity in obese African-American and Latino adolescents. Obesity (Silver Spring) 19:933–938

145. Frayn KN (2000) Visceral fat and insulin resistance—causative or correlative? Br J Nutr 83 Suppl 1:S71–S77

146. Cruz ML, Bergman RN, Goran MI (2002) Unique effect of visceral fat on insulin sensitivity in obese Hispanic children with a family history of type 2 diabetes. Diabetes Care 25:1631–1636
147. Going SB, Lohman TG, Cussler EC, Williams DP, Morrison JA, Horn PS (2011) Percent body fat and chronic disease risk factors in U.S. children and youth. Am J Prev Med 41:S77–S86
148. Indulekha K, Anjana RM, Surendar J, Mohan V (2011) Association of visceral and subcutaneous fat with glucose intolerance, insulin resistance, adipocytokines and inflammatory markers in Asian Indians (CURES-113). Clin Biochem 44:281–287
149. McLaughlin T, Lamendola C, Liu A, Abbasi F (2011) Preferential fat deposition in subcutaneous versus visceral depots is associated with insulin sensitivity. J Clin Endocrinol Metab 96:E1756–E1760
150. Patel P, Abate N (2013) Body fat distribution and insulin resistance. Nutrients 5:2019–2027
151. Rosenbaum M, Fennoy I, Accacha S, Altshuler L, Carey DE, Holleran S, Rapaport R, Shelov SP, Speiser PW, Ten S, Bhangoo A, Boucher-Berry C, Espinal Y, Gupta R, Hassoun AA, Iazetti L, Jean-Jacques F, Jean AM, Klein ML, Levine R, Lowell B, Michel L, Rosenfeld W (2013) Racial/Ethnic differences in clinical and biochemical type 2 diabetes mellitus risk factors in children. Obesity (Silver Spring) 21:2081–2090
152. Albu JB, Murphy L, Frager DH, Johnson JA, Pi-Sunyer FX (1997) Visceral fat and race-dependent health risks in obese nondiabetic premenopausal women. Diabetes 46:456–462
153. Wagenknecht LE, Langefeld CD, Scherzinger AL, Norris JM, Haffner SM, Saad MF, Bergman RN (2003) Insulin sensitivity, insulin secretion, and abdominal fat: the Insulin Resistance Atherosclerosis Study (IRAS) Family Study. Diabetes 52:2490–2496
154. Gautier JF, Milner MR, Elam E, Chen K, Ravussin E, Pratley RE (1999) Visceral adipose tissue is not increased in Pima Indians compared with equally obese Caucasians and is not related to insulin action or secretion. Diabetologia 42:28–34
155. Lear SA, Humphries KH, Kohli S, Birmingham CL (2007) The use of BMI and waist circumference as surrogates of body fat differs by ethnicity. Obesity (Silver Spring) 15:2817–2824
156. Misra A (2003) Revisions of cutoffs of body mass index to define overweight and obesity are needed for the Asian-ethnic groups. Int J Obes Relat Metab Disord 27:1294–1296
157. Razak F, Anand SS, Shannon H, Vuksan V, Davis B, Jacobs R, Teo KK, McQueen M, Yusuf S (2007) Defining obesity cut points in a multiethnic population. Circulation 115:2111–2118
158. Stevens J (2003) Ethnic-specific revisions of body mass index cutoffs to define overweight and obesity in Asians are not warranted. Int J Obes Relat Metab Disord 27:1297–1299
159. Carroll JF, Fulda KG, Chiapa AL, Rodriquez M, Phelps DR, Cardarelli KM, Vishwanatha JK, Cardarelli R (2009) Impact of race/ethnicity on the relationship between visceral fat and inflammatory biomarkers. Obesity (Silver Spring) 17:1420–1427
160. Conway JM, Yanovski SZ, Avila NA, Hubbard VS (1995) Visceral adipose tissue differences in black and white women. Am J Clin Nutr 61:765–771
161. Kanaley JA, Giannopoulou I, Tillapaugh-Fay G, Nappi JS, Ploutz-Snyder LL (2003) Racial differences in subcutaneous and visceral fat distribution in postmenopausal black and white women. Metabolism 52:186–191
162. Lovejoy JC, de la Bretonne JA, Klemperer M, Tulley R (1996) Abdominal fat distribution and metabolic risk factors: effects of race. Metabolism 45:1119–1124
163. Perry AC, Applegate EB, Jackson ML, Deprima S, Goldberg RB, Ross R, Kempner L, Feldman BB (2000) Racial differences in visceral adipose tissue but not anthropometric markers of health-related variables. J Appl Physiol (1985) 89:636–643
164. Fisher G, Hyatt TC, Hunter GR, Oster RA, Desmond RA, Gower BA (2012) Markers of inflammation and fat distribution following weight loss in African-American and white women. Obesity (Silver Spring) 20:715–720

165. Weinsier RL, Hunter GR, Gower BA, Schutz Y, Darnell BE, Zuckerman PA (2001) Body fat distribution in white and black women: different patterns of intraabdominal and subcutaneous abdominal adipose tissue utilization with weight loss. Am J Clin Nutr 74:631–636
166. Goran MI, Nagy TR, Treuth MS, Trowbridge C, Dezenberg C, McGloin A, Gower BA (1997) Visceral fat in white and African American prepubertal children. Am J Clin Nutr 65:1703–1708
167. Huang TT, Johnson MS, Figueroa-Colon R, Dwyer JH, Goran MI (2001) Growth of visceral fat, subcutaneous abdominal fat, and total body fat in children. Obes Res 9:283–289
168. Abate N, Chandalia M (2012) Role of subcutaneous adipose tissue in metabolic complications of obesity. Metab Syndr Relat Disord 10:319–320
169. Kelley DE, Thaete FL, Troost F, Huwe T, Goodpaster BH (2000) Subdivisions of subcutaneous abdominal adipose tissue and insulin resistance. Am J Physiol Endocrinol Metab 278: E941–E948
170. Tordjman J, Divoux A, Prifti E, Poitou C, Pelloux V, Hugol D, Basdevant A, Bouillot JL, Chevallier JM, Bedossa P, Guerre-Millo M, Clement K (2012) Structural and inflammatory heterogeneity in subcutaneous adipose tissue: relation with liver histopathology in morbid obesity. J Hepatol 56:1152–1158
171. Anand SS, Tarnopolsky MA, Rashid S, Schulze KM, Desai D, Mente A, Rao S, Yusuf S, Gerstein HC, Sharma AM (2011) Adipocyte hypertrophy, fatty liver and metabolic risk factors in South Asians: the Molecular Study of Health and Risk in Ethnic Groups (mol-SHARE). PLoS One 6:e22112
172. Kohli S, Lear SA (2013) Differences in subcutaneous abdominal adiposity regions in four ethnic groups. Obesity (Silver Springs) 21(11):2288–95
173. Kohli S, Sniderman AD, Tchernof A, Lear SA (2010) Ethnic-specific differences in abdominal subcutaneous adipose tissue compartments. Obesity (Silver Spring) 18:2177–2183
174. Nazare JA, Smith JD, Borel AL, Haffner SM, Balkau B, Ross R, Massien C, Almeras N, Despres JP (2012) Ethnic influences on the relations between abdominal subcutaneous and visceral adiposity, liver fat, and cardiometabolic risk profile: the International Study of Prediction of Intra-Abdominal Adiposity and Its Relationship With Cardiometabolic Risk/ Intra-Abdominal Adiposity. Am J Clin Nutr 96:714–726
175. Araneta MR, Barrett-Connor E (2005) Ethnic differences in visceral adipose tissue and type 2 diabetes: Filipino, African-American, and white women. Obes Res 13:1458–1465
176. Ashley MA, Buckley AJ, Criss AL, Ward JA, Kemp A, Garnett S, Cowell CT, Baur LA, Thompson CH (2002) Familial, anthropometric, and metabolic associations of intramyocellular lipid levels in prepubertal males. Pediatr Res 51:81–86
177. Bennett B, Larson-Meyer DE, Ravussin E, Volaufova J, Soros A, Cefalu WT, Chalew S, Gordon S, Smith SR, Newcomer BR, Goran M, Sothern M (2012) Impaired insulin sensitivity and elevated ectopic fat in healthy obese vs. nonobese prepubertal children. Obesity (Silver Spring) 20:371–375
178. Forouhi NG, Jenkinson G, Thomas EL, Mullick S, Mierisova S, Bhonsle U, McKeigue PM, Bell JD (1999) Relation of triglyceride stores in skeletal muscle cells to central obesity and insulin sensitivity in European and South Asian men. Diabetologia 42:932–935
179. Kelley DE, Goodpaster BH (2001) Skeletal muscle triglyceride. An aspect of regional adiposity and insulin resistance. Diabetes Care 24:933–941
180. Lee S, Kim Y, White DA, Kuk JL, Arslanian S (2012) Relationships between insulin sensitivity, skeletal muscle mass and muscle quality in obese adolescent boys. Eur J Clin Nutr 66:1366–1368
181. Maligie M, Crume T, Scherzinger A, Stamm E, Dabelea D (2012) Adiposity, fat patterning, and the metabolic syndrome among diverse youth: the EPOCH study. J Pediatr 161:875–880
182. Saukkonen T, Heikkinen S, Hakkarainen A, Hakkinen AM, van Leemput K, Lipsanen-Nyman M, Lundbom N (2010) Association of intramyocellular, intraperitoneal and liver fat with glucose tolerance in severely obese adolescents. Eur J Endocrinol 163:413–419

183. Sinha R, Dufour S, Petersen KF, LeBon V, Enoksson S, Ma YZ, Savoye M, Rothman DL, Shulman GI, Caprio S (2002) Assessment of skeletal muscle triglyceride content by (1)H nuclear magnetic resonance spectroscopy in lean and obese adolescents: relationships to insulin sensitivity, total body fat, and central adiposity. Diabetes 51:1022–1027

184. Kuhlmann J, Neumann-Haefelin C, Belz U, Kalisch J, Juretschke HP, Stein M, Kleinschmidt E, Kramer W, Herling AW (2003) Intramyocellular lipid and insulin resistance: a longitudinal in vivo 1H-spectroscopic study in Zucker diabetic fatty rats. Diabetes 52:138–144

185. Thamer C, Machann J, Bachmann O, Haap M, Dahl D, Wietek B, Tschritter O, Niess A, Brechtel K, Fritsche A, Claussen C, Jacob S, Schick F, Haring HU, Stumvoll M (2003) Intramyocellular lipids: anthropometric determinants and relationships with maximal aerobic capacity and insulin sensitivity. J Clin Endocrinol Metab 88:1785–1791

186. Lara-Castro C, Newcomer BR, Rowell J, Wallace P, Shaughnessy SM, Munoz AJ, Shiflett AM, Rigsby DY, Lawrence JC, Bohning DE, Buchthal S, Garvey WT (2008) Effects of short-term very low-calorie diet on intramyocellular lipid and insulin sensitivity in nondiabetic and type 2 diabetic subjects. Metabolism 57:1–8

187. Greco AV, Mingrone G, Giancaterini A, Manco M, Morroni M, Cinti S, Granzotto M, Vettor R, Camastra S, Ferrannini E (2002) Insulin resistance in morbid obesity: reversal with intramyocellular fat depletion. Diabetes 51:144–151

188. Ingram KH, Lara-Castro C, Gower BA, Makowsky R, Allison DB, Newcomer BR, Munoz AJ, Beasley TM, Lawrence JC, Lopez-Ben R, Rigsby DY, Garvey WT (2011) Intramyo-cellular lipid and insulin resistance: differential relationships in European and African Americans. Obesity (Silver Spring) 19:1469–1475

189. Koska J, Stefan N, Permana PA, Weyer C, Sonoda M, Bogardus C, Smith SR, Joanisse DR, Funahashi T, Krakoff J, Bunt JC (2008) Increased fat accumulation in liver may link insulin resistance with subcutaneous abdominal adipocyte enlargement, visceral adiposity, and hypoadiponectinemia in obese individuals. Am J Clin Nutr 87:295–302

190. Brumbaugh DE, Crume TL, Nadeau K, Scherzinger A, Dabelea D (2012) Intramyocellular lipid is associated with visceral adiposity, markers of insulin resistance, and cardiovascular risk in prepubertal children: the EPOCH study. J Clin Endocrinol Metab 97:E1099–E1105

191. Ou HY, Wang CY, Yang YC, Chen MF, Chang CJ (2013) The association between nonalcoholic fatty pancreas disease and diabetes. PLoS One 8:e62561

192. Kim JS, Le KA, Mahurkar S, Davis JN, Goran MI (2012) Influence of elevated liver fat on circulating adipocytokines and insulin resistance in obese Hispanic adolescents. Pediatr Obes 7:158–164

193. Singh GK, Vitola BE, Holland MR, Sekarski T, Patterson BW, Magkos F, Klein S (2013) Alterations in ventricular structure and function in obese adolescents with nonalcoholic fatty liver disease. J Pediatr 162:1160–1168, 1168.e1161

194. Targher G, Rossi AP, Zamboni GA, Fantin F, Antonioli A, Corzato F, Bambace C, Pozzi Mucelli R, Zamboni M (2012) Pancreatic fat accumulation and its relationship with liver fat content and other fat depots in obese individuals. J Endocrinol Invest 35:748–753

195. Toledo-Corral CM, Alderete TL, Hu HH, Nayak K, Esplana S, Liu T, Goran MI, Weigensberg MJ (2013) Ectopic fat deposition in prediabetic overweight and obese minority adolescents. J Clin Endocrinol Metab 98:1115–1121

196. Wicklow BA, Wittmeier KD, MacIntosh AC, Sellers EA, Ryner L, Serrai H, Dean HJ, McGavock JM (2012) Metabolic consequences of hepatic steatosis in overweight and obese adolescents. Diabetes Care 35:905–910

197. Wittmeier KD, Wicklow BA, MacIntosh AC, Sellers EA, Ryner LN, Serrai H, Gardiner PF, Dean HJ, McGavock JM (2012) Hepatic steatosis and low cardiorespiratory fitness in youth with type 2 diabetes. Obesity (Silver Spring) 20:1034–1040

198. Le KA, Ventura EE, Fisher JQ, Davis JN, Weigensberg MJ, Punyanitya M, Hu HH, Nayak KS, Goran MI (2011) Ethnic differences in pancreatic fat accumulation and its relationship with other fat depots and inflammatory markers. Diabetes Care 34:485–490

199. Szczepaniak LS, Victor RG, Mathur R, Nelson MD, Szczepaniak EW, Tyer N, Chen I, Unger
     RH, Bergman RN, Lingvay I (2012) Pancreatic steatosis and its relationship to beta-cell
     dysfunction in humans: racial and ethnic variations. Diabetes Care 35:2377–2383
200. Guerrero R, Vega GL, Grundy SM, Browning JD (2009) Ethnic differences in hepatic
     steatosis: an insulin resistance paradox? Hepatology 49:791–801
201. Hasson RE, Adam TC, Davis JN, Kelly LA, Ventura EE, Byrd-Williams CE, Toledo-Corral
     CM, Roberts CK, Lane CJ, Azen SP, Chou CP, Spruijt-Metz D, Weigensberg MJ, Berhane K,
     Goran MI (2012) Randomized controlled trial to improve adiposity, inflammation, and
     insulin resistance in obese African-American and Latino youth. Obesity (Silver Spring)
     20:811–818
202. Liska D, Dufour S, Zern TL, Taksali S, Cali AM, Dziura J, Shulman GI, Pierpont BM, Caprio
     S (2007) Interethnic differences in muscle, liver and abdominal fat partitioning in obese
     adolescents. PLoS One 2:e569
203. Maggio AB, Mueller P, Wacker J, Viallon M, Belli DC, Beghetti M, Farpour-Lambert NJ,
     McLin VA (2012) Increased pancreatic fat fraction is present in obese adolescents with
     metabolic syndrome. J Pediatr Gastroenterol Nutr 54:720–726
204. Ravussin E, Smith SR (2002) Increased fat intake, impaired fat oxidation, and failure of fat
     cell proliferation result in ectopic fat storage, insulin resistance, and type 2 diabetes mellitus.
     Ann N Y Acad Sci 967:363–378
205. Larson-Meyer DE, Heilbronn LK, Redman LM, Newcomer BR, Frisard MI, Anton S, Smith
     SR, Alfonso A, Ravussin E (2006) Effect of calorie restriction with or without exercise on
     insulin sensitivity, beta-cell function, fat cell size, and ectopic lipid in overweight subjects.
     Diabetes Care 29:1337–1344
206. Greenberg AS, Obin MS (2006) Obesity and the role of adipose tissue in inflammation and
     metabolism. Am J Clin Nutr 83:461S–465S
207. Strissel KJ, Stancheva Z, Miyoshi H, Perfield JW 2nd, DeFuria J, Jick Z, Greenberg AS, Obin
     MS (2007) Adipocyte death, adipose tissue remodeling, and obesity complications. Diabetes
     56:2910–2918
208. Dowling HJ, Fried SK, Pi-Sunyer FX (1995) Insulin resistance in adipocytes of obese
     women: effects of body fat distribution and race. Metabolism 44:987–995
209. Holly J, Sabin M, Perks C, Shield J (2006) Adipogenesis and IGF-1. Metab Syndr Relat
     Disord 4:43–50
210. Balas-Nakash M, Perichart-Perera O, Benitez-Arciniega A, Tolentino-Dolores M, Mier-
     Cabrera J, Vadillo-Ortega F (2013) Association between adiposity, inflammation and car-
     diovascular risk factors in school-aged Mexican children. Gac Med Mex 149:196–203
211. Spruijt-Metz D, Adar Emken B, Spruijt MR, Richey JM, Berman LJ, Belcher BR, Hsu YW,
     McClain AD, Lane CJ, Weigensberg MJ (2012) CRP is related to higher leptin levels in
     minority peripubertal females regardless of adiposity levels. Obesity (Silver Spring) 20:512–
     516
212. Utsal L, Tillmann V, Zilmer M, Maestu J, Purge P, Jurimae J, Saar M, Latt E, Maasalu K,
     Jurimae T (2012) Elevated serum IL-6, IL-8, MCP-1, CRP, and IFN-gamma levels in 10- to
     11-year-old boys with increased BMI. Horm Res Paediatr 78:31–39
213. Whincup PH, Nightingale CM, Owen CG, Rudnicka AR, Gibb I, McKay CM, Donin AS,
     Sattar N, Alberti KG, Cook DG (2010) Early emergence of ethnic differences in type
     2 diabetes precursors in the UK: the Child Heart and Health Study in England (CHASE
     Study). PLoS Med 7:e1000263
214. McArdle MA, Finucane OM, Connaughton RM, McMorrow AM, Roche HM (2013) Mech-
     anisms of obesity-induced inflammation and insulin resistance: insights into the emerging
     role of nutritional strategies. Front Endocrinol (Lausanne) 4:52
215. Gregor MF, Hotamisligil GS (2011) Inflammatory mechanisms in obesity. Annu Rev
     Immunol 29:415–445
216. Borst SE (2007) Nutrition and health: adipose tissue and adipokines in health and disease.
     Humana Press, Totowa

217. Lazar MA (2005) How obesity causes diabetes: not a tall tale. Science 307:373–375
218. Schwartz JL, Antoniades DZ, Zhao S (1993) Molecular and biochemical reprogramming of oncogenesis through the activity of prooxidants and antioxidants. Ann N Y Acad Sci 686:262–278, discussion 278-269
219. Boden G, Shulman GI (2002) Free fatty acids in obesity and type 2 diabetes: defining their role in the development of insulin resistance and beta-cell dysfunction. Eur J Clin Invest 32 Suppl 3:14–23
220. Fisher FM, McTernan PG, Valsamakis G, Chetty R, Harte AL, Anwar AJ, Starcynski J, Crocker J, Barnett AH, McTernan CL, Kumar S (2002) Differences in adiponectin protein expression: effect of fat depots and type 2 diabetic status. Horm Metab Res 34:650–654
221. Trujillo ME, Scherer PE (2005) Adiponectin—journey from an adipocyte secretory protein to biomarker of the metabolic syndrome. J Intern Med 257:167–175
222. Festa A, D'Agostino R Jr, Tracy RP, Haffner SM, Insulin Resistance Atherosclerosis S (2002) Elevated levels of acute-phase proteins and plasminogen activator inhibitor-1 predict the development of type 2 diabetes: the insulin resistance atherosclerosis study. Diabetes 51:1131–1137
223. Cai D, Yuan M, Frantz DF, Melendez PA, Hansen L, Lee J, Shoelson SE (2005) Local and systemic insulin resistance resulting from hepatic activation of IKK-beta and NF-kappaB. Nat Med 11:183–190
224. Fabbrini E, Cella M, McCartney SA, Fuchs A, Abumrad NA, Pietka TA, Chen Z, Finck BN, Han DH, Magkos F, Conte C, Bradley D, Fraterrigo G, Eagon JC, Patterson BW, Colonna M, Klein S (2013) Association between specific adipose tissue CD4+ T-cell populations and insulin resistance in obese individuals. Gastroenterology 145(366–374):366.e1–374.e3
225. He J, Le DS, Xu X, Scalise M, Ferrante AW, Krakoff J (2010) Circulating white blood cell count and measures of adipose tissue inflammation predict higher 24-h energy expenditure. Eur J Endocrinol 162:275–280
226. Le KA, Mahurkar S, Alderete TL, Hasson RE, Adam TC, Kim JS, Beale E, Xie C, Greenberg AS, Allayee H, Goran MI (2011) Subcutaneous adipose tissue macrophage infiltration is associated with hepatic and visceral fat deposition, hyperinsulinemia, and stimulation of NF-kappaB stress pathway. Diabetes 60:2802–2809
227. Spencer M, Unal R, Zhu B, Rasouli N, McGehee RE Jr, Peterson CA, Kern PA (2011) Adipose tissue extracellular matrix and vascular abnormalities in obesity and insulin resistance. J Clin Endocrinol Metab 96:E1990–E1998
228. Sbarbati A, Osculati F, Silvagni D, Benati D, Galie M, Camoglio FS, Rigotti G, Maffeis C (2006) Obesity and inflammation: evidence for an elementary lesion. Pediatrics 117:220–223
229. Tam CS, Tordjman J, Divoux A, Baur LA, Clement K (2012) Adipose tissue remodeling in children: the link between collagen deposition and age-related adipocyte growth. J Clin Endocrinol Metab 97:1320–1327
230. Kyrgios I, Galli-Tsinopoulou A, Stylianou C, Papakonstantinou E, Arvanitidou M, Haidich AB (2012) Elevated circulating levels of the serum acute-phase protein YKL-40 (chitinase 3-like protein 1) are a marker of obesity and insulin resistance in prepubertal children. Metabolism 61:562–568
231. Retnakaran R, Hanley AJ, Connelly PW, Harris SB, Zinman B (2006) Elevated C-reactive protein in Native Canadian children: an ominous early complication of childhood obesity. Diabetes Obes Metab 8:483–491
232. Albert MA, Glynn RJ, Buring J, Ridker PM (2004) C-reactive protein levels among women of various ethnic groups living in the United States (from the Women's Health Study). Am J Cardiol 93:1238–1242
233. Albert MA, Ridker PM (2004) Inflammatory biomarkers in African Americans: a potential link to accelerated atherosclerosis. Rev Cardiovasc Med 5 Suppl 3:S22–S27
234. Lin SX, Pi-Sunyer EX (2007) Prevalence of the metabolic syndrome among US middle-aged and older adults with and without diabetes—a preliminary analysis of the NHANES 1999–2002 data. Ethn Dis 17:35–39

235. Duncan BB, Schmidt MI, Chambless LE, Folsom AR, Carpenter M, Heiss G (2000) Fibrin-ogen, other putative markers of inflammation, and weight gain in middle-aged adults—the ARIC study. Atherosclerosis Risk in Communities. Obes Res 8:279–286
236. Patel DA, Srinivasan SR, Xu JH, Li S, Chen W, Berenson GS (2006) Distribution and metabolic syndrome correlates of plasma C-reactive protein in biracial (black-white) younger adults: the Bogalusa Heart Study. Metabolism 55:699–705
237. Ford ES, Giles WH, Mokdad AH, Myers GL (2004) Distribution and correlates of C-reactive protein concentrations among adult US women. Clin Chem 50:574–581
238. Liu S, Manson JE, Buring JE, Stampfer MJ, Willett WC, Ridker PM (2002) Relation between a diet with a high glycemic load and plasma concentrations of high-sensitivity C-reactive protein in middle-aged women. Am J Clin Nutr 75:492–498
239. Nettleton JA, Steffen LM, Mayer-Davis EJ, Jenny NS, Jiang R, Herrington DM, Jacobs DR Jr (2006) Dietary patterns are associated with biochemical markers of inflammation and endo-thelial activation in the Multi-Ethnic Study of Atherosclerosis (MESA). Am J Clin Nutr 83:1369–1379
240. King DE, Egan BM, Woolson RF, Mainous AG 3rd, Al-Solaiman Y, Jesri A (2007) Effect of a high-fiber diet vs a fiber-supplemented diet on C-reactive protein level. Arch Intern Med 167:502–506
241. Madan AK, Barden CB, Beech B, Fay K, Sintich M, Beech DJ (2002) Self-reported differences in daily raw vegetable intake by ethnicity in a breast screening program. J Natl Med Assoc 94:894–900
242. Despres JP, Nadeau A, Tremblay A, Ferland M, Moorjani S, Lupien PJ, Theriault G, Pinault S, Bouchard C (1989) Role of deep abdominal fat in the association between regional adipose tissue distribution and glucose tolerance in obese women. Diabetes 38:304–309
243. Kushner RF, Racette SB, Neil K, Schoeller DA (1995) Measurement of physical activity among black and white obese women. Obes Res 3 Suppl 2:261s–265s
244. Lovejoy JC, Smith SR, Rood JC (2001) Comparison of regional fat distribution and health risk factors in middle-aged white and African American women: The Healthy Transitions Study. Obes Res 9:10–16
245. Tuten C, Petosa R, Sargent R, Weston A (1995) Biracial differences in physical activity and body composition among women. Obes Res 3:313–318
246. Heilbronn L, Smith SR, Ravussin E (2004) Failure of fat cell proliferation, mitochondrial function and fat oxidation results in ectopic fat storage, insulin resistance and type II diabetes mellitus. Int J Obes Relat Metab Disord 28 Suppl 4:S12–S21
247. Bays H, Mandarino L, DeFronzo RA (2004) Role of the adipocyte, free fatty acids, and ectopic fat in pathogenesis of type 2 diabetes mellitus: peroxisomal proliferator-activated receptor agonists provide a rational therapeutic approach. J Clin Endocrinol Metab 89:463–478
248. Gaggini M, Morelli M, Buzzigoli E, DeFronzo RA, Bugianesi E, Gastaldelli A (2013) Non-alcoholic fatty liver disease (NAFLD) and its connection with insulin resistance, dyslipidemia, atherosclerosis and coronary heart disease. Nutrients 5:1544–1560
249. Manteiga S, Choi K, Jayaraman A, Lee K (2013) Systems biology of adipose tissue metab-olism: regulation of growth, signaling and inflammation. Wiley Interdiscip Rev Syst Biol Med 5:425–447
250. Salgin B, Ong KK, Thankamony A, Emmett P, Wareham NJ, Dunger DB (2012) Higher fasting plasma free fatty acid levels are associated with lower insulin secretion in children and adults and a higher incidence of type 2 diabetes. J Clin Endocrinol Metab 97:3302–3309
251. Burns SF, Kelsey SF, Arslanian SA (2009) Effects of an intravenous lipid challenge and free fatty acid elevation on in vivo insulin sensitivity in African American versus Caucasian adolescents. Diabetes Care 32:355–360
252. Goree LL, Darnell BE, Oster RA, Brown MA, Gower BA (2010) Associations of free fatty acids with insulin secretion and action among African-American and European-American girls and women. Obesity (Silver Spring) 18:247–253

253. Bashan N, Kovsan J, Kachko I, Ovadia H, Rudich A (2009) Positive and negative regulation of insulin signaling by reactive oxygen and nitrogen species. Physiol Rev 89:27–71
254. Emerit I (1994) Reactive oxygen species, chromosome mutation, and cancer: possible role of clastogenic factors in carcinogenesis. Free Radic Biol Med 16:99–109
255. Gago-Dominguez M, Castelao JE, Yuan JM, Ross RK, Yu MC (2002) Lipid peroxidation: a novel and unifying concept of the etiology of renal cell carcinoma (United States). Cancer Causes Control 13:287–293
256. Salganik RI (2001) The benefits and hazards of antioxidants: controlling apoptosis and other protective mechanisms in cancer patients and the human population. J Am Coll Nutr 20:464S–472S, discussion 473S–475S
257. Welsch CW (1995) Review of the effects of dietary fat on experimental mammary gland tumorigenesis: role of lipid peroxidation. Free Radic Biol Med 18:757–773
258. Niskanen LK, Salonen JT, Nyyssonen K, Uusitupa MI (1995) Plasma lipid peroxidation and hyperglycaemia: a connection through hyperinsulinaemia? Diabet Med 12:802–808
259. Vijayalingam S, Parthiban A, Shanmugasundaram KR, Mohan V (1996) Abnormal antioxidant status in impaired glucose tolerance and non-insulin-dependent diabetes mellitus. Diabet Med 13:715–719
260. Cederberg J, Basu S, Eriksson UJ (2001) Increased rate of lipid peroxidation and protein carbonylation in experimental diabetic pregnancy. Diabetologia 44:766–774
261. Collier A, Rumley A, Rumley AG, Paterson JR, Leach JP, Lowe GD, Small M (1992) Free radical activity and hemostatic factors in NIDDM patients with and without microalbuminuria. Diabetes 41:909–913
262. Griesmacher A, Kindhauser M, Andert SE, Schreiner W, Toma C, Knoebl P, Pietschmann P, Prager R, Schnack C, Schernthaner G et al (1995) Enhanced serum levels of thiobarbituric-acid-reactive substances in diabetes mellitus. Am J Med 98:469–475
263. Jain SK, McVie R, Jaramillo JJ, Palmer M, Smith T, Meachum ZD, Little RL (1996) The effect of modest vitamin E supplementation on lipid peroxidation products and other cardiovascular risk factors in diabetic patients. Lipids 31 Suppl:S87–S90
264. MacRury SM, Gordon D, Wilson R, Bradley H, Gemmell CG, Paterson JR, Rumley AG, MacCuish AC (1993) A comparison of different methods of assessing free radical activity in type 2 diabetes and peripheral vascular disease. Diabet Med 10:331–335
265. Neri S, Bruno CM, Raciti C, D'Angelo G, D'Amico R, Cristaldi R (1994) Alteration of oxide reductive and haemostatic factors in type 2 diabetics. J Intern Med 236:495–500
266. Sato Y, Hotta N, Sakamoto N, Matsuoka S, Ohishi N, Yagi K (1979) Lipid peroxide level in plasma of diabetic patients. Biochem Med 21:104–107
267. Velazquez E, Winocour PH, Kesteven P, Alberti KG, Laker MF (1991) Relation of lipid peroxides to macrovascular disease in type 2 diabetes. Diabet Med 8:752–758
268. Yaqoob M, McClelland P, Patrick AW, Stevenson A, Mason H, White MC, Bell GM (1994) Evidence of oxidant injury and tubular damage in early diabetic nephropathy. QJM 87:601–607
269. Santini SA, Marra G, Giardina B, Cotroneo P, Mordente A, Martorana GE, Manto A, Ghirlanda G (1997) Defective plasma antioxidant defenses and enhanced susceptibility to lipid peroxidation in uncomplicated IDDM. Diabetes 46:1853–1858
270. Wakatsuki A, Ikenoue N, Okatani Y, Fukaya T (2001) Estrogen-induced small low density lipoprotein particles may be atherogenic in postmenopausal women. J Am Coll Cardiol 37:425–430
271. Wakatsuki A, Okatani Y, Ikenoue N, Shinohara K, Watanabe K, Fukaya T (2003) Effect of lower dose of oral conjugated equine estrogen on size and oxidative susceptibility of low-density lipoprotein particles in postmenopausal women. Circulation 108:808–813
272. Evans JL, Goldfine ID, Maddux BA, Grodsky GM (2003) Are oxidative stress-activated signaling pathways mediators of insulin resistance and beta-cell dysfunction? Diabetes 52:1–8

273. Mehrotra S, Ling KL, Bekele Y, Gerbino E, Earle KA (2001) Lipid hydroperoxide and markers of renal disease susceptibility in African-Caribbean and Caucasian patients with Type 2 diabetes mellitus. Diabet Med 18:109–115

274. Haffner SM, Agil A, Mykkanen L, Stern MP, Jialal I (1995) Plasma oxidizability in subjects with normal glucose tolerance, impaired glucose tolerance, and NIDDM. Diabetes Care 18:646–653

275. Haffner SM, Miettinen H, Stern MP, Agil A, Jialal I (1996) Plasma oxidizability in Mexican-Americans and non-Hispanic whites. Metabolism 45:876–881

276. Lopes HF, Morrow JD, Stojiljkovic MP, Goodfriend TL, Egan BM (2003) Acute hyperlipidemia increases oxidative stress more in African Americans than in white Americans. Am J Hypertens 16:331–336

277. Il'yasova D, Spasojevic I, Base K, Zhang H, Wang F, Young SP, Millington DS, D'Agostino RB Jr, Wagenknecht LE (2012) Urinary F2-isoprostanes as a biomarker of reduced risk of type 2 diabetes. Diabetes Care 35:173–174

278. Il'yasova D, Wang F, Spasojevic I, Base K, D'Agostino RB Jr, Wagenknecht LE (2012) Racial differences in urinary F2-isoprostane levels and the cross-sectional association with BMI. Obesity (Silver Spring) 20:2147–2150

279. Lara-Castro C, Weinsier RL, Hunter GR, Desmond RE (2002) Visceral adipose tissue in women: longitudinal study of the effects of fat gain, time, and race. Obes Res 10:868–874

Chapter 9
Role of Ethnic Differences in Mediators of Energy Balance

Sarah S. Cohen and Loren Lipworth

Abstract The rapid increase in the prevalence of obesity in the USA in the past 30 years has many health implications. Obesity is strongly associated with incidence and mortality of several common cancers including post-menopausal breast cancer, renal cell cancer, colorectal cancer, prostate cancer, and endometrial cancer. Obesity patterns in the USA vary substantially by race, particularly for women, with black women having a higher prevalence of obesity than white women. Moreover, black men and women experience higher incidence and mortality for many common cancers and cardiovascular disease.

Keywords Racial differences—insulin IGF-1 • Racial differences—adiponectin • Racial differences—leptin • Racial differences—sex steroid hormones • Racial differences—inflammation

Definition and Measurement of Obesity

Obesity defined at the most basic level is an excess accumulation of body fat [1]. This excess results from a multifactorial process involving an imbalance in energy consumption and expenditure, resulting in enlarged fat cells (also known as adipose cells) as well as an increase in the number of adipose cells [2]. The major sites for adipose tissue storage are both within the abdominal cavity (abdominal or visceral fat) and just under the skin (subcutaneous fat). Visceral fat tends to

S.S. Cohen
EpidStat Institute, 2100 Commonwealth Blvd, Suite 203, Ann Arbor, MI 48105, USA
e-mail: sarah@epidstat.com

L. Lipworth (✉)
Division of Epidemiology, Department of Medicine, Vanderbilt-Ingram Cancer Center,
2525 West End Avenue, Suite 600, Nashville, TN 37203, USA
e-mail: loren.lipworth@vanderbilt.edu

D.J. Bowen et al. (eds.), *Impact of Energy Balance on Cancer Disparities*,
Energy Balance and Cancer 9, DOI 10.1007/978-3-319-06103-0_9,
© Springer International Publishing Switzerland 2014

accumulate with age and is more strongly associated with metabolic disorders and cardiovascular disease [3].

There is little agreement on the best way to measure obesity accurately in either clinical settings or in large-scale research studies. In epidemiologic studies, obesity is most often measured by BMI because the component measures (height and weight) are easily obtained via self-report from study participants or from inexpensive and easy-to-use tools [1, 4]. BMI is calculated as the weight in kilograms divided by the square of height in meters. Standard categories of BMI have been put forth by the World Health Organization and include underweight (BMI <18.5 kg/ m^2), healthy weight (BMI 18.5–24.9 kg/m^2), overweight (BMI 25.0–29.9 kg/m^2), obesity class I (BMI 30.0–34.9 kg/m^2), obesity class II (BMI 35.0–39.9 kg/m^2) and obesity class III or extreme obesity (BMI $\geq$40.0 kg/m^2) [5]. BMI has excellent validity as a measure of absolute fat mass adjusted for height [4], and the widely used, standardized cut-points established for BMI categories allow for ease of comparison across studies [5, 6]. However, because BMI includes body weight (which is made up of both lean body mass and fat tissue), it is a less valid measure for percent body fat than other measures that account for differences in the proportion of each type of body tissue [4].

Other measures to assess body composition include densitometry (underwater weighing) as well as newer techniques such as dual energy X-ray absorptiometry (DEXA), bioelectrical impedance analysis (BIA), and computed tomography (CT)/ magnetic resonance imaging (MRI). Densitometry requires that an individual be submerged in water. By measuring the ratio of body weight measured in air and body weight measured under water, an estimate of the proportion of fat in the total body mass can be calculated [4]. DEXA uses an X-ray with low- and high-energy peaks to distinguish fat mass, fat-free mass, and bone mineral mass in the whole body or by specific region (such as in the abdomen) [4]. DEXA is not able to distinguish visceral fat from subcutaneous fat [7]. BIA involves sending a weak electrical current through the body and measuring its impedance by muscle tissue; because muscle is composed mainly of water and fat tissue contains virtually no water, the impedance values can be used to estimate percentage body fat [4]. CT and MRI are considered to be the most accurate methods for assessing body composition including the quantification of visceral versus subcutaneous fat [7]. However, these three measures all require expensive equipment, specialized technicians, and can be time-consuming to perform on a large number of individuals, and thus are not widely used in epidemiologic research.

Distribution of body fat, not just the total amount, has also been shown to be related to health risks. While fat distribution can be measured using imaging tools such as DEXA and CT/MRI, waist circumference and waist-to-hip ratio (WHR) are also used to measure differences in fat tissue distribution and have been used frequently in epidemiologic studies because the required measurements are inexpensive and quick to obtain. One limitation of these measures is that many factors can affect the measurement of waist and hip circumferences including the degree of training of the individual making the measure, the time of day, and timing of the most recent meal [4].

Another level of complication arises when obesity is measured in individuals of different racial or ethnic backgrounds. Many studies have concluded that commonly used measures of obesity have different meanings for whites and blacks, likely due to differences in fat distribution. Several studies report that for a similar waist circumference and BMI, blacks have less visceral fat than whites [3, 8]. Hip circumferences of blacks have been found to be smaller than those in whites, resulting in an increased WHR for a given amount of central fat deposition [1]. The relationship between BMI and percent fat measured by DEXA has also been shown to differ by race with black women having lower body fatness than white women at the same BMI [9]. In contrast, Gallagher and colleagues found that BMI reflected the same level of fatness in black and white adults with BMI ≤ 35 kg/m^2 after age and sex adjustment [10]. Differences also exist between blacks and whites with respect to fat-free body mass with blacks generally having more bone mineral density and body protein than whites [11].

Racial Differences in Obesity Prevalence

Based on measured height and weight data from the National Health and Nutrition Examination Surveys (NHANES), the prevalence of obesity among American adults began increasing rapidly in the late 1970s and early 1980s [12] and started leveling off only in the most recent decade. The prevalence of obesity (BMI ≥ 30 kg/m^2) among all women in NHANES from 1988 to 1994 was 25.4 %, increased dramatically to 33.4 % in the NHANES data from 1999 to 2000 [13], then showed only a relatively small increase to 35.5 % in the 2007–2008 NHANES [12], and mostly recently was estimated at 36.1 % in the 2011–2012 NHANES [14]. For men, the prevalences of obesity were 20.2 %, 27.5 %, 32.2 %, and 33.5 % for 1988–1994, 1999–2000, 2007–2008, and 2011–2012, respectively [14, 15].

In addition to the rapid rate of increase in the obesity prevalence overall in the past 30 years, there is strong variation in the prevalence of obesity by race. NHANES data from 2011 to 2012 show that 32.8 % of white women were obese compared to 56.6 % of black women [14]; the difference for males is smaller with 32.4 % of white males being obese compared to 37.1 % of black males [14]. Differential increases in the prevalence of extreme obesity (BMI ≥ 40 kg/m^2) by race were even more pronounced with the prevalence increasing from 3.4 to 6.4 % among white women and from 7.9 to 14.2 % among black women between the NHANES surveys covering 1998–1994 and 2007–2008 [12, 13].

Environmental and Behavioral Determinants of Obesity

It is likely that genetic factors contribute to the ability of humans to store excess fat when food is abundant and to lose fat when food is scarce [15]. However, the recent increase in obesity in the USA, as well as in other populations around the globe, is unlikely to be explained solely by genetics because it has happened over such a short period of time. Thus, individual-level behavioral and environmental factors are also thought to be strong contributors to obesity, including physical activity levels, energy and nutrient intake, reproductive patterns, and socioeconomic status [16–19]. Population-level characteristics related to changes in occupations and infrastructure (such as changing modes of transportation) are also likely important influences on the development of obesity but are not reviewed here.

Physical Activity

The modern environment does not require nor encourage physical activity for most adults [20]. The Centers for Disease Control and the American College of Sports Medicine recommend that adults engage in at least 150 min per week of moderate-intensity physical activity [21] but data from the Behavioral Risk Factor Surveillance System (BRFSS) finds that more than half of US adults do not meet physical activity recommendations based on activity patterns in three domains (household work, transportation, and discretionary/leisure time) [21, 22]. Physical activity patterns by race have been extensively examined but there is inconsistency in the literature. Many studies have found blacks are less physically active than whites [23–25], in some cases, beginning as early as adolescence [26]. However, other studies have found no evidence for differences in physical activity levels across racial groups. Using data from the Health and Retirement Study, He and Baker found that leisure-time physical activity did not differ between blacks and whites after adjustment for education and health status [27]. Similarly, Marshall and colleagues found that within strata of social class (including education, income, employment status, and marital status), there were few differences in the prevalence of physical inactivity between white and black women [28].

Energy and Nutrient Intake

Energy intake that exceeds the energy needs of the body has been shown in controlled studies to cause weight gain in the form of stored fat [29]. However, the role of particular dietary factors as determinants of obesity is much less clear. Several methodological problems have been identified in studies of diet and obesity including short time periods of measurement, correlations between dietary factors and other determinants of obesity such as physical activity, and the validity and reliability of the tools used to measure dietary intakes [29]. Despite these limitations, ecologic, observational, and intervention studies have identified links between obesity and the consumption of fats, high-fructose corn syrup, fast food, and snack foods [30–33]. Recently, Drewnowski set forth a single explanation for these findings, namely that the consumption of low-cost foods which contain refined grains, added sugars, and added fats, explain the many links observed between weight gain and individual foods on a population level [34]. This hypothesis is consistent with the increased risk of overweight and obesity among black women who are disproportionately of lower socioeconomic status (SES) than white women.

Reproductive Factors

There may also be important racial differences in reproductive factors that affect the prevalence of obesity. While the role of parity in the development of obesity remains somewhat uncertain [35, 36], several studies have indicated that increasing parity is associated with an increase, albeit modest, in the risk of obesity [37–41]. In the few studies with sizable numbers of black women, it has been observed that black women may be more susceptible to weight gain following pregnancy than white women [42, 43]. For example, black women were found to retain more weight post-partum than white women at similar levels of gestational weight gain [42]. A recent analysis that stratified women by metropolitan status found that the effect of increased parity was significant only in black women living in metropolitan areas but not black women living in non-metropolitan areas [44]. A cross-sectional study using data from the Southern Community Cohort Study found a modest increase in the odds of obesity among both white and black women having five or more births compared to nulliparous women [45]. In addition, black women have more children on average than white women [46] and some studies have indicated that high levels of parity are most strongly associated with obesity [38, 39, 47]. Further, differences in the prevalence and length of breastfeeding exist with black women being less likely to breastfeed compared to white women [48, 49]. Some studies have reported that breastfeeding is associated with a small decrease in weight retention post-partum [50, 51] which in combination with a lower prevalence of breastfeeding among blacks could contribute to the disparity in the prevalence of obesity.

Socioeconomic Status

Underlying many of the observed associations between environmental and behavioral characteristics and obesity is the issue of SES. In a descriptive review of the literature regarding obesity and SES, McLaren reported that in resource-rich countries, such as the USA, lower SES was associated with larger body size among women in nearly two-thirds of the reviewed studies [52]. Racial differences in BMI have been found to be only partially explained by measures of SES such as education and income [53–56]. Wang and Beydoun [54] hypothesize that a major reason for this finding is that factors such as body image, lifestyle, social structure, and physical environment are responsible for much of the racial difference in body size and that these constructs are not adequately accounted for by adjustment for standard SES measures such as education and income.

Associations Between Obesity and Cancer

Obesity is one of the few modifiable risk factors for many cancer sites. The higher prevalence of overweight and obesity among blacks and observed disparities in cancer incidence and mortality for several common sites underscores the importance of studies of obesity in relation to cancer in diverse populations. For instance, the age-adjusted mortality rate for breast cancer in the USA in 2010 was 21.3 per 100,000 women among whites and 30.2 per 100,000 women among blacks [57]. Similarly for colorectal cancer, the age-adjusted mortality rate among whites (males and females) in 2010 was 18.1 per 100,000 compared to 27.5 per 100,000 among blacks [57, 58]. For prostate cancer, the racial disparity is very pronounced with the 2010 age-adjusted mortality rate being 20.1 per 100,000 for white males and 48.2 for black males [58].

All Cancers

Obesity is associated with incidence and mortality for several of the most common cancer sites [59–62]. In the Cancer Prevention Study II (CPS II), including 495,477 women followed for 16 years, the relative risk for all cancer mortality was 1.62 (1.40–1.87) comparing women with a BMI of ≥ 40 kg/m^2 to women with a BMI between 18.5 and 24.9 kg/m^2 [62]. Among the women in the CPS II, obesity was found to be associated with cancer of the esophagus, colon and rectum, liver, gallbladder, pancreas, kidney, breast, uterus, cervix, and ovaries with relative risks (RR) ranging from 1.46 (95 % confidence interval (CI) 0.94–2.24) for colorectal cancer to 6.25 (95 % CI 3.75–10.42) for the uterus [62]. The 2002 report on Cancer Prevention, Weight Control, and Physical Activity from the International Agency for Research on Cancer (IARC) concludes that, based on a comprehensive evaluation of the literature, there is sufficient evidence for a cancer-preventive effect of avoidance of weight gain for colon, post-menopausal breast cancer, endometrial cancer, kidney, and esophageal cancers [63]. This chapter will focus primarily on breast and kidney cancer in illustrating the role of ethnic differences in mediators of energy balance.

Breast Cancer

For breast cancer, existing epidemiologic studies indicate a complex relationship with obesity, which has generally been shown to be associated with increased risk among postmenopausal women but somewhat reduced risk among premenopausal women [64, 65]. For premenopausal women, in a meta-analysis, Ursin and colleagues found reductions in the RR for breast cancer in four cohort studies (RR for 8 kg/m^2 reduction in BMI $= 0.70$, 95 % CI $= 0.54$–0.91) and 19 case-control studies (RR for 8 kg/m^2 reduction in BMI $= 0.88$, 95 % CI $= 0.76$–1.02) although the individual study estimates were quite heterogeneous [66]. In the Pooling Project of Diet and Cancer, a pooled analysis of seven large prospective studies, the RR for premenopausal breast cancer among women with a BMI ≥ 33 kg/m^2 was 0.58 (95 % CI $= 0.34$–1.00) compared to women with a BMI <21 kg/m^2. A similar reduction in the RR was seen for women with a BMI between 31 and 33 but not for women with a BMI below 31 [67].

However, results are not entirely consistent for premenopausal women, and emerging evidence over the past several years suggests that obesity may in fact be associated with increased risk for breast cancer among premenopausal women [64, 68]. In particular, results have been noted to vary across studies by hormone receptor status or intrinsic subtype of breast cancer, and for triple negative breast cancer (TNBC) or basal-like breast cancers, an increased risk for breast cancer associated with obesity has been observed among premenopausal women in virtually all studies [68–72]. A recent review and meta-analysis focused on TNBC demonstrated, after stratification by menopausal status, a significant positive association between obesity and TNBC which was restricted to premenopausal women (odds ratio (OR) 1.43; 95 % CI 1.23–1.65 for those with BMI ≥ 30 versus <30), suggesting distinct molecular mechanisms involved in the onset of different subtypes of breast cancer during a woman's reproductive years.

The incidence of breast cancer is lower among black than white women greater than 40 years of age, but higher among black women at younger ages [73]. Despite a somewhat lower overall lifetime risk of breast cancer, black women are more likely to be diagnosed with aggressive and late-stage breast cancer, with lower survival rates, and breast cancers diagnosed among black women are more likely to have negative hormone receptor status. In fact, in every age group, white women have the highest rates of estrogen receptor-positive (ER+) breast cancer and black women have the highest rates of ER- breast cancer [73]. Despite the differential distribution of breast cancer subtypes among black compared to white women, and the higher prevalence of obesity among black women, limited and inconsistent information exists on the obesity-breast cancer association for premenopausal black women. Several studies among black premenopausal women [58, 74–79], but not all [58, 80], have indicated increased breast cancer risk associated with increased BMI, the standard index for assessing obesity, or with higher waist circumference or WHR, measures of central adiposity that are more prevalent among blacks than whites. In the Carolina Breast Cancer Study, a large case-control study of black and white women, BMI was inversely associated with breast cancer among white, but not black, premenopausal women [75], and an increased risk for basal-like tumors was associated with increased WHR among black and white pre- or postmenopausal women [69]. Few other studies have presented results for black women by hormone receptor status; in the Women's Circle of Health Study, elevated BMI, as well as waist circumference and WHR, were nonsignificantly positively associated with both ER+/PR+ and ER−/PR− breast cancer in black premenopausal women [74], while in the Multiethnic San Francisco Bay Area Breast Cancer Study, BMI was inversely associated only with ER+/PR + tumors but not with ER−/PR− tumors [81].

In contrast, among postmenopausal women, increased body size is positively associated with breast cancer risk [82, 83]. In the Pooling Project of Diet and Cancer, the RR for breast cancer among postmenopausal women with a BMI greater than 28 kg/m^2 was 1.26 (95 % CI = 1.09–1.46) compared to women with a BMI <21 kg/m^2, and a stronger positive association with obesity was seen among women who had never used hormone replacement therapy (HRT) [67]. In the Women's Health Initiative cohort, evidence of effect measure modification by HRT was also seen with obesity found to be a risk factor for breast cancer among nonusers of HRT but not among women who had ever used HRT [84]. Positive associations between adult weight gain and breast cancer risk as well as central adiposity and breast cancer risk have been consistently reported as well [82, 83]. While the positive association between obesity and postmenopausal breast cancer has been consistently observed among white women, the few studies of among black women have had more variable results; a 2011 review found only 8 studies conducted among black women, and the results were mixed with some studies showing a positive association with obesity and postmenopausal breast cancer and others being null [85]. In the Multiethnic Cohort Study, the hazard ratios (HR) for postmenopausal breast cancer for each five unit increase in BMI were very similar between white and black women (HR = 1.06 [95 % CI 1.00–1.14] and 1.08 [95 % CI 1.01–1.16], respectively)[86].

Renal Cell Carcinoma

An estimated 63,920 new cases of kidney cancer are expected in the USA in 2014, making it the sixth and eighth most commonly diagnosed primary cancer among men and women, respectively [87]. Approximately 85 % of kidney cancers are renal parenchyma (renal cell carcinoma, RCC) cancers, while the remainder are mainly urothelial cancers of the renal pelvis. Incidence rates for RCC have been steadily increasing over several decades; between 2001 and 2005, while the rate for all cancers combined dropped 1.8 % among men and 0.5 % among women in the USA, kidney cancer incidence rose 1.7 % and 2.2 % per year for men and women, respectively [88]. The most salient feature of these incidence trends has been the more rapid increase among blacks than whites [89, 90], leading to a pronounced shift in excess from among whites to among blacks beginning in the mid-1980s. Kidney cancer is now the 4th and 6th most common cancer among black men and women, respectively [91]. RCC is comprised of several histologic subtypes with distinct genetic and clinical features, the most common of which are clear cell, accounting for approximately 70 % of cases, papillary (10–15 %) and chromophobe (5 %); however, the proportional distribution of papillary RCC appears to be substantially higher among blacks than whites [92].

RCC is one of the malignancies most consistently and strongly associated with BMI among both men and women, regardless of study design or population [90, 93–95]. A recent quantitative summary analysis of the epidemiologic evidence reported RRs for RCC of 1.24 (95 % CI 1.15–1.34) among men and 1.34 (95 % CI 1.25–1.43) among women per 5 kg/m^2 increase in BMI [96]. Despite the higher and more rapidly increasing incidence of RCC among blacks, virtually no information exists on the obesity-RCC association specifically for blacks [97]. Limited evidence suggests that the association between obesity and RCC may differ by histologic subtype, with stronger associations observed for clear cell and chromophobe than papillary RCC among men [98], but this finding requires confirmation in larger studies.

Biological Mechanisms Linking Obesity to Cancer

Although mechanisms are not entirely clear, it is in general believed that obesity acts primarily by inducing insulin signaling and resistance, inflammation, and increased estrogen biosynthesis and signaling to increase the risk of cancer [99–101] (Fig. 9.1). The relative contribution of these mechanisms and their mediators to the complex association between obesity and cancer is likely to differ between blacks and whites, which may contribute to racial differences in risk and/or subtype distribution of cancer.

Insulin Resistance and IGF1 Pathway

The effects of obesity on cancer may be mediated by its effects on increased levels of insulin and related growth factors, in particular increased bioavailable concentrations of insulin-like growth factor 1 (IGF1) [102]. Over 90 % of IGF1 circulates bound to insulin-like growth factor-binding protein 3 (IGFBP3) [103], with less than 1 % of IGF circulating unbound [104]. IGF1 has important mitogenic, cell proliferative, and anti-apoptotic effects [102]. Insulin and IGFs induce signaling by binding to insulin receptor (IR) and IGF1 receptor (IGF-1R) and their hybrid receptors widely expressed on normal and neoplastic cells. These receptors may also play a role in the association between obesity and cancer [105, 106].

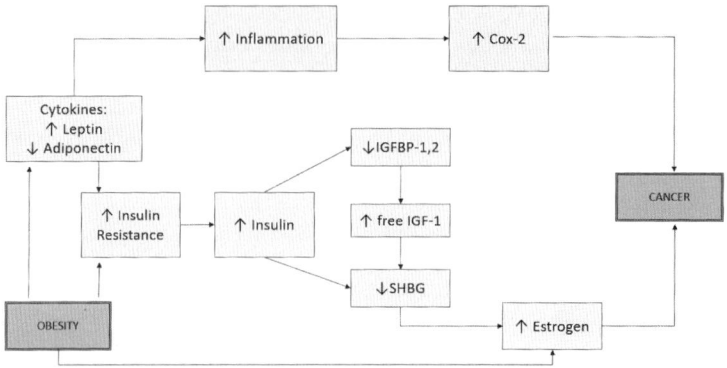

Figure adapted from Calle (2004), IARC (2002), and Kadowski (2005)

Fig. 9.1 Select obesity-mediated pathways leading to cancer. *IGF* insulin-like growth hormone, *IGFBP* insulin-like growth hormone binding protein, *SHBG* sex hormone-binding globulin. Figure adapted from [59], [63], and [165]

A recent study in a multiracial population demonstrated that black women had higher mean IGF1 and lower IGFBP3 levels than white women. IGF1 levels declined with rising BMI at age 21 among whites only, which led to increased racial differences in IGF1 among women who were obese in early adulthood [107]. Levels of insulin, IGF1 and IGFBP3 (and their ratio), and C-peptide, a surrogate of insulin secretion, have been linked to an increased risk of breast cancer in some, but not all, epidemiologic studies [108–117]. In a recent pooled analysis of 17 prospective studies, IGF1 was strongly positively associated with breast cancer risk the relative risk for women in the highest versus the lowest quintile of the distribution of IGF1 was 1.28 (95 % CI 1.14–1.44), with no significant difference according to menopausal status [116]. Although the association of both IGF1 and IGFBP3 appeared to be confined to ER+ tumors, the evidence is not entirely consistent, and whether IGF1 or IGF-binding protein 3 (IGFBP3) increase risk for breast cancer differentially for blacks and whites or for certain breast cancer subtypes, such as triple-negative breast cancer, is largely unknown [118, 119].

For RCC, in vitro studies have shown increased expression of insulin and IGF1 receptors in human RCC tissue, and IGF1 stimulates growth in human RCC cell lines [120–125]. An immunohistochemistry analysis of tissue from 180 RCC patients suggested differential IGF1 expression across RCC histologies, with stronger expression in clear cell tumors [125]. Some, but not all, in vitro studies have shown increased expression of insulin and IGF1 receptors in human RCC tissue, and IGF1 stimulates growth in human RCC cell lines [120–125]. An immunohistochemistry analysis of tissue from 180 RCC patients suggested differential IGF1 expression across RCC histologies, with stronger expression in clear cell tumors [125]. The only prospective analysis to date to investigate pre-diagnostic circulating levels of IGF1 in relation to RCC risk is a relatively small case-control study, nested in the Alpha-Tocopheral, Beta-Carotene (ATBC) Cancer Prevention Study cohort of Finnish male smokers; in this study, IGF1 was inversely associated with RCC risk [126, 127]. These results appear to contradict the observations from some experimental studies described above, in which IGF was demonstrated to stimulate renal carcinogenesis, but require confirmation.

Sex Hormones

The association between circulating sex steroid hormones and breast cancer risk among postmenopausal women has been studied extensively, with results consistently showing strong positive associations with estradiol (E2) and estrone (E1), and an inverse association with sex hormone-binding globulin (SHBG), which binds to estrogen to reduce its bioavailability [128–130]. SHBG also may act directly on breast cancer cells to inhibit E2-induced cell proliferation [131]. High BMI is consistently associated with higher levels of estrogens and lower levels of SHBG in postmenopausal women [132]. Among obese postmenopausal women, this reflects the higher rate of conversion of androgenic precursors to E2 through increased peripheral aromatase enzyme activity in adipose tissue.

Differences in endogenous steroid hormone levels between postmenopausal black and white women in the USA have been demonstrated in a small number of studies [133–135]. In a cross-sectional analysis including 240 black and 91 white postmenopausal women within the Multiethnic Cohort Study [135], black women had 20 % higher age-adjusted mean levels of E1 and total and bioavailable E2 compared with whites, but also the highest levels of SHBG, even after adjustment for BMI. However, few studies of postmenopausal breast cancer in relation to levels of steroid hormones have included black women [108, 136], and to our knowledge no study to date has evaluated these associations separately among black women. In a case-cohort analysis of incident postmenopausal breast cancer within the Women's Health Initiative Observational study, which included 56 black cases, E2 was associated with a significantly increased risk of breast cancer among women not using HRT [108].

Data among premenopausal women are more limited, primarily due to large variations in endogenous levels of estrogen throughout the menstrual cycle [137], but generally support a similar role for increased E2 and decreased sex hormone-binding globulin (SHBG) in the development of breast cancer among young women. SHBG binds to estrogen to reduce its bioavailability but may also act directly on breast cancer cells to inhibit E2-induced cell proliferation [131]. In a recent pooled analysis of data from seven prospective studies, including 767 cases of breast cancer, circulating concentrations of E2, calculated free E2 (a measure of bioavailable E2), and E1, as well as testosterone and androstenedione, were significantly positively associated with risk for breast cancer in premenopausal women, after adjustment for known breast cancer risk factors [138], with ORs ranging from 1.08 to 1.30 for a doubling in sex hormone concentrations. Sex hormones were more strongly associated with risk of ER+ breast cancer than ER− breast cancer, but the number of women with ER− disease was small. SHBG was not associated with breast cancer risk. Virtually all of the women included in the pooled analysis were of European ethnicity. There have been few studies of premenopausal breast cancer in relation to levels of steroid hormones among black women [108, 136], even though differences in endogenous steroid hormone levels between black and white women in the USA have been demonstrated [133, 139–141], including higher E2 levels throughout the menstrual cycle.

Obese premenopausal women have also been shown to have elevated levels of non-protein-bound and total estrogens and decreased levels of SHBG [142, 143]. Among controls in the pooled analysis described above, compared to normal weight women, those with a high BMI ($\geq$30 kg/m^2) had lower mean concentrations of E2 but higher concentrations of free E2 due to a strong inverse association of SHBG with BMI [138]. E1 was also positively associated with BMI, perhaps reflecting the higher rate of conversion of androgenic precursors through increased peripheral aromatase enzyme activity in adipose tissue, similar to what is observed among postmenopausal women. The few studies of sex hormone levels among premenopausal black women demonstrated a similar hormonal profile associated with obesity or increased WHR, with higher level of free E2 and a lower level of SHBG among obese compared to non-obese women in multivariate analyses [142, 144].

Inflammation

Most cancers develop in a background of chronic inflammation, and a tumor can be considered a chronic inflammatory state. It has been well documented that many types of cancer, including breast cancer, are heavily infiltrated by inflammatory cells. These cells express a large variety of cytokines and growth factors, some of which are known to function as regulators of tumor growth, metastasis, and angiogenesis. Adipose tissues produce not only estrogens, but also cytokines, including several major pro-inflammatory cytokines. A positive energy balance increases adipose tissue mass, but also induces hypoxia and cell necrosis in the fat depot. In response, there is a dramatic infiltration of pro-inflammatory macrophages. Similarly, there may be an increase in neutrophils and natural killer cells in adipose during the course of obesity. Obesity, and its consequent unbalanced inflammatory response, leads to uncontrolled chronic inflammation. Remarkably little is known about racial differences in inflammatory response in relation to breast cancer risk.

Many cytokines regulating immune system function and an inflammatory response, including tumor necrosis factor-α (TNFα), transforming growth factor β (TGFβ) and interleukins (IL), such as IL-6 and IL-1β, circulate at concentrations positively correlated with BMI, in total leading to a chronic pro-inflammatory state. IL-8 is an important pro-inflammatory CXC chemokine [145], and studies have demonstrated overexpression of IL-8 in tumor cells, as well as enrichment of normal breast tissue of obese women for markers of macrophage infiltration (CD68) and for genes associated with IL-8, IL-6 and other inflammation, or macrophage-associated pathways [146]. As a downstream biomarker, CRP provides functional integration of overall upstream cytokine activation and exerts several important pro-inflammatory effects. BMI, waist circumference, and WHR are significantly associated with higher serum CRP levels [147–149], while weight loss is associated with lower CRP levels [150, 151]. Ethnic differences in CRP between Eastern and Western populations have been shown to disappear after controlling for BMI and other metabolic factors [147]. However, population-based data to systematically assess the association of CRP with cancer risk are limited. In a large prospective study of blacks and whites in the southeastern USA, both the prevalence of elevated CRP and the magnitude of its association with BMI were markedly greater among blacks than whites, with the OR (95 % CI) for elevated CRP comparing obese with healthy-weight black women reaching 22.8 (7.1–73.8), compared with 4.6 (1.7–12.7) among white women [152].

Adipokines (Adiponectin and Leptin)

Associations between obesity and cancer may reflect levels of adiponectin, whose secretion from adipose tissue is down-regulated among obese individuals [153], and leptin, which is positively correlated with BMI. Leptin, which is positively correlated with obesity, may promote local or distant immunocyte differentiation, in addition to a proliferative signaling effect on cancer cells [154] and acts as a growth-promoting factor for cancer via the PI3K/Akt pathway [155]. Leptin also is thought to down-regulate the apoptotic response of tumor cells through as yet undetermined mechanisms [155]. In contrast, loss of adiponectin leads to an inflammatory response mediated via NF-κB activation [156]. Adiponectin exerts anti-cancer effects, including anti-angiogenesis and anti-proliferation via activation of its two known receptors, ADIPOR1 and ADIPOR2 [157–159], and may block effects of IGF1-stimulated PI3K/AKT signaling and cell proliferation [105, 160]. Since insulin exerts positive feedback on leptin gene expression and can suppress adiponectin secretion, these cytokines also may exert indirect effects on cancer through insulin-related mechanisms.

Adiponectin

Adiponectin is produced exclusively in adipose tissue and is found in a high concentration in the blood, accounting for approximately 0.01–0.05 % of total serum protein [161, 162]. Adiponectin levels appear to be relatively stable within individuals who do not undergo drastic changes in body weight. Circadian variation has been shown to be low overall with adiponectin levels varying less than 20 %

throughout the day, and slightly more variation is seen in females than in males [163]. Most evidence seems to indicate that adiponectin levels remain unchanged in relation to meal ingestion [163].

Two adiponectin receptors have been identified to date, AdipoR1 and AdipoR2 (6, 120). These receptors are found in the cell membrane and are located throughout the body in liver, muscle, and adipose tissue although AdipoR1 is found predominantly in muscle cells while AdipoR2 is primarily found in the liver [164]. The binding of adiponectin to these receptors mediates the activation of AMP kinase which leads to expression of peroxisome proliferators-activated receptor-alpha (PPAR-α) [165]. This activity is believed to increase gene expression of enzymes related to fatty acid oxidation and glucose uptake [161, 165]. This process is thought to be one of the main mechanisms linking adiponectin and insulin sensitivity. Additionally, increased obesity is thought to either directly decrease expression levels of adiponectin receptors or reduce the post-receptor signaling which may also contribute to insulin resistance [161].

Racial Differences in Adiponectin Levels

Racial differences in adiponectin levels have been examined in a relatively small number of studies most of which had low sample sizes; however, reports have been generally consistent that adiponectin levels are lower in blacks compared to whites and the differences emerge relatively early in life [166, 167]. Two small studies of children and early adolescents both reported that adiponectin levels were lower in blacks compared to whites in both genders [166] and among boys after matching on BMI percentile [167]. Two additional studies of middle-age adults reported lower adiponectin levels in American blacks ($N = 212$) [168] and in African blacks ($N = 27$) [169] compared to whites. Two larger cohorts of young (age 23–45) and middle-aged (age 48–58) adults also reported lower adiponectin levels in black participants compared to white participants [170, 171]. In a large study of older adults (age 70–79), adiponectin levels were found to be lower in blacks ($N = 1,044$) compared to whites ($N = 1,429$) [172]. Hulver and colleagues [173] reported mean adiponectin levels in strata of race and obesity status. They found that mean adiponectin levels were similar in obese white women, obese black women, and non-obese black women but higher in non-obese white women [173]. In a study of white and black South Africans, adiponectin levels were lower in normal weight blacks compared to whites but no differences were seen for overweight and obese women [174]. Both of these studies to examine adiponectin levels over categories of body size were limited by their small overall sample size ($N = 85$ and $N = 217$, respectively). Wassel Fyr and colleagues measured the percentage of European ancestry using 35 ancestry informative markers in a sample of 1,241 older adults (age 70–79) who self-reported as black. In models adjusted for adiposity, fasting glucose levels, insulin levels, blood pressure, and lipids, increasing adiponectin levels were found to increase as the percentage of European ancestry increased

[175]. This pattern is consistent with the previous reports described above that found lower adiponectin levels among blacks compared to whites using self-report.

Adiponectin Levels and Obesity

Serum adiponectin levels are negatively correlated with BMI and WHR [161, 162] which reflect overall adiposity and fat distribution, respectively. This is somewhat paradoxical given that most cytokines (such as leptin) increase directly in relation to body fat. It has been hypothesized that feedback loops exist between obesity, adiponectin expression, and regulation of the adiponectin receptors, resulting in the observed inverse association between obesity phenotypes and adiponectin levels in the blood [165, 176].

Most evidence to date regarding the obesity–adiponectin relationship has been observed in white or Japanese populations [177–180]. Despite the known differences in the prevalence of obesity and risk for obesity-related disease, a relatively low number of studies have examined the relationship between adiponectin and obesity in blacks and many have had very small sample sizes. In a study of adolescents including 40 white and 46 black participants, Degawa-Yamauchi et al. observed that adiponectin was negatively correlated with both BMI and percentiles of BMI [167]. In another small study, Hulver et al. found that adiponectin was negatively correlated with BMI only among whites ($N=48$) but not blacks ($N=37$) [173] while, in contrast, Araneta et al. found that adiponectin was negatively associated with increasing tertiles of BMI, waist circumference, and WHR in both blacks ($N=212$) and whites ($N=143$) [168]. Comparing black and white South Africans, adiponectin levels were found to be negatively correlated with BMI in each race group in univariate analysis although not in the final multivariate model [174]. In a genetically homogeneous sample of 431 individuals from 7 families living on the Caribbean island of Tobago, adiponectin was also found to be negatively correlated with BMI [181].

Reports from larger studies with sizeable black participant populations remain scant; among 522 black participants in the Insulin Resistance Atherosclerosis (IRAS) Family Study, visceral adipose tissue measured by CT was strongly negatively correlated with adiponectin levels [182]. In the Atherosclerosis Risk in Communities Study (ARIC), mean adiponectin levels were found to decrease over categories of BMI in 630 black and 523 white participants age 48–58. Further, the adjusted mean adiponectin values were lower for black women than for white women in each BMI category [171]. Waist circumference was found to be negatively associated with adiponectin levels in the CARDIA study of 1,615 white and 1,360 black young adults (age 23–45) [170]. Among 996 black and 996 white women age 40–79 enrolled in the Southern Community Cohort Study, which includes black and white participants from similar geographic and socioeconomic situations, black women had significantly lower adiponectin levels than white women even after adjustment for BMI. Both race groups demonstrated a strong inverse association between adiponectin and BMI although the trend was mono-tonic in white women but leveled off for black women with severe obesity [183].

Leptin

Leptin was first discovered in 1994 [184], and like adiponectin, is a protein produced and secreted by adipose tissue. The leptin protein is translated as a 167 amino acid polypeptide including a signal peptide consisting of a cleaved strand of the first 21 amino acids [185]. Leptin plays a critical role in regulating energy intake, energy expenditure, and overall adiposity [183]. Leptin secretion shows clear circadian variation with basal levels observed between 08:00 and 12:00, then rising progressively to a peak between 24:00 and 04:00 and finally receding steadily to a low point again by 12:00 [186].

Rodent models with leptin deficiency (homozygous for a mutant *ob* gene) have morbid obesity, and enough evidence has accumulated to implicate leptin as having a critical role in behavior, metabolism, and endocrinology [187]. Two general types of mutation affect the *ob* gene and alter the structure of protein that is produced. The first is a nonsense mutation that produces a stop codon prematurely in the nucleo-tide sequence, resulting in a shortened and nonfunctional protein. The other muta-tion affects the promoter region, resulting in no transcription and thus no protein is produced at all [185].

Leptin Levels and Obesity

Several studies including individuals of different ethnic backgrounds have shown that leptin levels are universally positively associated with BMI. This positive leptin-BMI association was demonstrated in all subpopulations within a relatively large study of European, Chinese, South Asian, and Aboriginal Canadians [188]. In the Multi-ethnic Study of Atherosclerosis (MESA), leptin levels were positively associated with BMI in white, Chinese, black, and Hispanic individuals although the magnitude of association varied with blacks having the weakest association compared to the other groups [186]. In the Southern Community Cohort Study, adjusted leptin levels also increased as body size increased in both black and white women. In contrast to the MESA results, however, leptin levels were similar for both race groups at BMI <18.5 and 18.5–24.9 but were higher in overweight and obese black women compared to white women at the same BMI [189].

Racial Differences in Leptin Levels

Initial studies examining leptin levels across race groups generally utilized conve-nience samples, and conflicting results have been reported, with some studies finding no difference in leptin levels according to race [12, 59] and at least one finding lower levels in black women compared to white women [190]. Larger population-based studies, however, have consistently found leptin levels to be higher in black women than in white women. In the Multiethnic Cohort Study, mean leptin levels were 27.9 ng/ml among 73 black women versus 21.4 ng/ml among 71 white women [191]. The Health, Aging, and Body Composition Study ($n = 718$ black and 840 white women aged 70–79 years) also reported higher leptin levels among the black women (geometric mean 20.2 ng/ml) versus white women (13.9 ng/ml) [1] as did the ARIC study ($n = 305$ blacks and $n = 388$ whites) with median leptin levels of 20.3 ng/ml in black versus 9.8 ng/ml in white participants [2]. In the third National Health and Nutrition Examination Survey ($n = 957$ black women and $n = 1,441$ white women), mean leptin levels were found to be 16.4 ng/ml among the black women and 12.2 ng/ml among the white women [3]. Also, within the Southern Community Cohort Study, mean adjusted leptin levels were again higher in black women ($n = 829$) than white women ($n = 915$), 22.7 vs. 18.8 ng/ml [4].

Differences in leptin levels between black and white women may reflect actual physiological or genetic differences in the production of leptin in adipose tissue between race groups. But a methodological consideration is that incomplete adjustment for fat distribution could be responsible for the consistently observed higher leptin levels in black women compared with white women in large population-based studies. It has been established that black women have an overall higher proportion of subcutaneous fat than white women [5, 6], and leptin secretion is known to be higher in subcutaneous fat than in visceral fat [7, 8]. Thus, incomplete adjustment for fat distribution could bias estimates of differences of leptin levels between blacks and whites. However, at least two studies have adjusted for fat distribution, either percentage fat and visceral fat [1] or skinfold thicknesses and waist and hip circumferences [3], and both still found that black women had modestly higher leptin levels than white women.

Adiponectin and Leptin Levels and Cancer

In vitro studies have examined the effects of adiponectin on epithelial breast tissue and on breast cancer cell lines. MCF-7 breast cancer cells were found to express functional adiponectin receptors in several in vitro studies [192–195]. Conflicting evidence exists as to whether breast cancer cell proliferation is inhibited by adiponectin with some groups finding evidence for this activity in vitro [192, 195, 196] while others have been unable to replicate this finding [193, 194]. Additionally, several tumor cell lines have been shown to express the adiponectin receptors AdipoR1 and AdipoR2 indicating that adiponectin could act directly on cancer cells through signaling of its receptors [197].

To date, at least seven studies in human populations have examined associations between adiponectin and breast cancer risk. Five relatively small case-control studies conducted in women residing in Japan, Greece, and Taiwan found a reduced risk of breast cancer at the highest levels of adiponectin compared to the lowest levels [198–202]. In two studies [198, 201], the results were consistent between pre- and post-menopausal women while two others found an association only among post-menopausal women [200, 202]. The Japanese study also found that lower adiponectin levels were associated with larger tumors and higher grade tumors but these results have yet to be replicated [201]. A fifth case-control study, conducted in Korea, found no association between tertiles of adiponectin and breast cancer risk (OR = 0.92, 95 % CI = 0.46–1.81) [203]. The Nurses' Health Study used pre-diagnosis blood samples for a prospective case-control study including 1,477 cases and 2,196 controls [204]. These authors found that breast cancer risk was reduced when comparing the highest quartile of adiponectin to the lowest among post-menopausal women (OR = 0.73, 95 % CI = 0.55–0.98) but not among pre-menopausal women [204]. However, a recent meta-analysis reported an inverse association between circulating adiponectin and breast cancer risk among premenopausal women [205]. With the exception of the Nurses' Health Study, the remaining case-control studies to have examined adiponectin levels in relation to breast cancer used blood samples collected post-diagnosis. Without a clear understanding of the determinants of adiponectin levels, the measurement of adiponectin in blood samples in women after a cancer diagnosis has been made has serious implications for potential bias and is an important limitation. Notably, none of these studies included any appreciable numbers of women of African descent.

For leptin, a positive association has been reported with breast cancer, albeit markedly less consistently than the association with adiponectin [206]. Among women with breast cancer, blood levels of leptin are reported to increase concomitantly with E2 levels, and it is hypothesized that the role of leptin may be specific to postmenopausal ER+ breast cancer among overweight women [207].

Several case-control studies have demonstrated associations between levels of adiponectin or leptin and risk of RCC, or between adiponectin (inverse) or leptin (positive) levels and markers of tumor aggressiveness [208–215]. However, reverse causation cannot be ruled out in these studies which were based on post-diagnostic blood samples. The only prospective analyses to date to investigate pre-diagnostic circulating levels of adipokines in relation to RCC risk reported an inverse association between adiponectin and RCC risk among Finnish male smokers [126, 127].

Crosstalk

While each of the pathways of insulin resistance, estrogen biosynthesis/signaling, and inflammation/adipokines may play an independent role in the association between obesity and breast cancer development [99], there is evidence for crosstalk between the insulin/IGF1 and ER signaling pathways [101, 216, 217]. Chronic hyperinsulinemia is associated with increased ovarian estrogen production and reduced secretion of SHBG [218, 219], leading to increased estrogen bioavailability. Moreover, estrogen induces expression of IGF1 and IGF-1R in ER+ breast cancer cells [220, 221], resulting in enhanced activation of signaling pathways downstream insulin and IGF1 receptors [106]. Despite this crosstalk, not all studies of circulating IGF1 levels have evaluated associations according to tumor ER status or controlled for endogenous sex hormone levels [108, 116]. A better understanding of the joint effects of hormones is needed in order to assess the independent associations of IGF1 and estrogen with breast cancer risk. Similarly, increased TNFα expression in adipose tissue blocks insulin signaling, thereby inducing insulin resistance [222], and also regulates IL-6 synthesis and aromatase expression, thus stimulating estrogen production [223]. Serum concentrations and adipose tissue expression of IL-6 also are positively associated with insulin resistance [224, 225], and expression of IL-8 is regulated in part by steroid hormones [145]. Moreover, estrogens can suppress secretion of adiponectin by adipocytes [226], and studies of postmenopausal women show strong associations between adiponectin and SHBG (positive) or free E2 (inverse) [108, 227–229].

Conclusion

Several types of cancer are strongly and consistently associated with obesity. There are complex biologic mechanisms underlying these associations, and it is possible that racial differences in the contribution of these biological mechanisms and their mediators play an important role in observed racial differences in cancer incidence and mortality patterns. In particular, accumulating evidence demonstrates racial differences in biomarkers related to obesity, lending support to the hypothesis that these biomarkers act differently in their association with obesity and risk of breast cancer between racial groups. The examination of racial differences in metabolic biomarkers of obesity is important, because although BMI is the standard index for assessing general obesity in epidemiologic studies, it is not a biologic trait and thus does not capture body fat distribution or distinguish between adipose tissue, fat-free mass, and skeletal muscle mass [3, 8, 230], which vary widely across multiethnic populations for a given BMI value [231] and are crucial for characterizing the "multifaceted" obese phenotype and its physiological and pathological risks [232].

Identification of mechanisms through which obesity increases risk for cancer will contribute to delineation of persons at high (and low) risk, specific to race, and pave the way for developing clinically useful cancer prediction biomarkers. This may provide novel targets for development of mechanism-based prevention or treatment approaches in obese patients with cancer, including strategies for cancer prevention linked to insulin resistance or inflammation pathways (e.g., via agents that block the IGF-1 receptor to decrease IGF signaling, or nonsteroidal anti-inflammatory drugs (NSAID), to disrupt the inflammatory process and thereby reduce risk of cancer).

References

1. Pi-Sunyer FX (2000) Obesity: criteria and classification. Proc Nutr Soc 59(4):505–509
2. Bray GA (2004) Medical consequences of obesity. J Clin Endocrinol Metab 89(6):2583–2589
3. Snijder MB, van Dam RM, Visser M, Seidell JC (2006) What aspects of body fat are particularly hazardous and how do we measure them? Int J Epidemiol 35(1):83–92
4. Willett WC (1998) Nutritional epidemiology, 2nd edn. Oxford University Press, New York, NY
5. World Health Organization (1997) Obesity: preventing and managing the global epidemic. Report of a WHO consultation on obesity. WHO, Geneva
6. (1998) Executive summary of the clinical guidelines on the identification, evaluation, and treatment of overweight and obesity in adults. Arch Intern Med 158(17):1855–1867
7. Hu FB (2008) Obesity epidemiology, 1st edn. Oxford University Press, New York, NY
8. Lovejoy JC, de la Bretonne JA, Klemperer M, Tulley R (1996) Abdominal fat distribution and metabolic risk factors: effects of race. Metabolism 45(9):1119–1124
9. Evans EM, Rowe DA, Racette SB, Ross KM, McAuley E (2006) Is the current BMI obesity classification appropriate for black and white postmenopausal women? Int J Obes (Lond) 30(5):837–843
10. Gallagher D, Visser M, Sepulveda D, Pierson RN, Harris T, Heymsfield SB (1996) How useful is body mass index for comparison of body fatness across age, sex, and ethnic groups? Am J Epidemiol 143(3):228–239
11. Wagner DR, Heyward VH (2000) Measures of body composition in blacks and whites: a comparative review. Am J Clin Nutr 71(6):1392–1402
12. Flegal KM, Carroll MD, Ogden CL, Curtin LR (2010) Prevalence and trends in obesity among US adults, 1999–2008. JAMA 303(3):235–241
13. Flegal KM, Carroll MD, Ogden CL, Johnson CL (2002) Prevalence and trends in obesity among US adults, 1999–2000. JAMA 288(14):1723–1727
14. Ogden CL, Carroll MD, Kit BK, Flegal KM (2013) Prevalence of obesity among adults: United States, 2011–2012. NCHS Data Brief No. 131. Hyattsville, MD
15. Ogden CL, Yanovski SZ, Carroll MD, Flegal KM (2007) The epidemiology of obesity. Gastroenterology 132(6):2087–2102
16. Hill JO, Melanson EL (1999) Overview of the determinants of overweight and obesity: current evidence and research issues. Med Sci Sports Exerc 31(11 Suppl):S515–S521
17. Kumanyika S (1987) Obesity in black women. Epidemiol Rev 9:31–50
18. Parker JD, Abrams B (1993) Differences in postpartum weight retention between black and white mothers. Obstet Gynecol 81(5 Pt 1):768–774
19. Patel KA, Schlundt DG (2001) Impact of moods and social context on eating behavior. Appetite 36(2):111–118
20. Hill JO, Wyatt HR, Reed GW, Peters JC (2003) Obesity and the environment: where do we go from here? Science 299:853–855
21. Department of Health and Human Services (2008) Physical activity guidelines advisory committee physical activity guidelines advisory committee report, Department of Health and Human Services, Washington, DC
22. Centers for Disease Control and Prevention (2005) Adult participation in recommended levels of physical activity-United States, 2001 and 2003. Morb Mortal Wkly Rep 54: 1208–1212
23. Ahmed NU, Smith GL, Flores AM, Pamies RJ, Mason HR, Woods KF, Stain SC (2005) Racial/ethnic disparity and predictors of leisure-time physical activity among U.S. men. Ethn Dis 15(1):40–52
24. Crespo CJ, Smit E, Andersen RE, Carter-Pokras O, Ainsworth BE (2000) Race/ethnicity, social class and their relation to physical inactivity during leisure time: results from the Third National Health and Nutrition Examination Survey, 1988–1994. Am J Prev Med 18(1):46–53

25. Macera CA, Ham SA, Yore MM, Jones DA, Ainsworth BE, Kimsey CD, Kohl HW 3rd (2005) Prevalence of physical activity in the United States: behavioral risk factor surveillance system, 2001. Prev Chronic Dis 2(2):A17

26. Kimm SY, Glynn NW, Kriska AM, Barton BA, Kronsberg SS, Daniels SR, Crawford PB, Sabry ZI, Liu K (2002) Decline in physical activity in black girls and white girls during adolescence. N Engl J Med 347(10):709–715

27. He XZ, Baker DW (2005) Differences in leisure-time, household, and work-related physical activity by race, ethnicity, and education. J Gen Intern Med 20(3):259–266

28. Marshall SJ, Jones DA, Ainsworth BE, Reis JP, Levy SS, Macera CA (2007) Race/ethnicity, social class, and leisure-time physical inactivity. Med Sci Sports Exerc 39(1):44–51

29. Jebb SA (2007) Dietary determinants of obesity. Obes Rev 8(Suppl 1):93–97

30. Bowman SA, Vinyard BT (2004) Fast food consumption of U.S. adults: impact on energy and nutrient intakes and overweight status. J Am Coll Nutr 23(2):163–168

31. Bray GA, Nielsen SJ, Popkin BM (2004) Consumption of high-fructose corn syrup in beverages may play a role in the epidemic of obesity. Am J Clin Nutr 79(4):537–543

32. Bray GA, Paeratakul S, Popkin BM (2004) Dietary fat and obesity: a review of animal, clinical and epidemiological studies. Physiol Behav 83(4):549–555

33. Isganaitis E, Lustig RH (2005) Fast food, central nervous system insulin resistance, and obesity. Arterioscler Thromb Vasc Biol 25(12):2451–2462

34. Drewnowski A (2007) The real contribution of added sugars and fats to obesity. Epidemiol Rev 29:160–171

35. Lewis CE, Smith DE, Wallace DD, Williams OD, Bild DE, Jacobs DR Jr (1997) Seven-year trends in body weight and associations with lifestyle and behavioral characteristics in black and white young adults: the CARDIA study. Am J Public Health 87(4):635–642

36. Linne Y, Dye L, Barkeling B, Rossner S (2003) Weight development over time in parous women—the SPAWN study—15 years follow-up. Int J Obes Relat Metab Disord 27(12):1516–1522

37. Brown JE, Kaye SA, Folsom AR (1992) Parity-related weight change in women. Int J Obes Relat Metab Disord 16(9):627–631

38. Williamson DF, Madans J, Pamuk E, Flegal KM, Kendrick JS, Serdula MK (1994) A prospective study of childbearing and 10-year weight gain in US white women 25 to 45 years of age. Int J Obes Relat Metab Disord 18(8):561–569

39. Harris HE, Ellison GT, Holliday M (1997) Is there an independent association between parity and maternal weight gain? Ann Hum Biol 24(6):507–519

40. Lahmann PH, Lissner L, Gullberg B, Berglund G (2000) Sociodemographic factors associated with long-term weight gain, current body fatness and central adiposity in Swedish women. Int J Obes Relat Metab Disord 24(6):685–694

41. Bastian LA, West NA, Corcoran C, Munger RG (2005) Number of children and the risk of obesity in older women. Prev Med 40(1):99–104

42. Keppel KG, Taffel SM (1993) Pregnancy-related weight gain and retention: implications of the 1990 Institute of Medicine guidelines. Am J Public Health 83(8):1100–1103

43. Smith DE, Lewis CE, Caveny JL, Perkins LL, Burke GL, Bild DE (1994) Longitudinal changes in adiposity associated with pregnancy. The CARDIA Study. Coronary Artery Risk Development in Young Adults Study. JAMA 271(22):1747–1751

44. Lee SK, Sobal J, Frongillo EA, Olson CM, Wolfe WS (2005) Parity and body weight in the United States: differences by race and size of place of residence. Obes Res 13(7):1263–1269

45. Cohen SS, Larson CO, Matthews CE, Buchowski MS, Signorello LB, Hargreaves MK, Blot WJ (2009) Parity and breastfeeding in relation to obesity among black and white women in the southern community cohort study. J Womens Health (Larchmt) 18(9):1323–1332

46. National Center for Health Statistics (2006) Health, United States 2006, with Chartbook on Trends in the Health of Americans. Hyattsville, MD

47. Szklarska A, Jankowska EA (2003) Independent effects of social position and parity on body mass index among Polish adult women. J Biosoc Sci 35(4):575–583

48. Bernstein L, Teal CR, Joslyn S, Wilson J (2003) Ethnicity-related variation in breast cancer risk factors. Cancer 97(1 Suppl):222–229
49. Ludington-Hoe SM, McDonald PE, Satyshur R (2002) Breastfeeding in African-American women. J Natl Black Nurses Assoc 13(1):56–64
50. Ohlin A, Rossner S (1996) Factors related to body weight changes during and after pregnancy: the Stockholm Pregnancy and Weight Development Study. Obes Res 4(3):271–276
51. Rooney BL, Schauberger CW (2002) Excess pregnancy weight gain and long-term obesity: one decade later. Obstet Gynecol 100(2):245–252
52. McLaren L (2007) Socioeconomic status and obesity. Epidemiol Rev 29:29–48
53. Robert SA, Reither EN (2004) A multilevel analysis of race, community disadvantage, and body mass index among adults in the US. Soc Sci Med 59(12):2421–2434
54. Wang Y, Beydoun MA (2007) The obesity epidemic in the United States—gender, age, socioeconomic, racial/ethnic, and geographic characteristics: a systematic review and meta-regression analysis. Epidemiol Rev 29:6–28
55. Zhang Q, Wang Y (2004) Trends in the association between obesity and socioeconomic status in U.S. adults: 1971 to 2000. Obes Res 12(10):1622–1632
56. Zhang Q, Wang Y (2004) Socioeconomic inequality of obesity in the United States: do gender, age, and ethnicity matter? Soc Sci Med 58(6):1171–1180
57. U.S. Cancer Statistics Working Group (2009) United States cancer statistics: 1999–2005 incidence and mortality web-based report, US Department of Health and Human Services, Centers for Disease Control and Prevention and National Cancer Institute, Atlanta, GA
58. U.S. Cancer Statistics Working Group (2013) United States cancer statistics: 1999–2010 incidence and mortality web-based report, U.S. Department of Health and Human Services, Centers for Disease Control and Prevention and National Cancer Institute, Atlanta, GA
59. Calle EE, Thun MJ (2004) Obesity and cancer. Oncogene 23(38):6365–6378
60. Abu-Abid S, Szold A, Klausner J (2002) Obesity and cancer. J Med 33(1–4):73–86
61. Bianchini F, Kaaks R, Vainio H (2002) Overweight, obesity, and cancer risk. Lancet Oncol 3(9):565–574
62. Calle EE, Rodriguez C, Walker-Thurmond K, Thun MJ (2003) Overweight, obesity, and mortality from cancer in a prospectively studied cohort of U.S. adults. N Engl J Med 348(17):1625–1638
63. IARC (2002) IARC handbooks of cancer prevention. Weight control and physical activity. International Agency for Research on Cancer, Lyon
64. Anderson GL, Neuhouser ML (2012) Obesity and the risk for premenopausal and postmenopausal breast cancer. Cancer Prev Res (Phila) 5(4):515–521
65. Suzuki R, Orsini N, Saji S, Key TJ, Wolk A (2009) Body weight and incidence of breast cancer defined by estrogen and progesterone receptor status—a meta-analysis. Int J Cancer 124(3):698–712
66. Ursin G, Longnecker MP, Haile RW, Greenland S (1995) A meta-analysis of body mass index and risk of premenopausal breast cancer. Epidemiology 6(2):137–141
67. van den Brandt PA, Spiegelman D, Yaun SS, Adami HO, Beeson L, Folsom AR, Fraser G, Goldbohm RA, Graham S, Kushi L, Marshall JR, Miller AB, Rohan T, Smith-Warner SA, Speizer FE, Willett WC, Wolk A, Hunter DJ (2000) Pooled analysis of prospective cohort studies on height, weight, and breast cancer risk. Am J Epidemiol 152(6):514–527
68. Cecchini RS, Costantino JP, Cauley JA, Cronin WM, Wickerham DL, Land SR, Weissfeld JL, Wolmark N (2012) Body mass index and the risk for developing invasive breast cancer among high-risk women in NSABP P-1 and STAR breast cancer prevention trials. Cancer Prev Res (Phila) 5(4):583–592
69. Millikan RC, Newman B, Tse CK, Moorman PG, Conway K, Dressler LG, Smith LV, Labbok MH, Geradts J, Bensen JT, Jackson S, Nyante S, Livasy C, Carey L, Earp HS, Perou CM (2008) Epidemiology of basal-like breast cancer. Breast Cancer Res Treat 109(1):123–139

70. Yang XR, Chang-Claude J, Goode EL, Couch FJ, Nevanlinna H, Milne RL, Gaudet M, Schmidt MK, Broeks A, Cox A, Fasching PA, Hein R, Spurdle AB, Blows F, Driver K, Flesch-Janys D, Heinz J, Sinn P, Vrieling A, Heikkinen T, Aittomaki K, Heikkila P, Blomqvist C, Lissowska J, Peplonska B, Chanock S, Figueroa J, Brinton L, Hall P, Czene K, Humphreys K, Darabi H, Liu J, Veer LJ V 't, van Leeuwen FE, Andrulis IL, Glendon G, Knight JA, Mulligan AM, O'Malley FP, Weerasooriya N, John EM, Beckmann MW, Hartmann A, Weihbrecht SB, Wachter DL, Jud SM, Loehberg CR, Baglietto L, English DR, Giles GG, McLean CA, Severi G, Lambrechts D, Vandorpe T, Weltens C, Paridaens R, Smeets A, Neven P, Wildiers H, Wang X, Olson JE, Cafourek V, Fredericksen Z, Kosel M, Vachon C, Cramp HE, Connley D, Cross SS, Balasubramanian SP, Reed MW, Dork T, Bremer M, Meyer A, Karstens JH, Ay A, Park-Simon TW, Hillemanns P, Arias Perez JI, Menendez Rodriguez P, Zamora P, Benitez J, Ko YD, Fischer HP, Hamann U, Pesch B, Bruning T, Justenhoven C, Brauch H, Eccles DM, Tapper WJ, Gerty SM, Sawyer EJ, Tomlinson IP, Jones A, Kerin M, Miller N, McInerney N, Anton-Culver H, Ziogas A, Shen CY, Hsiung CN, Wu PE, Yang SL, Yu JC, Chen ST, Hsu GC, Haiman CA, Henderson BE, Le Marchand L, Kolonel LN, Lindblom A, Margolin S, Jakubowska A, Lubinski J, Huzarski T, Byrski T, Gorski B, Gronwald J, Hooning MJ, Hollestelle A, van den Ouweland AM, Jager A, Kriege M, Tilanus-Linthorst MM, Collee M, Wang-Gohrke S, Pylkas K, Jukkola-Vuorinen A, Mononen K, Grip M, Hirvikoski P, Winqvist R, Mannermaa A, Kosma VM, Kauppinen J, Kataja V, Auvinen P, Soini Y, Sironen R, Bojesen SE, Orsted DD, Kaur-Knudsen D, Flyger H, Nordestgaard BG, Holland H, Chenevix-Trench G, Manoukian S, Barile M, Radice P, Hankinson SE, Hunter DJ, Tamimi R, Sangrajrang S, Brennan P, McKay J, Odefrey F, Gaborieau V, Devilee P, Huijts PE, Tollenaar RA, Seynaeve C, Dite GS, Apicella C, Hopper JL, Hammet F, Tsimiklis H, Smith LD, Southey MC, Humphreys MK, Easton D, Pharoah P, Sherman ME, Garcia-Closas M (2011) Associations of breast cancer risk factors with tumor subtypes: a pooled analysis from the Breast Cancer Association Consortium studies. J Natl Cancer Inst 103(3):250–263
71. Turkoz FP, Solak M, Petekkaya I, Keskin O, Kertmen N, Sarici F, Arik Z, Babacan T, Ozisik Y, Altundag K (2013) Association between common risk factors and molecular subtypes in breast cancer patients. Breast 22(3):344–350
72. Petekkaya I, Sahin U, Gezgen G, Solak M, Yuce D, Dizdar O, Arslan C, Ayyildiz V, Altundag K (2013) Association of breast cancer subtypes and body mass index. Tumori 99(2).129–133
73. Desantis C, Ma J, Bryan L, Jemal A (2013) Breast cancer statistics, 2013. CA Cancer J Clin 64:52–62
74. Bandera EV, Chandran U, Zirpoli G, Gong Z, McCann SE, Hong CC, Ciupak G, Pawlish K, Ambrosone CB (2013) Body fatness and breast cancer risk in women of African ancestry. BMC Cancer 13:475
75. Hall IJ, Newman B, Millikan RC, Moorman PG (2000) Body size and breast cancer risk in black women and white women: the Carolina Breast Cancer Study. Am J Epidemiol 151(8): 754–764
76. Harvie M, Hooper L, Howell AH (2003) Central obesity and breast cancer risk: a systematic review. Obes Rev 4(3):157–173
77. Mayberry RM (1994) Age-specific patterns of association between breast cancer and risk factors in black women, ages 20 to 39 and 40 to 54. Ann Epidemiol 4(3):205–213
78. Schatzkin A, Palmer JR, Rosenberg L, Helmrich SP, Miller DR, Kaufman DW, Lesko SM, Shapiro S (1987) Risk factors for breast cancer in black women. J Natl Cancer Inst 78(2): 213–217
79. Zhu K, Caulfield J, Hunter S, Roland CL, Payne-Wilks K, Texter L (2005) Body mass index and breast cancer risk in African American women. Ann Epidemiol 15(2):123–128
80. Palmer JR, Adams-Campbell LL, Boggs DA, Wise LA, Rosenberg L (2007) A prospective study of body size and breast cancer in black women. Cancer Epidemiol Biomarkers Prev 16(9):1795–1802

81. John EM, Sangaramoorthy M, Phipps AI, Koo J, Horn-Ross PL (2011) Adult body size, hormone receptor status, and premenopausal breast cancer risk in a multiethnic population: the San Francisco Bay Area breast cancer study. Am J Epidemiol 173(2):201–216
82. Stephenson GD, Rose DP (2003) Breast cancer and obesity: an update. Nutr Cancer 45(1): 1–16
83. Carmichael AR, Bates T (2004) Obesity and breast cancer: a review of the literature. Breast 13(2):85–92
84. Morimoto LM, White E, Chen Z, Chlebowski RT, Hays J, Kuller L, Lopez AM, Manson J, Margolis KL, Muti PC, Stefanick ML, McTiernan A (2002) Obesity, body size, and risk of postmenopausal breast cancer: the Women's Health Initiative (United States). Cancer Causes Control 13(8):741–751
85. Halaas JL, Gajiwala KS, Maffei M, Cohen SL, Chait BT, Rabinowitz D, Lallone RL, Burley SK, Friedman JM (1995) Weight-reducing effects of the plasma protein encoded by the obese gene. Science 269(5223):543–546
86. White KK, Park SY, Kolonel LN, Henderson BE, Wilkens LR (2012) Body size and breast cancer risk: the Multiethnic Cohort. Int J Cancer 131(5):E705–E716
87. Siegel R, Ma J, Zou Z, Jemal A (2014) Cancer statistics, 2014. CA Cancer J Clin 64(1):9–29
88. Jemal A, Thun MJ, Ries LA, Howe HL, Weir HK, Center MM, Ward E, Wu XC, Eheman C, Anderson R, Ajani UA, Kohler B, Edwards BK (2008) Annual report to the nation on the status of cancer, 1975–2005, featuring trends in lung cancer, tobacco use, and tobacco control. J Natl Cancer Inst 100(23):1672–1694
89. Lipworth L, McLaughlin JK, Tarone RE, Blot WJ (2011) Renal cancer paradox: higher incidence but not higher mortality among African-Americans. Eur J Cancer Prev 20(4): 331–333
90. Lipworth L, Tarone RE, McLaughlin JK (2011) Renal cell cancer among African Americans: an epidemiologic review. BMC Cancer 11:133
91. Desantis C, Naishadham D, Jemal A (2013) Cancer statistics for African Americans, 2013. CA Cancer J Clin 63:151–166
92. Sankin A, Cohen J, Wang H, Macchia RJ, Karanikolas N (2011) Rate of renal cell carcinoma subtypes in different races. Int Braz J Urol 37(1):29–32, discussion 33–24
93. Ildaphonse G, George PS, Mathew A (2009) Obesity and kidney cancer risk in men: a meta-analysis (1992–2008). Asian Pac J Cancer Prev 10(2):279–286
94. Mathew A, George PS, Ildaphonse G (2009) Obesity and kidney cancer risk in women: a meta-analysis (1992–2008). Asian Pac J Cancer Prev 10(3):471–478
95. Klinghoffer Z, Yang B, Kapoor A, Pinthus JH (2009) Obesity and renal cell carcinoma: epidemiology, underlying mechanisms and management considerations. Expert Rev Anticancer Ther 9(7):975–987
96. Renehan AG, Tyson M, Egger M, Heller RF, Zwahlen M (2008) Body-mass index and incidence of cancer: a systematic review and meta-analysis of prospective observational studies. Lancet 371(9612):569–578
97. Samanic C, Gridley G, Chow WH, Lubin J, Hoover RN, Fraumeni JF Jr (2004) Obesity and cancer risk among white and black United States veterans. Cancer Causes Control 15(1):35–43
98. Purdue MP, Moore LE, Merino MJ, Boffetta P, Colt JS, Schwartz KL, Bencko V, Davis FG, Graubard BI, Janout V, Ruterbusch JJ, Beebe-Dimmer J, Cote ML, Shuch B, Mates D, Hofmann JN, Foretova L, Rothman N, Szeszenia-Dabrowska N, Matveev V, Wacholder S, Zaridze D, Linehan WM, Brennan P, Chow WH (2012) An investigation of risk factors for renal cell carcinoma by histologic subtype in two case-control studies. Int J Cancer 132:2640–2647
99. Arcidiacono B, Iiritano S, Nocera A, Possidente K, Nevolo MT, Ventura V, Foti D, Chiefari E, Brunetti A (2012) Insulin resistance and cancer risk: an overview of the pathogenetic mechanisms. Exp Diabetes Res 2012:789174

100. Morris PG, Hudis CA, Giri D, Morrow M, Falcone DJ, Zhou XK, Du B, Brogi E, Crawford CB, Kopelovich L, Subbaramaiah K, Dannenberg AJ (2011) Inflammation and increased aromatase expression occur in the breast tissue of obese women with breast cancer. Cancer Prev Res (Phila) 4(7):1021–1029
101. Hursting SD, Digiovanni J, Dannenberg AJ, Azrad M, Leroith D, Demark-Wahnefried W, Kakarala M, Brodie A, Berger NA (2012) Obesity, energy balance, and cancer: new opportunities for prevention. Cancer Prev Res (Phila) 5(11):1260–1272
102. Renehan AG, Roberts DL, Dive C (2008) Obesity and cancer: pathophysiological and biological mechanisms. Arch Physiol Biochem 114(1):71–83
103. Pollak M (2008) Insulin and insulin-like growth factor signalling in neoplasia. Nat Rev Cancer 8(12):915–928
104. Douglas JB, Silverman DT, Pollak MN, Tao Y, Soliman AS, Stolzenberg-Solomon RZ (2010) Serum IGF-I, IGF-II, IGFBP-3, and IGF-I/IGFBP-3 molar ratio and risk of pancreatic cancer in the prostate, lung, colorectal, and ovarian cancer screening trial. Cancer Epidemiol Biomarkers Prev 19(9):2298–2306
105. Braun S, Bitton-Worms K, LeRoith D (2011) The link between the metabolic syndrome and cancer. Int J Biol Sci 7(7):1003–1015
106. Wysocki PJ, Wierusz-Wysocka B (2010) Obesity, hyperinsulinemia and breast cancer: novel targets and a novel role for metformin. Expert Rev Mol Diagn 10(4):509–519
107. Fowke JH, Matthews CE, Yu H, Cai Q, Cohen S, Buchowski MS, Zheng W, Blot WJ (2010) Racial differences in the association between body mass index and serum IGF1, IGF2, and IGFBP3. Endocr Relat Cancer 17(1):51–60
108. Gunter MJ, Hoover DR, Yu H, Wassertheil-Smoller S, Rohan TE, Manson JE, Li J, Ho GY, Xue X, Anderson GL, Kaplan RC, Harris TG, Howard BV, Wylie-Rosett J, Burk RD, Strickler HD (2009) Insulin, insulin-like growth factor-I, and risk of breast cancer in postmenopausal women. J Natl Cancer Inst 101(1):48–60
109. Kaaks R, Lundin E, Rinaldi S, Manjer J, Biessy C, Soderberg S, Lenner P, Janzon L, Riboli E, Berglund G, Hallmans G (2002) Prospective study of IGF-I, IGF-binding proteins, and breast cancer risk, in northern and southern Sweden. Cancer Causes Control 13(4):307–316
110. Keinan-Boker L, Bueno De Mesquita HB, Kaaks R, Van Gils CH, Van Noord PA, Rinaldi S, Riboli E, Seidell JC, Grobbee DE, Peeters PH (2003) Circulating levels of insulin-like growth factor I, its binding proteins -1,-2, -3, C-peptide and risk of postmenopausal breast cancer. Int J Cancer 106(1):90–95
111. Renehan AG, Egger M, Minder C, O'Dwyer ST, Shalet SM, Zwahlen M (2005) IGF-I, IGF binding protein-3 and breast cancer risk: comparison of 3 meta-analyses. Int J Cancer 115(6): 1006–1007, author reply 1008
112. Renehan AG, Zwahlen M, Minder C, O'Dwyer ST, Shalet SM, Egger M (2004) Insulin-like growth factor (IGF)-I, IGF binding protein-3, and cancer risk: systematic review and meta-regression analysis. Lancet 363(9418):1346–1353
113. Shi R, Yu H, McLarty J, Glass J (2004) IGF-I and breast cancer: a meta-analysis. Int J Cancer 111(3):418–423
114. Sugumar A, Liu YC, Xia Q, Koh YS, Matsuo K (2004) Insulin-like growth factor (IGF)-I and IGF-binding protein 3 and the risk of premenopausal breast cancer: a meta-analysis of literature. Int J Cancer 111(2):293–297
115. Verheus M, Peeters PH, Rinaldi S, Dossus L, Biessy C, Olsen A, Tjonneland A, Overvad K, Jeppesen M, Clavel-Chapelon F, Tehard B, Nagel G, Linseisen J, Boeing H, Lahmann PH, Arvaniti A, Psaltopoulou T, Trichopoulou A, Palli D, Tumino R, Panico S, Sacerdote C, Sieri S, van Gils CH, Bueno-de-Mesquita BH, Gonzalez CA, Ardanaz E, Larranaga N, Garcia CM, Navarro C, Quiros JR, Key T, Allen N, Bingham S, Khaw KT, Slimani N, Riboli E, Kaaks R (2006) Serum C-peptide levels and breast cancer risk: results from the European Prospective Investigation into Cancer and Nutrition (EPIC). Int J Cancer 119(3):659–667

116. Key TJ, Appleby PN, Reeves GK, Roddam AW (2010) Insulin-like growth factor 1 (IGF1), IGF binding protein 3 (IGFBP3), and breast cancer risk: pooled individual data analysis of 17 prospective studies. Lancet Oncol 11(6):530–542

117. Schernhammer ES, Holly JM, Pollak MN, Hankinson SE (2005) Circulating levels of insulin-like growth factors, their binding proteins, and breast cancer risk. Cancer Epidemiol Biomarkers Prev 14(3):699–704

118. Maiti B, Kundranda MN, Spiro TP, Daw HA (2010) The association of metabolic syndrome with triple-negative breast cancer. Breast Cancer Res Treat 121(2):479–483

119. Pichard C, Plu-Bureau G, Neves ECM, Gompel A (2008) Insulin resistance, obesity and breast cancer risk. Maturitas 60(1):19–30

120. Kellerer M, von Eye CH, Muhlhofer A, Capp E, Mosthaf L, Bock S, Petrides PE, Haring HU (1995) Insulin- and insulin-like growth-factor-I receptor tyrosine-kinase activities in human renal carcinoma. Int J Cancer 62(5):501–507

121. Rosendahl A, Forsberg G (2004) Influence of IGF-IR stimulation or blockade on proliferation of human renal cell carcinoma cell lines. Int J Oncol 25(5):1327–1336

122. Rosendahl AH, Forsberg G (2006) IGF-I and IGFBP-3 augment transforming growth factor-beta actions in human renal carcinoma cells. Kidney Int 70(9):1584–1590

123. Cheung CW, Vesey DA, Nicol DL, Johnson DW (2004) The roles of IGF-I and IGFBP-3 in the regulation of proximal tubule, and renal cell carcinoma cell proliferation. Kidney Int 65 (4):1272–1279

124. Ramp U, Jaquet K, Reinecke P, Schardt C, Friebe U, Nitsch T, Marx N, Gabbert HE, Gerharz CD (1997) Functional intactness of stimulatory and inhibitory autocrine loops in human renal carcinoma cell lines of the clear cell type. J Urol 157(6):2345–2350

125. Schips L, Zigeuner R, Ratschek M, Rehak P, Ruschoff J, Langner C (2004) Analysis of insulin-like growth factors and insulin-like growth factor I receptor expression in renal cell carcinoma. Am J Clin Pathol 122(6):931–937

126. Liao LM, Weinstein SJ, Pollak M, Li Z, Virtamo J, Albanes D, Chow WH, Purdue MP (2013) Prediagnostic circulating adipokine concentrations and risk of renal cell carcinoma in male smokers. Carcinogenesis 34(1):109–112

127. Major JM, Pollak MN, Snyder K, Virtamo J, Albanes D (2010) Insulin-like growth factors and risk of kidney cancer in men. Br J Cancer 103(1):132–135

128. Kaaks R, Rinaldi S, Key TJ, Berrino F, Peeters PH, Biessy C, Dossus L, Lukanova A, Bingham S, Khaw KT, Allen NE, Bueno-de-Mesquita HB, van Gils CH, Grobbee D, Boeing H, Lahmann PH, Nagel G, Chang-Claude J, Clavel-Chapelon F, Fournier A, Thiebaut A, Gonzalez CA, Quiros JR, Tormo MJ, Ardanaz E, Amiano P, Krogh V, Palli D, Panico S, Tumino R, Vineis P, Trichopoulou A, Kalapothaki V, Trichopoulos D, Ferrari P, Norat T, Saracci R, Riboli E (2005) Postmenopausal serum androgens, oestrogens and breast cancer risk: the European prospective investigation into cancer and nutrition. Endocr Relat Cancer 12(4):1071–1082

129. Key T, Appleby P, Barnes I, Reeves G (2002) Endogenous sex hormones and breast cancer in postmenopausal women: reanalysis of nine prospective studies. J Natl Cancer Inst 94(8): 606–616

130. Missmer SA, Eliassen AH, Barbieri RL, Hankinson SE (2004) Endogenous estrogen, androgen, and progesterone concentrations and breast cancer risk among postmenopausal women. J Natl Cancer Inst 96(24):1856–1865

131. Catalano MG, Frairia R, Boccuzzi G, Fortunati N (2005) Sex hormone-binding globulin antagonizes the anti-apoptotic effect of estradiol in breast cancer cells. Mol Cell Endocrinol 230(1–2):31–37

132. Bezemer ID, Rinaldi S, Dossus L, Gils CH, Peeters PH, Noord PA, Bueno-de-Mesquita HB, Johnsen SP, Overvad K, Olsen A, Tjonneland A, Boeing H, Lahmann PH, Linseisen J, Nagel G, Allen N, Roddam A, Bingham S, Khaw KT, Kesse E, Tehard B, Clavel-Chapelon F, Agudo A, Ardanaz E, Quiros JR, Amiano P, Martinez-Garcia C, Tormo MJ, Pala V, Panico S, Vineis P, Palli D, Tumino R, Trichopoulou A, Baibas N, Zilis D, Hemon B, Norat T, Riboli E,

Kaaks R (2005) C-peptide, IGF-I, sex-steroid hormones and adiposity: a cross-sectional study in healthy women within the European Prospective Investigation into Cancer and Nutrition (EPIC). Cancer Causes Control 16(5):561–572

133. Pinheiro SP, Holmes MD, Pollak MN, Barbieri RL, Hankinson SE (2005) Racial differences in premenopausal endogenous hormones. Cancer Epidemiol Biomarkers Prev 14(9): 2147–2153

134. Probst-Hensch NM, Pike MC, McKean-Cowdin R, Stanczyk FZ, Kolonel LN, Henderson BE (2000) Ethnic differences in post-menopausal plasma oestrogen levels: high oestrone levels in Japanese-American women despite low weight. Br J Cancer 82(11):1867–1870

135. Setiawan VW, Haiman CA, Stanczyk FZ, Le Marchand L, Henderson BE (2006) Racial/ethnic differences in postmenopausal endogenous hormones: the multiethnic cohort study. Cancer Epidemiol Biomarkers Prev 15(10):1849–1855

136. Woolcott CG, Shvetsov YB, Stanczyk FZ, Wilkens LR, White KK, Caberto C, Henderson BE, Le Marchand L, Kolonel LN, Goodman MT (2010) Plasma sex hormone concentrations and breast cancer risk in an ethnically diverse population of postmenopausal women: the Multiethnic Cohort Study. Endocr Relat Cancer 17(1):125–134

137. Key TJ (2011) Endogenous oestrogens and breast cancer risk in premenopausal and post-menopausal women. Steroids 76(8):812–815

138. Endogenous Hormones and Breast Cancer Collaborative Group (2013) Sex hormones and risk of breast cancer in premenopausal women: a collaborative reanalysis of individual participant data from seven prospective studies. Lancet Oncol 14(10):1009–1019

139. Lamon-Fava S, Barnett JB, Woods MN, McCormack C, McNamara JR, Schaefer EJ, Longcope C, Rosner B, Gorbach SL (2005) Differences in serum sex hormone and plasma lipid levels in Caucasian and African-American premenopausal women. J Clin Endocrinol Metab 90(8):4516–4520

140. Randolph JF Jr, Sowers M, Gold EB, Mohr BA, Luborsky J, Santoro N, McConnell DS, Finkelstein JS, Korenman SG, Matthews KA, Sternfeld B, Lasley BL (2003) Reproductive hormones in the early menopausal transition: relationship to ethnicity, body size, and menopausal status. J Clin Endocrinol Metab 88(4):1516–1522

141. Marsh EE, Shaw ND, Klingman KM, Tiamfook-Morgan TO, Yialamas MA, Sluss PM, Hall JE (2011) Estrogen levels are higher across the menstrual cycle in African-American women compared with Caucasian women. J Clin Endocrinol Metab 96(10):3199–3206

142. Barnett JB, Woods MN, Rosner B, McCormack C, Longcope C, Houser RF Jr, Gorbach SL (2001) Sex hormone levels in premenopausal African-American women with upper and lower body fat phenotypes. Nutr Cancer 41(1–2):47–56

143. Key TJ (1999) Serum oestradiol and breast cancer risk. Endocr Relat Cancer 6(2):175–180

144. Paxton RJ, King DW, Garcia-Prieto C, Connors SK, Hernandez M, Gor BJ, Jones LA (2013) Associations between body size and serum estradiol and sex hormone-binding globulin levels in premenopausal African American women. J Clin Endocrinol Metab 98(3):E485–E490

145. Waugh DJ, Wilson C (2008) The interleukin-8 pathway in cancer. Clin Cancer Res 14(21): 6735–6741

146. Sun X, Casbas-Hernandez P, Bigelow C, Makowski L, Joseph Jerry D, Smith Schneider S, Troester MA (2012) Normal breast tissue of obese women is enriched for macrophage markers and macrophage-associated gene expression. Breast Cancer Res Treat 131(3): 1003–1012

147. Anand SS, Razak F, Yi Q, Davis B, Jacobs R, Vuksan V, Lonn E, Teo K, McQueen M, Yusuf S (2004) C-reactive protein as a screening test for cardiovascular risk in a multiethnic population. Arterioscler Thromb Vasc Biol 24(8):1509–1515

148. Festa A, D'Agostino R Jr, Williams K, Karter AJ, Mayer-Davis EJ, Tracy RP, Haffner SM (2001) The relation of body fat mass and distribution to markers of chronic inflammation. Int J Obes Relat Metab Disord 25(10):1407–1415

149. Ford ES (1999) Body mass index, diabetes, and C-reactive protein among U.S. adults. Diabetes Care 22(12):1971–1977

150. Esposito K, Pontillo A, Di Palo C, Giugliano G, Masella M, Marfella R, Giugliano D (2003) Effect of weight loss and lifestyle changes on vascular inflammatory markers in obese women: a randomized trial. JAMA 289(14):1799–1804
151. Puglisi MJ, Fernandez ML (2008) Modulation of C-reactive protein, tumor necrosis factor-alpha, and adiponectin by diet, exercise, and weight loss. J Nutr 138(12):2293–2296
152. Zhang X, Shu XO, Signorello LB, Hargreaves MK, Cai Q, Linton MF, Fazio S, Zheng W, Blot WJ (2008) Correlates of high serum C-reactive protein levels in a socioeconomically disadvantaged population. Dis Markers 24(6):351–359
153. Vona-Davis L, Rose DP (2007) Adipokines as endocrine, paracrine, and autocrine factors in breast cancer risk and progression. Endocr Relat Cancer 14(2):189–206
154. Fernandez-Riejos P, Najib S, Santos-Alvarez J, Martin-Romero C, Perez-Perez A, Gonzalez-Yanes C, Sanchez-Margalet V (2010) Role of leptin in the activation of immune cells. Mediators Inflamm 2010:568343
155. Jarde T, Perrier S, Vasson MP, Caldefie-Chezet F (2011) Molecular mechanisms of leptin and adiponectin in breast cancer. Eur J Cancer 47(1):33–43
156. Devaraj S, Torok N, Dasu MR, Samols D, Jialal I (2008) Adiponectin decreases C-reactive protein synthesis and secretion from endothelial cells: evidence for an adipose tissue-vascular loop. Arterioscler Thromb Vasc Biol 28(7):1368–1374
157. Crimmins NA, Martin LJ (2007) Polymorphisms in adiponectin receptor genes ADIPOR1 and ADIPOR2 and insulin resistance. Obes Rev 8(5):419–423
158. Takahashi M, Arita Y, Yamagata K, Matsukawa Y, Okutomi K, Horie M, Shimomura I, Hotta K, Kuriyama H, Kihara S, Nakamura T, Yamashita S, Funahashi T, Matsuzawa Y (2000) Genomic structure and mutations in adipose-specific gene, adiponectin. Int J Obes Relat Metab Disord 24(7):861–868
159. Jeong YJ, Bong JG, Park SH, Choi JH, Oh HK (2011) Expression of leptin, leptin receptor, adiponectin, and adiponectin receptor in ductal carcinoma in situ and invasive breast cancer. J Breast Cancer 14(2):96–103
160. Grossmann ME, Cleary MP (2012) The balance between leptin and adiponectin in the control of carcinogenesis—focus on mammary tumorigenesis. Biochimie 94:2164–2171
161. Koerner A, Kratzsch J, Kiess W (2005) Adipocytokines: leptin—the classical, resistin—the controversial, adiponectin—the promising, and more to come. Best Pract Res Clin Endocrinol Metab 19(4):525–546
162. Ronti T, Lupattelli G, Mannarino E (2006) The endocrine function of adipose tissue: an update. Clin Endocrinol (Oxf) 64(4):355–365
163. Swarbrick MM, Havel PJ (2008) Physiological, pharmacological, and nutritional regulation of circulating adiponectin concentrations in humans. Metab Syndr Relat Disord 6(2):87–102
164. Takeuchi T, Adachi Y, Ohtsuki Y, Furihata M (2007) Adiponectin receptors, with special focus on the role of the third receptor, T-cadherin, in vascular disease. Med Mol Morphol 40(3):115–120
165. Kadowaki T, Yamauchi T (2005) Adiponectin and adiponectin receptors. Endocr Rev 26(3):439–451
166. Bush NC, Darnell BE, Oster RA, Goran MI, Gower BA (2005) Adiponectin is lower among African Americans and is independently related to insulin sensitivity in children and adolescents. Diabetes 54(9):2772–2778
167. Degawa-Yamauchi M, Dilts JR, Bovenkerk JE, Saha C, Pratt JH, Considine RV (2003) Lower serum adiponectin levels in African-American boys. Obes Res 11(11):1384–1390
168. Araneta MR, Barrett-Connor E (2007) Adiponectin and ghrelin levels and body size in normoglycemic Filipino, African-American, and white women. Obesity (Silver Spring) 15(10):2454–2462
169. Ferris WF, Naran NH, Crowther NJ, Rheeder P, van der Merwe L, Chetty N (2005) The relationship between insulin sensitivity and serum adiponectin levels in three population groups. Horm Metab Res 37(11):695–701

170. Steffes MW, Gross MD, Schreiner PJ, Yu X, Hilner JE, Gingerich R, Jacobs DR Jr (2004) Serum adiponectin in young adults—interactions with central adiposity, circulating levels of glucose, and insulin resistance: the CARDIA study. Ann Epidemiol 14(7):492–498

171. Duncan BB, Schmidt MI, Pankow JS, Bang H, Couper D, Ballantyne CM, Hoogeveen RC, Heiss G (2004) Adiponectin and the development of type 2 diabetes: the atherosclerosis risk in communities study. Diabetes 53(9):2473–2478

172. Kanaya AM, Wassel Fyr C, Vittinghoff E, Havel PJ, Cesari M, Nicklas B, Harris T, Newman AB, Satterfield S, Cummings SR (2006) Serum adiponectin and coronary heart disease risk in older Black and White Americans. J Clin Endocrinol Metab 91(12):5044–5050

173. Hulver MW, Saleh O, MacDonald KG, Pories WJ, Barakat HA (2004) Ethnic differences in adiponectin levels. Metabolism 53(1):1–3

174. Schutte AE, Huisman HW, Schutte R, Malan L, van Rooyen JM, Malan NT, Schwarz PE (2007) Differences and similarities regarding adiponectin investigated in African and Caucasian women. Eur J Endocrinol 157(2):181–188

175. Wassel Fyr CL, Kanaya AM, Cummings SR, Reich D, Hsueh WC, Reiner AP, Harris TB, Moffett S, Li R, Ding J, Miljkovic-Gacic I, Ziv E (2007) Genetic admixture, adipocytokines, and adiposity in Black Americans: the health, aging, and body composition study. Hum Genet 121(5):615–624

176. Chandran M, Phillips SA, Ciaraldi T, Henry RR (2003) Adiponectin: more than just another fat cell hormone? Diabetes Care 26(8):2442–2450

177. Hotta K, Funahashi T, Arita Y, Takahashi M, Matsuda M, Okamoto Y, Iwahashi H, Kuriyama H, Ouchi N, Maeda K, Nishida M, Kihara S, Sakai N, Nakajima T, Hasegawa K, Muraguchi M, Ohmoto Y, Nakamura T, Yamashita S, Hanafusa T, Matsuzawa Y (2000) Plasma concentrations of a novel, adipose-specific protein, adiponectin, in type 2 diabetic patients. Arterioscler Thromb Vasc Biol 20(6):1595–1599

178. Arita Y, Kihara S, Ouchi N, Takahashi M, Maeda K, Miyagawa J, Hotta K, Shimomura I, Nakamura T, Miyaoka K, Kuriyama H, Nishida M, Yamashita S, Okubo K, Matsubara K, Muraguchi M, Ohmoto Y, Funahashi T, Matsuzawa Y (1999) Paradoxical decrease of an adipose-specific protein, adiponectin, in obesity. Biochem Biophys Res Commun 257(1):79–83

179. Cnop M, Havel PJ, Utzschneider KM, Carr DB, Sinha MK, Boyko EJ, Retzlaff BM, Knopp RH, Brunzell JD, Kahn SE (2003) Relationship of adiponectin to body fat distribution, insulin sensitivity and plasma lipoproteins: evidence for independent roles of age and sex. Diabetologia 46(4):459–469

180. Staiger H, Tschritter O, Machann J, Thamer C, Fritsche A, Maerker E, Schick F, Haring HU, Stumvoll M (2003) Relationship of serum adiponectin and leptin concentrations with body fat distribution in humans. Obes Res 11(3):368–372

181. Miljkovic-Gacic I, Wang X, Kammerer CM, Bunker CH, Wheeler VW, Patrick AL, Kuller LH, Evans RW, Zmuda JM (2007) Genetic determination of adiponectin and its relationship with body fat topography in multigenerational families of African heritage. Metabolism 56(2):234–238

182. Hanley AJ, Bowden D, Wagenknecht LE, Balasubramanyam A, Langfeld C, Saad MF, Rotter JI, Guo X, Chen YD, Bryer-Ash M, Norris JM, Haffner SM (2007) Associations of adiponectin with body fat distribution and insulin sensitivity in nondiabetic hispanics and African-Americans. J Clin Endocrinol Metab 92(7):2665–2671

183. Anubhuti AS (2008) Leptin and its metabolic interactions: an update. Diabetes Obes Metab 10(11):973–993

184. Zhang Y, Proenca R, Maffei M, Barone M, Leopold L, Friedman JM (1994) Positional cloning of the mouse obese gene and its human homologue. Nature 372(6505):425–432

185. Auwerx J, Staels B (1998) Leptin. Lancet 351(9104):737–742

186. Sinha MK, Sturis J, Ohannesian J, Magosin S, Stephens T, Heiman ML, Polonsky KS, Caro JF (1996) Ultradian oscillations of leptin secretion in humans. Biochem Biophys Res Commun 228(3):733–738

187. Gautron L, Elmquist JK (2011) Sixteen years and counting: an update on leptin in energy balance. J Clin Invest 121(6):2087–2093
188. Mente A, Razak F, Blankenberg S, Vuksan V, Davis AD, Miller R, Teo K, Gerstein H, Sharma AM, Yusuf S, Anand SS (2010) Ethnic variation in adiponectin and leptin levels and their association with adiposity and insulin resistance. Diabetes Care 33(7):1629–1634
189. Cohen SS, Fowke JH, Cai Q, Buchowski MS, Signorello LB, Hargreaves MK, Zheng W, Blot WJ, Matthews CE (2012) Differences in the association between serum leptin levels and body mass index in black and white women: a report from the Southern Community Cohort Study. Ann Nutr Metab 60(2):90–97
190. Ries LAG, Harkins D, Krapcho M, Mariotto A, Miller BA, Feuer EJ, Clegg L, Eisner MP, Horner MJ, Howlader N, Hayat M, Hankey BF, Edwards BK (2006) SEER cancer statistics review, 1975–2003. National Cancer Institute Bethesda, MD. http://seer.cancer.gov/csr/1975_2003/, based on November 2005 SEER data submission, posted to the SEER web site, 2006
191. Gibbons GH (2004) Physiology, genetics, and cardiovascular disease: focus on African Americans. J Clin Hypertens (Greenwich) 6(4 Suppl 1):11–18
192. Dieudonne MN, Bussiere M, Dos Santos E, Leneveu MC, Giudicelli Y, Pecquery R (2006) Adiponectin mediates antiproliferative and apoptotic responses in human MCF7 breast cancer cells. Biochem Biophys Res Commun 345(1):271–279
193. Treeck O, Lattrich C, Juhasz-Boess I, Buchholz S, Pfeiler G, Ortmann O (2008) Adiponectin differentially affects gene expression in human mammary epithelial and breast cancer cells. Br J Cancer 99(8):1246–1250
194. Arditi JD, Venihaki M, Karalis KP, Chrousos GP (2007) Antiproliferative effect of adiponectin on MCF7 breast cancer cells: a potential hormonal link between obesity and cancer. Horm Metab Res 39(1):9–13
195. Jarde T, Caldefie-Chezet F, Goncalves-Mendes N, Mishellany F, Buechler C, Penault-Llorca-F, Vasson MP (2009) Involvement of adiponectin and leptin in breast cancer: clinical and in vitro studies. Endocr Relat Cancer 16(4):1197–1210
196. Kang JH, Lee YY, Yu BY, Yang BS, Cho KH, Yoon DK, Roh YK (2005) Adiponectin induces growth arrest and apoptosis of MDA-MB-231 breast cancer cell. Arch Pharm Res 28(11):1263–1269
197. Barb D, Williams CJ, Neuwirth AK, Mantzoros CS (2007) Adiponectin in relation to malignancies: a review of existing basic research and clinical evidence. Am J Clin Nutr 86(3):s858–s866
198. Chen DC, Chung YF, Yeh YT, Chaung HC, Kuo FC, Fu OY, Chen HY, Hou MF, Yuan SS (2006) Serum adiponectin and leptin levels in Taiwanese breast cancer patients. Cancer Lett 237(1):109–114
199. Korner A, Pazaitou-Panayiotou K, Kelesidis T, Kelesidis I, Williams CJ, Kaprara A, Bullen J, Neuwirth A, Tseleni S, Mitsiades N, Kiess W, Mantzoros CS (2006) Total and high molecular weight adiponectin in breast cancer: in vitro and in vivo studies. J Clin Endocrinol Metab 92:1041–1048
200. Mantzoros C, Petridou E, Dessypris N, Chavelas C, Dalamaga M, Alexe DM, Papadiamantis Y, Markopoulos C, Spanos E, Chrousos G, Trichopoulos D (2004) Adiponectin and breast cancer risk. J Clin Endocrinol Metab 89(3):1102–1107
201. Miyoshi Y, Funahashi T, Kihara S, Taguchi T, Tamaki Y, Matsuzawa Y, Noguchi S (2003) Association of serum adiponectin levels with breast cancer risk. Clin Cancer Res 9(15): 5699–5704
202. Tian YF, Chu CH, Wu MH, Chang CL, Yang T, Chou YC, Hsu GC, Yu CP, Yu JC, Sun CA (2007) Anthropometric measures, plasma adiponectin, and breast cancer risk. Endocr Relat Cancer 14(3):669–677
203. Kang JH, Yu BY, Youn DS (2007) Relationship of serum adiponectin and resistin levels with breast cancer risk. J Korean Med Sci 22(1):117–121

204. Tworoger SS, Eliassen AH, Kelesidis T, Colditz GA, Willett WC, Mantzoros C, Hankinson SE (2007) Plasma adiponectin concentrations and risk of incident breast cancer. J Clin Endocrinol Metab 92:1510–1516
205. Liu LY, Wang M, Ma ZB, Yu LX, Zhang Q, Gao DZ, Wang F, Yu ZG (2013) The role of adiponectin in breast cancer: a meta-analysis. PLoS One 8(8):e73183
206. Vona-Davis L, Howard-McNatt M, Rose DP (2007) Adiposity, type 2 diabetes and the metabolic syndrome in breast cancer. Obes Rev 8(5):395–408
207. Maccio A, Madeddu C (2011) Obesity, inflammation, and postmenopausal breast cancer: therapeutic implications. ScientificWorldJournal 11:2020–2036
208. Spyridopoulos TN, Dessypris N, Antoniadis AG, Gialamas S, Antonopoulos CN, Katsifoti K, Adami HO, Chrousos GP, Petridou ET (2012) Insulin resistance and risk of renal cell cancer: a case-control study. Hormones (Athens) 11(3):308–315
209. Spyridopoulos TN, Petridou ET, Dessypris N, Terzidis A, Skalkidou A, Deliveliotis C, Chrousos GP (2009) Inverse association of leptin levels with renal cell carcinoma: results from a case-control study. Hormones (Athens) 8(1):39–46
210. Horiguchi A, Ito K, Sumitomo M, Kimura F, Asano T, Hayakawa M (2008) Decreased serum adiponectin levels in patients with metastatic renal cell carcinoma. Jpn J Clin Oncol 38(2): 106–111
211. Horiguchi A, Sumitomo M, Asakuma J, Asano T, Zheng R, Nanus DM, Hayakawa M (2006) Increased serum leptin levels and over expression of leptin receptors are associated with the invasion and progression of renal cell carcinoma. J Urol 176(4 Pt 1):1631–1635
212. Pinthus JH, Kleinmann N, Tisdale B, Chatterjee S, Lu JP, Gillis A, Hamlet T, Singh G, Farrokhyar F, Kapoor A (2008) Lower plasma adiponectin levels are associated with larger tumor size and metastasis in clear-cell carcinoma of the kidney. Eur Urol 54(4):866–873
213. Rasmuson T, Grankvist K, Jacobsen J, Olsson T, Ljungberg B (2004) Serum insulin-like growth factor-1 is an independent predictor of prognosis in patients with renal cell carcinoma. Acta Oncol 43(8):744–748
214. Spyridopoulos TN, Petridou ET, Skalkidou A, Dessypris N, Chrousos GP, Mantzoros CS (2007) Low adiponectin levels are associated with renal cell carcinoma: a case-control study. Int J Cancer 120(7):1573–1578
215. Iimura Y, Saito K, Fujii Y, Kumagai J, Kawakami S, Komai Y, Yonese J, Fukui I, Kihara K (2009) Development and external validation of a new outcome prediction model for patients with clear cell renal cell carcinoma treated with nephrectomy based on preoperative serum C-reactive protein and TNM classification: the TNM-C score. J Urol 181(3):1004–1012, discussion 1012
216. Dupont J, Le Roith D (2001) Insulin-like growth factor 1 and oestradiol promote cell proliferation of MCF-7 breast cancer cells: new insights into their synergistic effects. Mol Pathol 54(3):149–154
217. Hamelers IH, Steenbergh PH (2003) Interactions between estrogen and insulin-like growth factor signaling pathways in human breast tumor cells. Endocr Relat Cancer 10(2):331–345
218. Poretsky L, Kalin MF (1987) The gonadotropic function of insulin. Endocr Rev 8(2):132–141
219. Pugeat M, Crave JC, Elmidani M, Nicolas MH, Garoscio-Cholet M, Lejeune H, Dechaud H, Tourniaire J (1991) Pathophysiology of sex hormone binding globulin (SHBG): relation to insulin. J Steroid Biochem Mol Biol 40(4–6):841–849
220. LeRoith D, Werner H, Beitner-Johnson D, Roberts CT Jr (1995) Molecular and cellular aspects of the insulin-like growth factor I receptor. Endocr Rev 16(2):143–163
221. Maor S, Papa MZ, Yarden RI, Friedman E, Lerenthal Y, Lee SW, Mayer D, Werner H (2007) Insulin-like growth factor-I controls BRCA1 gene expression through activation of transcription factor Sp1. Horm Metab Res 39(3):179–185
222. Hotamisligil GS, Shargill NS, Spiegelman BM (1993) Adipose expression of tumor necrosis factor-alpha: direct role in obesity-linked insulin resistance. Science 259(5091):87–91
223. Purohit A, Newman SP, Reed MJ (2002) The role of cytokines in regulating estrogen synthesis: implications for the etiology of breast cancer. Breast Cancer Res 4(2):65–69

224. Gonullu G, Ersoy C, Ersoy A, Evrensel T, Basturk B, Kurt E, Oral B, Gokgoz S, Manavoglu O (2005) Relation between insulin resistance and serum concentrations of IL-6 and TNF-alpha in overweight or obese women with early stage breast cancer. Cytokine 31(4): 264–269

225. Vozarova B, Weyer C, Hanson K, Tataranni PA, Bogardus C, Pratley RE (2001) Circulating interleukin-6 in relation to adiposity, insulin action, and insulin secretion. Obes Res 9(7): 414–417

226. Fasshauer M, Klein J, Neumann S, Eszlinger M, Paschke R (2002) Hormonal regulation of adiponectin gene expression in 3T3-L1 adipocytes. Biochem Biophys Res Commun 290(3): 1084–1089

227. Gaudet MM, Falk RT, Gierach GL, Lacey JV Jr, Graubard BI, Dorgan JF, Brinton LA (2010) Do adipokines underlie the association between known risk factors and breast cancer among a cohort of United States women? Cancer Epidemiol 34(5):580–586

228. Tworoger SS, Mantzoros C, Hankinson SE (2007) Relationship of plasma adiponectin with sex hormone and insulin-like growth factor levels. Obesity (Silver Spring) 15(9):2217–2224

229. Wildman RP, Wang D, Fernandez I, Mancuso P, Santoro N, Scherer PE, Sowers MR (2012) Associations of testosterone and sex hormone binding globulin with adipose tissue hormones in midlife women. Obesity (Silver Spring) 21:629–636

230. Szymanska E, Bouwman J, Strassburg K, Vervoort J, Kangas AJ, Soininen P, Ala-Korpela M, Westerhuis J, van Duynhoven JP, Mela DJ, Macdonald IA, Vreeken RJ, Smilde AK, Jacobs DM (2012) Gender-dependent associations of metabolite profiles and body fat distribution in a healthy population with central obesity: towards metabolomics diagnostics. OMICS 16(12): 652–667

231. Wells JC (2012) Ethnic variability in adiposity, thrifty phenotypes and cardiometabolic risk: addressing the full range of ethnicity, including those of mixed ethnicity. Obes Rev 13(Suppl 2):14–29

232. Muller MJ, Lagerpusch M, Enderle J, Schautz B, Heller M, Bosy-Westphal A (2012) Beyond the body mass index: tracking body composition in the pathogenesis of obesity and the metabolic syndrome. Obes Rev 13(Suppl 2):6–13

Chapter 10
Community-Based Strategies to Alter Energy Balance in Underserved Breast Cancer Survivors

Melinda Stolley

Abstract Breast cancer survival rates are significantly lower for African-American women compared to white women. Additionally, African-American women with breast cancer are more likely than other women to die from comorbid conditions including diabetes and hypertension. Such disparities are not easily explained and likely involve complex issues related to social injustices. However, obesity and behavioral factors may be additional contributors. Seventy-eight percent of African-American women are overweight or obese, and data suggest that many do not engage in regular physical activity and tend to have diets high in fat and low in vegetables and whole grains. The combined effects of obesity, unhealthy diet, and inactivity may contribute to the disparity in breast cancer survival between African-American and white women and may be the easiest modifiable factors to address in the near term. Although several weight loss interventions have reported beneficial results for breast cancer survivors, the inclusion of AA women has been extremely limited. This chapter presents a review of health behaviors among African-American breast cancer survivors, followed by a discussion of qualitative work exploring the beliefs, attitudes, barriers, and facilitators related to health behaviors and weight loss. A summary of interventions to date is provided as well as an in-depth look at one particular community-based intervention, Moving Forward.

Keywords Breast cancer • African-American women • Moving Forward • Weight loss • Health behaviors • Economically stressed neighborhoods • Weight loss barriers and facilitators • Exercise barriers and facilities • Community-based interventions • Socio-ecological model • Social cognitive therapy • Self-efficacy

M. Stolley (✉)
Division of Health Promotion Research, Department of Medicine, University of Illinois at Chicago (MC 275), 456 Westside Research Office Bldg., 1747 West Roosevelt Road, Chicago, IL 60608, USA
e-mail: mstolley@uic.edu

D.J. Bowen et al. (eds.), *Impact of Energy Balance on Cancer Disparities*,
Energy Balance and Cancer 9, DOI 10.1007/978-3-319-06103-0_10,
© Springer International Publishing Switzerland 2014

Breast cancer is the second leading cause of cancer death among African-American women [1]. Despite lower incidence, breast cancer mortality rates for Black women are higher than those for women of other races even after controlling for age, SES, tumor stage and histology, hormone receptor status, and menopausal status [2–4]. Additionally, African-American women with breast cancer are more likely to die from comorbid conditions including diabetes and hypertension [5, 6]. Energy balance contributes to breast cancer progression as well as the development and exacerbation of many comorbid conditions [7–12]. This association remains after adjusting for stage at diagnosis, nodal status, treatment type, and menopausal status prior to diagnosis [9, 13–15]. Obesity is thought to promote tumor progression by three primary mechanisms: (1) producing higher concentrations of estrogen and testosterone [3, 16, 17], contributing to (2) insulin resistance leading to increased levels of insulin-like growth factor-I (IGF-1) and insulin-like growth factor-binding protein-3 (IGFBP-3) [11, 18, 19], and (3) chronic inflammation [20]. Obesity is also related to increased risk for all-cause and cardiovascular mortality among women with breast cancer [8, 12]. Seventy-eight percent of African-American women are overweight or obese [21]. Given the associations between obesity, BC prognosis, and all-cause mortality, the prevalence of high body mass index (BMI) could be a significant factor in the lower survival rates in AA women [8, 22].

Although weight loss or prevention of further weight gain is important to many breast cancer patients [9, 23–25], it is often an elusive goal. Studies of white women show that rather than lose weight, most gain weight over the course of chemotherapy treatments [26–29]. This weight gain is related to time since diagnosis, post-menopausal status, adjuvant chemotherapy, current energy intake, and physical activity [26–30]. Little is known about posttreatment weight gain in Black women. However, the Women's Healthy Eating and Living Study Group (WHELS) (91 % white, 9 % African-American) showed that posttreatment weight gain was positively related to African-American ethnicity [30]. Moreover, average weight gain was significantly greater in African-American (13 lb) compared to white participants (6 lb) [30]. These results are concerning given the likelihood that many Black women are overweight or obese at the time of diagnosis [31].

Changing behaviors that contribute to obesity could benefit breast cancer survivors [9, 32, 33]. In particular, efforts to eat a higher quality, lower calorie diet and increase physical activity may help to reduce the risk for breast cancer recurrence [9, 34, 35], secondary cancers [36], and comorbid conditions [37] as well as improve quality of life [38–41]. Anecdotal evidence supports that being diagnosed with cancer can promote self-initiated changes in health behaviors [42–44]. For example, in a study of 126 breast cancer patients nearly 60 % made dietary changes, 30 % began a new physical activity, and 64 % took new dietary supplements since their diagnosis [45]. In another study of 250 breast cancer patients, 41 % reported dietary changes, with decreases in meat and increases in fruit/vegetable intake being the most frequently identified changes [46]. Participation in behavioral interventions may also promote healthful behavior changes among breast cancer survivors [47–52]. The Women's Intervention Nutrition Study (WINS) [48] reported decreases in dietary fat, while the WHELS reported decreases in dietary fat as well as increases in fruits and vegetables [49]. Other interventions that addressed sedentary lifestyles report increased physical activity [47, 50–58]. These results are encouraging, but efforts thus far have primarily targeted white women.

In this chapter, we focus on efforts to address weight loss and health behavior change among African-American breast cancer survivors. We begin with a discussion of health behaviors in survivors and qualitative work conducted to understand more about the beliefs, attitudes, barriers, and facilitators related to health behaviors and weight loss. We proceed to provide a review of community-based efforts including a description of the "Moving Forward" program followed by recommendations for future directions.

Health Behaviors, Weight, and Breast Cancer

Few studies have examined the diet and physical activity patterns of African-American breast cancer survivors. Those that have show high caloric and low fruit and vegetable consumption and low levels of physical activity [59–62]. In an analysis of dietary intake at baseline among minority women participating in the WHELS trial, Paxton and colleagues reported that African-Americans ($N = 118$) consumed significantly more calories from fat and less fruit than Asians or whites [62]. Dennis-Parker and colleagues examined compliance with national nutrition recommendations in 31 overweight African-American breast cancer survivors enrolled in a weight loss program. Although the majority of survivors were consuming the recommended daily servings of fruits and vegetables, they exceeded recommendations for energy intake from fat, saturated fat, and added sugars. Additionally, most did not meet the recommendations for whole grains and fiber intake [60]. Both studies reflect the dietary behaviors of women enrolled in intervention trials and thus are not generalizable to the general survivor population. Population-based studies are needed.

Low physical activity levels have been reported in two studies. The first analyzed baseline data of minority women participating in the WHELS trial. African-American women were less likely than other women to meet the guidelines for physical activity. Additionally, health-related quality of life was positively associated with physical activity [62]. The second study surveyed 468 African-American breast cancer survivors and reported similar results. The majority of women did not exercise regularly, and median television viewing was over 5 h daily [61].

Qualitative studies provide further insight into African-American survivors' thoughts and experiences regarding health behaviors, weight, and breast cancer [63–65]. Data support anecdotal evidence that many African-American women make behavioral changes following their diagnosis [63, 64]. For some, this stems from their beliefs that their eating and exercise behaviors contributed to their cancer.

> I'm becoming more conscious. I just know that all the fat I used to eat in Haagen Dazs probably contributed to my cancer. I just have to forget about all of that sweet stuff. It's not going to help me in the long run. So I am trying to give it up. Now I try to eat sherbet and keep up with my exercise—jazzercise 4 days a week. I know I should do more though.

Additional reasons for changing behavior include wanting to improve overall health, to feel better physically and emotionally, and to lose weight. Increasing fruit and vegetable intake, decreasing meat intake, choosing lower fat foods, and increasing physical activity are the most commonly adopted behaviors [63, 64]. The focus of dietary changes may differ by socioeconomic status. In one study, survivors with more education and higher incomes spoke of increasing food items that they had heard might reduce cancer risk [44]. Such foods included green tea, flaxseeds, foods with omega-3 fatty acids, soy, and whole grains.

> You might want to grind the flax seeds and sprinkle them on your yogurt. The dieticians that I've spoken to suggested that the seeds themselves rather than the oil are what help you.

Survivors with lower incomes were more focused on getting rid of specific "bad foods" including high-salt foods, fried foods, or sugar.

> I know that your immune system will fight whatever problems are in your body and what we have to do is to watch what we eat, what will make our immune system weaker, the chocolates, sweets, no raw sugar at all. I found that out. It depresses your immune system for like three hours.

In terms of exercise, qualitative data suggest that many African-American breast cancer survivors are aware of the multiple health benefits of regular physical activity. Those with a history of exercising also report on their improved quality of life. However, many report difficulty with initiating and/or maintaining a regular exercise program [63].

> I know that exercise is very important. I exercised a lot for weeks before my surgery (mastectomy) because I had to get my blood pressure down so I had to get the weight off. I did that and I kept doing that even when I was going through radiation. But now, the regular physical exercise that we all know we need to have, I don't do that. I am having trouble getting going again. I know that it is important to do to maintain good health especially if being a survivor of an illness.

Despite their efforts to make behavioral changes following their diagnosis, survivors relate feeling that they are not doing enough. For many, this frustration is related to the challenge of losing weight. Weight gain is common after breast cancer and serves as an important source of concern and distress [63–65].

> Now I find that I don't eat a lot, but the least little thing I am just picking up weight. And I thought, is it just me or is this something we all of us survivors go through you know this weight gain thing? And that sort of bothers me because I am not used to the extra weight I have gained, and it makes me feel so sluggish. It's miserable with this weight.

In addition to feeling "weighed down," survivors report being concerned about their appearance, not being able to fit into their clothes, needing to afford to buy new wardrobes, and, for some, about how the weight impacts their risk for a recurrence or comorbid conditions [63, 64]. Overall, the weight gain impacts quality of life leaving women feeling frustrated and helpless. Their helplessness may be due in part to not expecting the weight gain but also due to believing that the weight is a result of their hormone treatments.

> I've gained 25 pounds on hormone treatments. I mean 25 pounds. And my eating habits really haven't changed that much, I'm just tired of feeling tired all the time. I don't want to do anything. I work and come home. I'm too tired to go out with friends.

At the same time, survivors recognize that personal behaviors influence weight status and that making lifestyle changes will impact their weight as well as their overall health.

> . . . The only thing I want to say about hormone treatment is the jury with me is still out. Once I started exercising and working weight watchers I've lost 26 pounds and I'm still on the hormone treatment. So for me, I know now it was my high caloric intake and lack of exercise. Simple.

> I was on the hormone treatment for five years, I completed it and I gained the weight maybe the first couple of years I was on it. Then I was ok. When I started to exercise that I found out that no matter what you eat you've gotta have some kind of exercise to balance out what you eat.

As evidenced by their efforts to make behavioral changes after their cancer diagnosis, African-American breast cancer survivors are interested and motivated to adopt healthier eating and exercise patterns and lose weight. However, urban African-American women face multiple barriers in their quest to practice healthy lifestyles. For example, a higher percentage of African-Americans in the general population live in economically stressed neighborhoods where access to fresh fruits and vegetables may be limited or cost prohibitive [66, 67]. Alternatively, cheaper high-fat foods are easily accessible. In addition, opportunities for physical activity in disadvantaged communities are frequently limited by a lack of safe open spaces, sidewalks in disrepair, gang violence, poor lighting, and insufficient police [68]. Focus group data with African-American breast cancer survivors highlight other important barriers as well as facilitators. Barriers include pain, family, mood, and confusion [63, 64]. Facilitators include faith and spirituality, family and friend support, desire to reduce overall health risks, and risk of recurrence (Table 10.1).

Pain is mentioned in several qualitative studies with African-American breast cancer survivors [63–65]. Restricted range of motion as a result of surgery, joint and bone pain due to hormone treatments, and/or arthritis are common and interfere with physical activity.

> I just started really exercising because at first I didn't feel like doing it. And I was in a lot of pain too. So what I do now I just walk around the plaza about three miles you know, as much as I can. When I get tired I just stop. But you know my bones hurt. My body hurts and there's only so much you can do with your arms and I don't force myself to do anything that might hurt me.

Negative emotions such as anxiety, depression, and sadness are identified both as instigators for unhealthy eating and inhibitors of exercise. This may be particularly true for women who are retired or live alone. Additionally, for some women, the breast cancer experience is still quite upsetting.

Table 10.1 Barriers and facilitators to healthy lifestyle changes

Barriers	Facilitators
Access (easy access to cheap high-fat foods; difficult access to fresh fruits and vegetables and safe places to exercise	Faith and spirituality
Pain due to surgery, hormone treatments, arthritis	Family support
Negative emotions (anxiety, depression, sadness)	Friend support
Family food preferences	Group participation in exercise
Confusion about which dietary and physical activity recommendations to follow	Desire to reduce the risk of breast cancer recurrence and comorbid health condition risk

> I have fish and I steam my vegetables … but when I am not feeling so well about things, especially before my doctor's appointments, I grab a bag of Lays. It's like a comfort. I recognize that my eating habits tend to change with my emotions.

An additional barrier is confusion. Survivors relate feeling bombarded with information about the "right" way to eat to lose weight and/or stave off a cancer recurrence. Specific topics about which many have questions are organics, soy, supplements, dietary fat, and timing of meals. "When I got diagnosed, I was told by a dietician here to stay away from soy." While another survivor offers, "I work with a lot of Asian women who say you have to put soy in your diet and eat a lot of tofu." Survivors also express confusion about the best way to lose weight and what types of exercises are most helpful in terms of weight loss and reducing health risk.

Family is another significant barrier that is also a facilitator. As a barrier, family members' food preferences make it difficult to avoid eating high-fat and sweet foods. Also challenging is accommodating the traditional value of preparing preferred foods as an expression of love for friends and family. "I have grandkids and they really like fried food, so I cook it for them and then I eat it with them." The pervasive linking of food and family within African-American culture is illustrated in this comment, "Any celebration is a food celebration. Think of soul food, think of my family. Come over Sunday, FOOD." As a facilitator, survivors acknowledge that making good choices is easier when other family members and friends are making the same good choices. "It's easier for me to eat right because my husband eats right. He is not a junk food fanatic, he eats soy, vegetables, fruits and fish and he looks like a million." Providing motivation and company for exercise is an important enabler. "At first I didn't like it. I mean they were, "C'mon, c'mon." And I was, "No, no." But once I started keeping pace, I enjoyed it, I really did."

In keeping with the importance of the support of friends and family, African-American survivors make special note of the value of group participation when attempting to initiate or maintain lifestyle changes. This is particularly true for exercise.

> I think group activities are a help to me. When I can be a part of a group and be accountable or responsible to being there, it's a big motivation to me to have somebody other than myself. If I miss my group and they say where were you last week? That kind of thing helps you keep it going. It's very helpful to me in keeping the commitment to the exercise program.

Other facilitators such as faith and spirituality and the desire to reduce overall health risks and risk of recurrence also support efforts to change behaviors and lose weight. Historically, the African-American culture has relied on faith and spirituality to cope with difficult and painful experiences [69–71]. Breast cancer survivors relate that their faith enabled them to get through their breast cancer treatments and will now help them to lose the weight to improve their health.

> I took myself to the Lord and He lost my desires. Everything through Christ. I changed my life, I changed my desires. The Lord still has some more dealings with me, but I'll get there, we'll get there.

Concerns about comorbid health conditions such as diabetes and hypertension, along with worry about recurrence, are often the strongest motivator for initiating behavior change.

> For the first time my sugar level rose . . . so that was something that shocked me even more than the cancer. I have friends that are on dialysis and I don't want to go that route. So . . . now, I walk and I eat small meals.

These health concerns are equally important during slips or lapses.

> I think it's the realization that there are consequences and there is no avoiding that thought when I peer into the ice cream section at the grocery store . . .. I can't even enjoy it the way I used to because it's looming over me that I might be doing something to bring it back. And maybe this is what brought it the first time. There is no way for me to know.

Weight loss intervention efforts that acknowledge and address barriers while supporting facilitators are needed. Further, African-American breast cancer survivors want programs that provide holistic information on how to make realistic changes that can be incorporated into their lives [57]. They are not interested in diet or exercise patterns that are not sustainable. Survivors also want to know more about "psychological" strategies, such as mindful eating, controlling their environ ments (stimulus control), and relapse prevention.

Weight Loss Interventions for Breast Cancer Survivors

Several weight loss interventions have reported beneficial results for breast cancer survivors, including weight loss [54, 55, 57, 58, 72]; prevention of weight gain [47]; improved body composition and lipids [53, 55]; decreases in sex hormones [53]; decreases in dietary fat intake [47]; increases in fruit, vegetable, and/or fiber intake [47]; increased physical activity [47, 57]; and improved psychological status [47]. Until recently, the inclusion of African-American women was limited and no intervention had targeted African-American women [72–75].

Considering the high rates of breast cancer mortality, comorbidities, and obesity among African-American breast cancer survivors, weight loss is an important goal for women in this group. However, due to a complex interaction of behavioral, cultural, and societal factors, data suggest that African-American women are less likely to participate in traditional weight loss programs, more apt to drop out, and lose less weight than white women [76, 77]. To meet the needs of African-American breast cancer survivors, weight loss programs must consider and address the personal (e.g., preferred tastes, body image, hair concerns), interpersonal (social support, roles, and responsibilities), and environmental factors (access to and availability of food and physical activity resources) that influence their behaviors [63, 64, 78–80]. In recent years, several studies have examined the feasibility and efficacy of weight loss interventions for African-American breast cancer survivors [57, 72–74].

One such program examined the effects of an 8-week walking intervention based on tenets of the Health Belief Model [73]. Groups met weekly at a community location to discuss the benefits of and barriers to exercise, the relationship of exercise to health and cancer risk, and self-monitoring/problem solving related to motivation. Twenty-two of 24 African-American breast cancer survivors completed the program, and 95 % reported satisfaction with the number of sessions. Significant improvements were noted for steps per day (+3,506), body weight (−2.0 lb), percent body fat (−3.4), diastolic and systolic blood pressure, and exercise attitude. Seventeen participants completed a 3-month follow-up. The increase in steps per day was maintained; however weight and percent body fat had increased, but not significantly [73].

A second intervention trial explored the effects of the commercial *Curves* program on the weight of 42 survivors, 9 of whom were African-American [74]. *Curves* is a commercial exercise facility that offers a comprehensive weight loss program. The 6-month weight loss program included a 30-min exercise circuit and a high-vegetable/low-fat/calorie-restricted diet. Weekly dietary counseling was provided, but sessions took place at a university hospital, as opposed to the community-based *Curves*. Women in the treatment arm lost an average of 3.3 % (± 3.5 %) of their body weight, compared to 1.8 % (±2.9 %) of the waitlist control group. At a 6-month follow-up, the treatment group had regained some of their weight, such that there were no longer group differences. Participants noted that weight loss maintenance was challenging due to the costs associated with maintaining their membership at *Curves*.

Djuric and colleagues examined a spirituality-based weight loss maintenance intervention following a standard weight loss program [72]. Thirty-one obese African-American breast cancer survivors participated. The weight loss program included dietitian-led counseling by telephone, weight watchers coupons, and the recommendation that participants exercise at least 30 min most days each week. At the end of 6 months, 24 participants were randomized to either a dietitian-led maintenance program or one that included weekly spirituality counseling. The goal of the spirituality counseling was to address barriers to weight loss maintenance: (1) dealing with a crisis, (2) setting priorities, (3) coping with emotions that might lead to relapse, and (4) developing accountability for adhering to healthy diet and exercise patterns. Twenty-three women completed the program. Results were modest; 2.5 % and 2.6 kg weight loss was reported for the standard weight loss program. There was a slight regain at the 18-month follow-up, and no differences were noted between the two maintenance groups [72].

These studies are important in that they establish the feasibility and efficacy of weight loss interventions for African-American breast cancer survivors and address the needs of an underserved group. However, further work is needed to develop comprehensive programs that address diet and physical activity patterns as well as cognitive-behavioral strategies related to lifestyle changes. Basing the program within a community setting will promote sustainability. Attention to the psychosocial needs of breast cancer survivors is also important. For many African-American women, breast cancer is a topic not easily shared and thus incorporating support within the context of a weight loss program facilitates discussion. Finally, a significant limitation of the research to date is the lack of any data on the biological impact of weight loss for African-American breast cancer survivors. Weight loss trials with white breast cancer survivors support the positive impact of weight loss on intermediate markers of breast cancer including sex hormones (estrogen, estradiol, testosterone, sex hormone-binding globulin), chronic inflammation (C-reactive protein [CRP], interleukin-6 [IL-6], and TNF-α), and hyperinsulinemia (leptin, IGF-1, IGFBP3). These data, along with those for body composition (percent body fat vs. lean mass), are particularly important for African-American survivors given the historically low levels of weight loss observed in interventions. Furthermore, results from a prospective study of 278 overweight/obese postmenopausal women (38 %/105 African-American) not affected by breast cancer within the Weight Loss Maintenance Trial showed that African-American women exhibited higher levels of estrogen and testosterone concentrations, independent of adiposity [81]. Gathering body composition and biological data will enhance our understanding of how weight loss, even small amounts, impacts breast cancer recurrence risk and overall health risk among African-American women.

Moving Forward: A Community-Based Weight Loss Intervention

To address the limitations of intervention efforts to date, researchers in collaboration with African-American breast cancer survivors developed "Moving Forward," a weight loss program and intervention trial for African-American breast cancer survivors [57]. Preliminarily, focus groups informed the development of the intervention, after which an advisory board of African-American survivors reviewed and adapted the program. Twenty-three African-American survivors participated in a pilot study. Attendance and satisfaction data supported the acceptability and feasibility of the program with the majority of the women attending at least 75 % of the 52 classes and 87 % completing the 6-month program. Post-intervention differences were significant for weight (-5.57 lb), BMI (-1.0 kg/m^2), dietary fat intake (-23.6 g), vegetable consumption ($+1.6$ svgs/day), vigorous physical activity ($+23.6$ min/day), and social support. Although a nonsignificant increase of 20-min moderate physical activity was observed post-intervention, the combination of vigorous and moderate physical activity (>40 min/day) represented an important shift in behavior. Quality of life, as measured by the FACT-B and ES, was high at baseline, and thus no significant change was noted post-intervention.

Based on feedback from the pilot study, the investigators and advisory board once again reviewed and adapted the intervention. Primary changes included adding more information and activities related to how lifestyle and weight status impact breast cancer recurrence and comorbidity risk; integrating more practical advice related to grocery shopping and meal planning; and basing the program within the community at a location with affordable exercise facilities. Following we provide a description of the *Moving Forward* program and a description of the methods for the intervention trial.

Program Structure

Moving Forward is a 6-month program that meets twice weekly. The program is conducted in partnership with a city park district. The program is conducted in city park district facilities where participants enjoy reduced-fee or free memberships, ongoing access to classes and fitness rooms, and the opportunity to maintain contact with program participants once the program concludes. The content of the intervention is designed to provide participants with the tools they need to make independent changes in important health behaviors. Table 10.2 provides a list of weekly curriculum topics. The first meeting each week includes a 60-min class that addresses *knowledge* (e.g., relationship between obesity and breast cancer; food label reading; portions; available healthy living community resources), *attitudes* (e.g., pros and cons of weight loss; understanding the roles that food plays in one's life; the concept of fail to plan, plan to fail), and *cognitive behavioral strategies* including self-monitoring of weight, food, and physical activity; realistic goal setting; stimulus control; problem solving; mindfulness; cognitive restructuring; and relapse prevention. Pilot data showed that many women entered the program with low levels of knowledge about healthy eating and exercise. Thus, the first weeks are devoted to teaching core concepts (e.g., concept of calories in/out; food label reading; measuring heart rate). Other class activities include weekly weigh-in; completing a food and activity self-monitoring record for the current day; increasing awareness of portions by weighing and measuring foods according to one's typical portions and then according to recommended portions; creating stimulus control plans for home, car, and work; identifying barriers to healthy eating and/or exercise and problem solving within small groups; going on a field trip to a local grocery store to practice reading food labels; creating an eating out management plan; and identifying high-risk situations and brainstorming ways to manage them. The first weekly meeting also includes a support "icebreaker" (share the funniest moment of your breast cancer journey; what has been the most frustrating; etc.) and a 60-min exercise class taught by a certified cancer exercise trainer.

Table 10.2 Moving Forward—weekly curriculum topics

Week 1	Introduction to program
Week 2	Self-monitoring and goal setting
Week 3	Using self-monitoring tools to make better choices
Week 4	Energy requirements
Week 5	Reading food labels and monitoring heart rate
Week 6	Measuring portions
Week 7	Breakfast and water—two key tools to losing weight
Week 8	Healthy grocery shopping
Week 9	Meal planning
Week 10	Holiday eating (moved according to when holiday falls)
Week 11	Stimulus control
Week 12	Mindful eating
Week 13	Eating away from home—restaurant and party strategies
Week 14	Program review—where were you, where are you now
Week 15	Building movement into your daily life
Week 16	Barriers to healthy eating and exercise
Week 17	Problem solving
Week 18	The power of habit
Week 19	Strategies to increase fruits and vegetables
Week 20	Where you were, where you are, and where you plan to go
Week 21	Relapse prevention I—what is a lapse vs. relapse
Week 22	Relapse prevention II—identifying high-risk situations
Week 23	Relapse prevention III—maintaining a physically active lifestyle
Week 24	Relapse prevention IV—motivation to maintain changes
Week 25	Transitioning from Moving Forward to being on your own
Week 26	Graduation

The second meeting each week is a stand alone 60-min exercise class taught by a certified cancer exercise trainer. The exercise classes incorporate a variety of activities, including traditional aerobics, line dancing, African dance, salsa, yoga, Pilates, and strength and flexibility training. Class time is also spent learning to use the park district fitness facility equipment to ensure that women feel comfortable and competent on the equipment, thus promoting enhanced self-efficacy and mastery of new skills. Many participants enter the program at very low levels of fitness; therefore, physical activity levels are increased gradually with special attention to concerns such as lymphedema and balance. Increased physical activity outside of class is encouraged by suggesting enrollment in additional exercise classes, providing safe outdoor walking routes, and alerting women to activity resources online and on Fit TV or other television channels that provide regular fitness and exercise programs.

Participants often need further support and reinforcement of lifestyle changes outside of class as well as timely information related to healthy eating and exercise resources. To do this, Moving Forward uses text messaging, a strategy successfully used in previous weight loss interventions with low-income African-American women [82, 83]. All participants receive three text messages each week. Messages are 200 characters in length and are written to be brief, clear, and motivational. The intent is to reinforce concepts covered in class while also supporting self-efficacy, social support, and perceived access.

Intervention Goals

The overall goal of Moving Forward is to make independent changes in health behaviors to promote a healthy weight. The weight loss goal is 7 % of baseline body weight (1–2 lb/week) consistent with the recommendations of an expert panel at NIH [84]. Dietary goals aimed at producing weight loss, decreasing BC recurrence risk, and improving overall health include (1) a decrease in daily caloric intake (based on weight in pounds × 12 kcal/day with 500–750 cal subtracted to create an energy deficit); (2) a decrease in dietary fat consumption to 20 % of total calories; (3) an increase in fruit and vegetable consumption to seven daily servings; and (4) an increase in fiber to 25 g/day. For exercise, participants will gradually increase their activity to a minimum of 180 min/week at 55–65 % maximal heart rate.

Theoretical Framework

The Moving Forward intervention integrates concepts from social cognitive theory (SCT) [85] and the socio-ecological model (SEM) [86, 87] to promote independent behavior change. SCT suggests that behavior can be explained by the dynamic interaction between behavior, personal factors (e.g., self-efficacy), and the environment (e.g., social support). Self-efficacy is a person's confidence in performing a particular behavior and overcoming barriers to that behavior. A number of studies have supported the mediating role of self-efficacy in making independent health behavior changes [88–92]. The intervention also incorporates tenets of the SEM [86, 87], a model that goes beyond individual-level variables and emphasizes that support from the larger social context is needed for long-term behavior change [93]. Accordingly, SEM posits that weight status, diet, and physical activity are influenced by individual (e.g., beliefs, taste preferences), interpersonal (e.g., social support, traditions, and role expectations), and community factors (e.g., access to resources that support health promotion) [94]. Interventions hoping to promote long-term behavior change must address these three levels of influence [95, 96]. Moving Forward accomplishes this by addressing (1) *individual* factors—acknowledging heavier body image ideals and the importance of hairstyles; (2) *interpersonal*—the importance of food in the African-American culture and finding ways to integrate this value with healthful eating; providing low-fat versions of culturally traditional "soul food" recipes; acknowledging and addressing family roles and family resistance/support to change; providing information on integrating healthful lifestyle practices for the family; facilitating social support for making changes in diet, physical activity, and weight; and understanding the important role of religion and worship in the women's lives and how it affects their health perspectives; and (3) *community*—incorporating a sustainable link to a community physical activity resource that can address barriers to regular physical activity (i.e., safety, weather, access); problem solving around cost and availability of healthy food; and introducing participants to unfamiliar community resources. Interestingly, a positive sense of community (e.g., social bonds between individuals and between individuals and their community) is associated with self-efficacy for physical activity among African-American women [97].

Based on these theoretical models, the content and structure of Moving Forward were developed to address mediators of long-term behavior change by enhancing self-efficacy, increasing social support, and addressing perceived access to healthy eating and exercise resources. Self-efficacy is enhanced by (1) teaching participants the information and cognitive behavioral skills necessary to make healthy changes; (2) modeling the process of structuring behavior change into incremental steps that participants can realistically achieve; (3) providing opportunities for repeated practice and mastery of skills and behaviors that support weight loss; (4) having participants share testimonials of their achievements; and (5) fostering social support.

Increased social support is facilitated by (1) creating a cohesive community of African-American breast cancer survivors who are working toward a common goal; (2) providing opportunities to tell stories related to their breast cancer and weight loss experiences; (3) sharing potluck dinners in honor of holidays and significant events with group participants and their family members; (4) participating regularly as a group (with family members) in local breast cancer events such as walks; (5) creating a "Moving Forward" Facebook page that will continue to be maintained and updated once the program is completed; (6) providing contact information (with permission) to enable participants to text or call each other; (7) exercise buddies; (8) CPD sites devoting a section of their bulletin boards to address BC information; and (9) sending motivating text messages.

Perceived access to environmental resources is improved by (1) showing participants strategies for learning about healthy eating and exercise resources in their home and work communities via computer searches and media ads; (2) providing participants with a list and description of local healthy eating and exercise resources; (3) problem solving around access issues to healthy eating and exercise resources; (4) creating individual and group plans to overcome barriers to resources (including identifying safe, reliable transportation); (5) increasing familiarity with resources by doing class visits or having community resource staff come to class; (6) Park district postings on local resources, store specials; and (7) text messages to announce/remind participants about sales/events at local sites.

Intervention Trial Methodology: Details of the Moving Forward Trial Are Provided Below

Study Design

We are conducting a randomized study with 240 African-American diagnosed with stage I, II, or III breast cancer. Study aims include evaluating the effects of a guided weight loss program (Moving Forward) on the BMI, body composition, and waist: hip circumference of overweight/obese African-American breast cancer survivors. Diet and physical activity patterns, intermediate markers of breast cancer recurrence (i.e., estradiol, estrone, testosterone, leptin, C-peptide, IGF-1, IGFBP-3, SHBG, and CRP), fatigue, and quality of life will also be examined. The study will be based in six predominantly African-American communities in Chicago, and the intervention will be conducted at Park district facilities. Forty African-American breast cancer survivors will be recruited from each community area (20 guided, 20 self-guided). Figure 10.1 provides an overview of the study design, and Fig. 10.2 provides an overview of the conceptual framework for the study.

Procedure

Women who respond to recruitment efforts will complete a brief telephone interview to verify eligibility. Once eligibility is established, all participants are asked to complete a 75-min pre-intervention interview plus a blood draw, blood pressure measurement, and height and weight measurements. Subsequently, participants are randomly assigned to one of the two 6-month interventions: (1) Moving Forward guided weight loss intervention (MF) or (2) Moving Forward self-guided weight loss intervention (SG) that includes a program binder with all information, activities, and supplies, but no participation in classes or exercise sessions. Following the intervention and at a 6-month follow-up, participants from both groups complete a 75-min post-intervention interview plus a blood draw, blood pressure measurement, and weight measurement.

Recruitment

Recruitment centers around a number of community and institutional partners including breast cancer support organizations, hospital cancer registries, local churches, community leaders such as aldermen, block clubs, and community centers. Thus far, the most effective mode has been direct contact with survivors through support groups or hospital registries. An equally effective strategy is

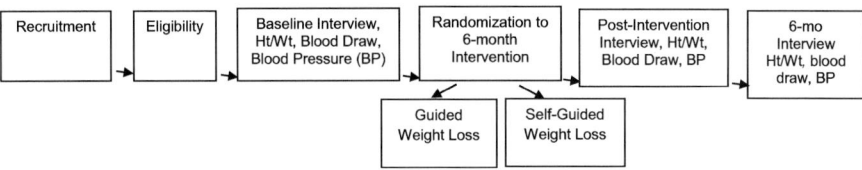

Fig. 10.1 Study design

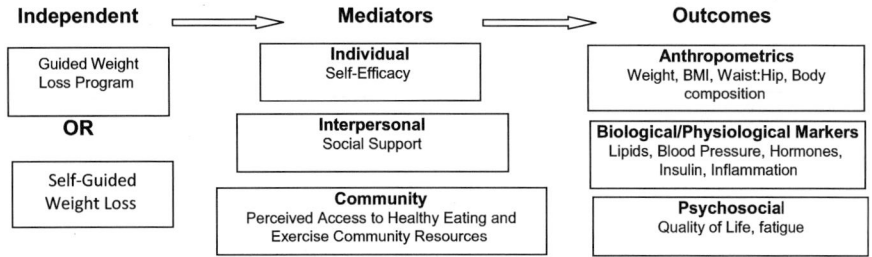

Fig. 10.2 Study conceptual framework

hosting educational events about breast cancer at community venues such as police stations, churches, and community centers. Community leaders and stakeholders are invited, and lunch is served. Presentations address breast cancer screening (presented by a radiologist from a community hospital), diagnosis, what happens during treatment, life after treatment, and, finally, a brief review on Moving Forward and survivorship.

Eligibility

Inclusion Criteria

Inclusion criteria include the following: (1) self-identification as Black or AA (including individuals who are biracial but identify themselves as Black or AA); (2) female; (3) stage I, II, and III invasive breast carcinoma; (4) treatment (surgery, chemotherapy, and/or radiation) completed at least 6 months *prior to* recruitment (ongoing treatment with tamoxifen or aromatase inhibitors is acceptable); (5) age 18 or above at the time of diagnosis; (6) BMI at least 25 kg/m^2—chosen because this includes only those participants who are overweight and would not be harmed by a 7 % weight loss; (7) physically able to participate in a moderate physical activity program as assessed by a screening questionnaire and PCP approval; (8) agreeable to random assignment and data collection including blood draw; and (9) able to attend twice-weekly classes for 6 months.

Exclusion Criteria

Exclusion criteria include the following: (1) plans to move from the community during the study; (2) medical condition limiting adherence as assessed by PCP; (3) history of significant mental illness; (4) currently pregnant, less than 3 months postpartum, or pregnancy anticipated during the study; (5) current/planned use of an FDA-approved or over-the-counter weight loss medication; or (6) participation in another structured weight loss program.

Measures

Demographics

Demographic data will include name, address, date of birth, marital status, number of children, education, occupational status, annual income, and insurance status.

Breast Cancer Data

Diagnosis and treatment history are collected from the treating oncologist.

Comorbid Condition Rating Scale

The Modified Cumulative Illness Rating Scale [98, 99] classifies comorbidities by 14 organ systems that may be affected and rates them according to severity from 0 to 4. This measure can generate four ratings including total score, number of categories endorsed, severity index (total score/number of categories endorsed), and number of categories at level 3.

Self-Efficacy for Eating and Exercise Behaviors [100]

This measure consists of 12 items that assess exercise self-efficacy (e.g., beliefs about one's ability to exercise five times a week despite barriers) and 20 items that assess healthy eating self-efficacy (e.g., eating low-fat foods and healthy portions despite high-fat temptations).

Social Support for Eating and Exercise [101]

This questionnaire asks respondents to rate on a five-point scale (1 = none, 5 = very often) the frequency that friends and family have done or said certain things related to the respondents' efforts to change dietary or exercise habits. The social support for eating survey includes ten items and two subscales (i.e., encouragement and discouragement) each for friends and family.

Perceived Access to Healthy Eating and Exercise

Respondents rate their level of agreement to statements related to access to physical activity resources and five items related to healthy eating resources. There are also five items asking about the availability of activity-related facilities that require a yes/no response. These questionnaires were used in two studies with urban minority populations [102, 103].

Body Mass Index

Height (baseline only) is assessed using a portable stadiometer. Weight will be assessed using a Seca company digital scale with participants wearing light clothes and no shoes. BMI is calculated as weight (kg)/height (m)^2.

Waist-to-Hip Ratio

Waist-to-hip ratio (WHR) is measured with participants standing without outer garments and with empty pockets. Waist is measured at the level midway between the lower rib margin and the iliac crest, with the participant breathing out gently. Hip is recorded as the maximum circumference over the buttocks.

Diet: Brief Block 98 Food Frequency Measure

While there is no gold standard for dietary assessment and all dietary assessments are prone to underreporting [104], a combination of 24-h recalls and a food frequency questionnaire (FFQ) may be sufficient to address the concern of underreporting. However, this method is costly, burdensome for participants, and not necessary for determining mean consumption of dietary components for a group of participants. Our goal is to determine group means for consumption of energy, fruits and vegetables, fat, and fiber. A semiquantitative FFQ is the most appropriate tool in this case [105–107]. The Block 2005 Food Frequency Questionnaire [108] estimates the usual intake of a wide array of nutrients and food groups. Reliability and validity have been established for the measure in a wide range of age, gender, income, and ethnic groups [109, 110].

Physical Activity (Self-Report and Objective)

The Modified Activity Questionnaire (MAQ) [111] assesses self-reported leisure and occupational activity, television viewing, and inactivity due to disability. For leisure activity, respondents review a list of 29 popular activities (e.g., walking, gardening) and select those that they performed on at least ten occasions in the last year. Respondents then provide information on average frequency and duration for each activity. For occupational activity, respondents provide information on common activities performed at work and transportation to/from work. The MAQ has been used in many large studies with diverse samples, including cancer survivors [112], and has well-established reliability and validity [111].

Accelerometer

The limitations of self-reported physical activity are well established [113]. Therefore, the ActiGraph GT1M activity monitor is used to obtain an objective measure of physical activity. The ActiGraph is a small, lightweight accelerometer designed to detect normal body motion. Participants are asked to wear the ActiGraph during waking hours for 7 days. Only days on which the participant wore the accelerometers for at least 10 h are included; participants with fewer than four valid days are excluded. Thresholds suggested by Troiano and colleagues will be used to calculate the amount of time spent in moderate and vigorous physical activity [114].

Biological/Physiological Markers of BC Progression

Markers for three proposed mechanisms by which obesity may contribute to BC progression are being measured. These include levels of sex hormones (markers: estradiol, estrogen, sex hormone-binding globulin, testosterone), hyperinsulinemia (markers: leptin, IGF-I, IGBP3, C-peptide), and chronic inflammation (markers: IL-6, CRP) [115].

Biological/Physiological Markers of Comorbidities

A fasting blood sample is also drawn (at the same time as sample above) for lipid profile analysis (HDL, LDL, triglycerides) as a marker of dyslipidemia and HbA1c as a marker of diabetes. Diastolic and systolic blood pressure are also measured using a standard protocol.

Psychosocial (Quality of Life and Fatigue)

The SF-36-item short-form health survey [116] assesses eight health concepts: (1) limitation in physical activities due to health problems, (2) limitations in social activities due to physical or emotional problems, (3) limitations in usual role activities due to physical health problems, (4) bodily pain, (5) general mental health (psychological distress and well-being), (6) limitations in usual role activities because of emotional problems, (7) vitality (energy and fatigue), and (8) general health perceptions. This instrument has been widely used with diverse healthy and clinical populations and has good reliability and validity [117–119]. Specific scales within the Functional Assessment of Cancer Therapy series are also administered: the FACT-B (assesses effects of breast cancer treatment), the FACT-ES (assesses the side effects and putative benefits of hormonal treatments for breast cancer), and FACT-F (assesses cancer-related fatigue) [120, 121].

In sum, the Moving Forward intervention trial seeks to address the limitations of the literature to date by examining the impact of a community-based weight loss intervention on physical, biological, and psychosocial outcomes.

Conclusions

African-American women with breast cancer are more likely to die from breast cancer and comorbid conditions than women with breast cancer of other races. The combined effects of obesity, diet, and physical inactivity may contribute to this disparity. Developing interventions that address these risk factors in African-American women is an important public health goal. Although such interventions have been shown to be feasible and effective with white women [47, 54–56, 58], limited efforts have been initiated for African-American women [72]. Four interventions have established the feasibility and efficacy of weight loss program for African-American women. Results reinforce the importance of offering comprehensive, community-based programs that can be sustained once the evaluation is completed. Further efforts to address the needs of African-American survivors are warranted, as are data that provide insight into the biological and physiological impact of energy balance interventions.

References

1. American Cancer Society (2004) Cancer facts and figures 2004. American Cancer Society, Atlanta, GA
2. Joslyn SA, West MM (2000) Racial differences in breast carcinoma survival. Cancer 88:114–123

3. Newman LA, Griffith KA, Jatoi I et al (2006) Meta-analysis of survival in African American and White American patients with breast cancer: ethnicity compared with socioeconomic status. J Clin Oncol 24:1342–1349
4. Surveillance Epidemiology and End Results (SEER) Program
5. Tammemagi CM, Nerenz D, Neslund-Dudas C et al (2005) Comorbidity and survival disparities among Black and White patients with breast cancer. JAMA 294:1765–1772
6. Eley JW, Hill HA, Chen VW et al (1994) Racial differences in survival from breast cancer. Results of the National Cancer Institute Black/White Cancer Survival Study. JAMA 272:947–954
7. McCullough ML, Feigelson HS, Diver WR et al (2005) Risk factors for fatal breast cancer in African-American women and White women in a large US prospective cohort. Am J Epidemiol 162:734–742
8. McKenzie F, Jeffreys M (2009) Do lifestyle or social factors explain ethnic/racial inequalities in breast cancer survival? Epidemiol Rev 31:52–66
9. Chlebowski RT, Aiello E, McTiernan A (2002) Weight loss in breast cancer patient management. J Clin Oncol 20:1128–1143
10. Dignam JJ, Wieand K, Johnson KA et al (2003) Obesity, tamoxifen use, and outcomes in women with estrogen receptor-positive early-stage breast cancer. J Natl Cancer Inst 95:1467–1476
11. Stephenson GD, Rose DP (2003) Breast cancer and obesity: an update. Nutr Canc 45:1–16
12. Nichols HB, Trentham-Dietz A, Egan KM et al (2009) Body mass index before and after breast cancer diagnosis: associations with all-cause, breast cancer, and cardiovascular disease mortality. Cancer Epidemiol Biomarkers Prev 18:1403–1409
13. Reeves GK, Patterson J, Vessey MP et al (2000) Hormonal and other factors in relation to survival among breast cancer patients. Int J Cancer 89:293–299
14. Vatten LJ, Foss OP, Kvinnsland S (1991) Overall survival of breast cancer patients in relation to preclinically determined total serum cholesterol, body mass index, height and cigarette smoking: a population-based study. Eur J Cancer 27:641–646
15. Kyogoku S, Hirohata T, Takeshita S et al (1990) Survival of breast-cancer patients and body size indicators. Int J Cancer 46:824–831
16. Endogenous Hormones Breast Cancer Collaborative G (2003) Body mass index, serum sex hormones, and breast cancer risk in postmenopausal women. J Nat Canc Inst 95:1218–1226
17. McTiernan A, Rajan KB, Tworoger SS et al (2003) Adiposity and sex hormones in post-menopausal breast cancer survivors. J Clin Oncol 21:1961–1966
18. Goodwin PJ, Ennis M, Pritchard KI et al (2002) Fasting insulin and outcome in early-stage breast cancer: results of a prospective cohort study. J Clin Oncol 20:42–51
19. Blackburn GL, Wang KA (2007) Dietary fat reduction and breast cancer outcome: results from the Women's Intervention Nutrition Study (WINS). Am J Clin Nutr 86:878S–881S
20. Pierce BL, Ballard-Barbash R, Bernstein L et al (2009) Elevated biomarkers of inflammation are associated with reduced survival among breast cancer patients. J Clin Oncol 27:3437–3444
21. Flegal KM, Graubard BI, Williamson DF et al (2010) Sources of differences in estimates of obesity-associated deaths from first National Health and Nutrition Examination Survey (NHANES I) hazard ratios. Am J Clin Nutr 91:519–527
22. Chlebowski RT, Chen Z, Anderson GL et al (2005) Ethnicity and breast cancer: factors influencing differences in incidence and outcome. J Natl Cancer Inst 97:439–448
23. Newman S, Miller A, Howe G (1986) A study of the effect of weight and dietary fat on breast cancer survival time. Am J Epidemiol 123:767–774
24. Senie RT, Rosen PP, Rhodes P et al (1992) Obesity at diagnosis of breast carcinoma influences duration of disease-free survival. Ann Intern Med 116:26–32
25. Boyd N, Campbell J, Germanson T et al (1981) Body weight and prognosis in breast cancer. J Natl Cancer Inst 67:785–789

26. Camoriano JK, Loprinzi CL, Ingle JN et al (1990) Weight change in women treated with adjuvant therapy or observed following mastectomy for node-positive breast cancer. J Clin Oncol 8:1327–1334
27. Goodwin PJ, Panzarella T, Boyd NF (1988) Weight gain in women with localized breast cancer – a descriptive study. Breast Cancer Res Treat 11:59–66
28. Heasman KZ, Sutherland HJ, Campbell JA et al (1985) Weight gain during adjuvant chemotherapy for breast cancer. Breast Canc Res Treat 5:195–200
29. Demark-Wahnefried W, Rimer BK, Winer EP (1997) Weight gain in women diagnosed with breast cancer. J Am Diet Assoc 97:519–526, 529; quiz 527–8
30. Rock CL, Flatt SW, Newman V et al (1999) Factors associated with weight gain in women after diagnosis of breast cancer. Women's Healthy Eating and Living Study Group. J Am Diet Assoc 99:1212–1221
31. Flegal KM, Carroll MD, Ogden CL et al (2002) Prevalence and trends in obesity among US adults, 1999–2000 [comment]. JAMA 288:1723–1727
32. Demark-Wahnefried W, Morey MC, Clipp EC et al (2003) Leading the Way in Exercise and Diet (Project LEAD): intervening to improve function among older breast and prostate cancer survivors. Control Clin Trials 24:206–223
33. Rock CL, Demark-Wahnefried W (2002) Can lifestyle modification increase survival in women diagnosed with breast cancer? J Nutr 132:3504S–3507S
34. Adams-Campbell LL, Rosenberg L, Rao RS et al (2001) Strenuous physical activity and breast cancer risk in African-American women. J Natl Med Assoc 93:267–275
35. Hoffman-Goetz L, Apter D, Demark-Wahnefried W et al (1998) Possible mechanisms mediating an association between physical activity and breast cancer. Cancer 83:621–628
36. American Cancer Society (2003) Cancer facts & figures for African Americans 2003–2004. American Cancer Society, Atlanta, GA
37. Mitchell BD, Stern MP, Haffner SM et al (1990) Risk factors for cardiovascular mortality in Mexican Americans and non-Hispanic whites. Am J Epidemiol 131:423–432
38. Tangney CC, Young JA, Murtaugh MA et al (2002) Self-reported dietary habits, overall dietary quality and symptomatology of breast cancer survivors: a cross-sectional examination. Breast Canc Res Treat 71:113–123
39. Bowen DJ, Kestin M, McTiernan A et al (1995) Effects of dietary fat intervention on mental health in women. Cancer Epidemiol Biomarkers Prev 4:555–559
40. Courneya KS, Keats MR, Turner AR (2000) Physical exercise and quality of life in cancer patients following high dose chemotherapy and autologous bone marrow transplantation. Psychooncology 9:127–136
41. Courneya KS, Friedenreich CM (1997) Relationship between exercise during cancer treatment and current quality of life in survivors of breast cancer. J Psychol Oncol 15:35–57
42. Spencer SM, Carver CS, Price AA (1998) Psychological and social factors in adaptation. In: Holland JC (ed) Psycho-oncology. Oxford University Press, New York, NY, pp 211–222
43. Reardon KK, Aydin CE (1993) Changes in lifestyle initiated by breast cancer patients: who does and who doesn't? Health Commun 5:263–282
44. Taylor SE, Lichtman RR, Wood JV (1984) Attributions, beliefs about control, and adjustment to breast cancer. J Pers Soc Psychol 46:489–502
45. Patterson RE, Neuhouser ML, Hedderson MM et al (2003) Changes in diet, physical activity, and supplement use among adults diagnosed with cancer. J Am Diet Assoc 103:323–328
46. Maunsell E, Drolet M, Brisson J et al (2002) Dietary change after breast cancer: extent, predictors, and relation with psychological distress. J Clin Oncol 20:1017–1025
47. Goodwin P, Esplen MJ, Butler K et al (1998) Multidisciplinary weight management in locoregional breast cancer: results of a phase II study. Breast Cancer Res Treat 48:53–64
48. Chlebowski RT, Blackburn GL, Buzzard IM et al (1993) Adherence to a dietary fat intake reduction program in postmenopausal women receiving therapy for early breast cancer. The Women's Intervention Nutrition Study [see comments]. J Clin Oncol 11:2072–2080

49. Pierce JP, Faerber S, Wright FA et al (1997) Feasibility of a randomized trial of a high-vegetable diet to prevent breast cancer recurrence. Nutr Canc 28:282–288
50. Berglund G, Bolund C, Gustafsson UL et al (1994) One-year follow-up of the 'Starting Again' group rehabilitation programme for cancer patients. Eur J Cancer 30A:1744–1751
51. Mock V, Dow KH, Meares CJ et al (1997) Effects of exercise on fatigue, physical functioning, and emotional distress during radiation therapy for breast cancer. Oncol Nurs Forum 24:991–1000
52. Segar ML, Katch VL, Roth RS et al (1998) The effect of aerobic exercise on self-esteem and depressive and anxiety symptoms among breast cancer survivors. Oncol Nurs Forum 25:107–113
53. McTiernan A, Ulrich C, Kumai C et al (1998) Anthropometric and hormone effects of an eight-week exercise-diet intervention in breast cancer patients: results of a pilot study. Cancer Epidemiol Biomarkers Prev 7:477–481
54. Djuric Z, DiLaura NM, Jenkins I et al (2002) Combining weight-loss counseling with the weight watchers plan for obese breast cancer survivors. Obes Res 10:657–665
55. Mefferd K, Nichols J, Pakiz B et al (2007) A cognitive behavioral therapy intervention to promote weight loss improves body composition and blood lipid profiles among overweight breast cancer survivors. Breast Cancer Res Treat 104:145–152
56. Saxton JM, Daley A, Woodroofe N et al (2006) Study protocol to investigate the effect of a lifestyle intervention on body weight, psychological health status and risk factors associated with disease recurrence in women recovering from breast cancer treatment. BMC Cancer 6:35
57. Stolley MR, Sharp LK, Oh A et al (2009) A weight loss intervention for African American breast cancer survivors, 2006. Prev Chronic Dis 6:A22
58. de Waard F, Ramlau R, Mulders Y et al (1993) A feasibility study on weight reduction in obese postmenopausal breast cancer patients. Eur J Cancer Prev 2:233–238
59. Paxton RJ, Phillips KL, Jones LA et al (2012) Associations among physical activity, body mass index, and health-related quality of life by race/ethnicity in a diverse sample of breast cancer survivors. Cancer 118:4024–4031
60. Dennis Parker EA, Sheppard VB, Adams-Campbell L (2013) Compliance with national nutrition recommendations among breast cancer survivors in "Stepping Stone". Integr Cancer Ther 12(2):114–120
61. Paxton RJ, Taylor WC, Chang S et al (2013) Lifestyle behaviors of African American breast cancer survivors: a Sisters Network, Inc. study. PLoS One 8:e61854
62. Paxton RJ, Jones LA, Chang S et al (2011) Was race a factor in the outcomes of the women's health eating and living study? Cancer 117:3805–3813
63. Stolley MR, Sharp LK, Wells AM et al (2006) Health behaviors and breast cancer: experiences of urban African American women. Health Educ Behav 33:604–624
64. Weathers B, Frances K, Barg A et al (2006) Perceptions of changes in weight among African American breast cancer survivors. Psychooncology 15:174–179
65. Halbert CH, Weathers B, Esteve R et al (2008) Experiences with weight change in African-American breast cancer survivors. Breast J 14:182–187
66. Fitzgibbon ML, Stolley MR (2004) Environmental changes may be needed for prevention of overweight in minority children. Pediatr Ann 33:45–49
67. Burdette HL, Whitaker RC (2004) Neighborhood playgrounds, fast food restaurants, and crime: relationships to overweight in low-income preschool children. Prev Med 38:57–63
68. Ross CE, Mirowsky J (2001) Neighborhood disadvantage, disorder, and health. J Health Soc Behav 42:258–276
69. Lackey NR, Gates MF, Brown G (2001) African American women's experiences with the initial discovery, diagnosis, and treatment of breast cancer. Oncol Nurs Forum 28:519–527
70. Potts RG (1996) Spirituality and the experience of cancer in an African-American community: implications for psychosocial oncology. J Psychosoc Oncol 14:1–19

71. Post-White J, Ceronsky C, Kreitzer MJ et al (1996) Hope, spirituality, sense of coherence, and quality of life in patients with cancer. Oncol Nurs Forum 23:1571–1586

72. Djuric Z, Mirasolo J, Kimbrough LV et al (2009) A pilot trial of spirituality counseling for weight loss maintenance in African American breast cancer survivors. J Natl Med Assoc 101:552

73. Wilson DB, Porter JS, Parker G et al (2005) Anthropometric changes using a walking intervention in African American breast cancer survivors: a pilot study. Prev Chronic Dis 2:A16

74. Greenlee HA, Crew KD, Mata JM et al (2013) A pilot randomized controlled trial of a commercial diet and exercise weight loss program in minority breast cancer survivors. Obesity (Silver Spring) 21:65–76

75. Spector D, Deal AM, Amos KD et al (2013) A pilot study of a home-based motivational exercise program for African American breast cancer survivors: clinical and quality-of-life outcomes. Integr Cancer Ther 13(2):121–132

76. Foster GD, Wadden TA, Swain RM et al (1999) Changes in resting energy expenditure after weight loss in obese African American and white women. Am J Clin Nutr 69:13–17

77. Kumanyika S (2002) Obesity treatment in minorities. In: Wadden T, Stunkard A (eds) Handbook of obesity treatment. Guilford Press, New York, NY, pp 416–446

78. Kumanyika SK, Whitt-Glover MC, Gary TL et al (2007) Expanding the obesity research paradigm to reach African American communities. Prev Chronic Dis 4:A112

79. Davis EM, Clark JM, Carrese JA et al (2005) Racial and socioeconomic differences in the weight-loss experiences of obese women. Am J Public Health 95:1539–1543

80. Bronner Y, Boyington JE (2002) Developing weight loss interventions for African-American women: elements of successful models. J Natl Med Assoc 94:224–235

81. Stolzenberg-Solomon RZ, Falk RT, Stanczyk F et al (2012) Sex hormone changes during weight loss and maintenance in overweight and obese postmenopausal African-American and non-African-American women. Breast Cancer Res 14:R141

82. Gerber BS, Stolley MR, Thompson AL et al (2009) Mobile phone text messaging to promote healthy behaviors and weight loss maintenance: a feasibility study. Health Informatics J 15:17–25

83. Stolley MR, Fitzgibbon ML, Schiffer L et al (2009) Obesity Reduction Black Intervention Trial (ORBIT): six-month results. Obesity 17:100–106

84. NIH/NHLBI (1998) Clinical guidelines on the identification, evaluation, and treatment of overweight and obesity in adults – the evidence report. Obes Res 6:51S–209S

85. Bandura A (1986) Social foundations of thought and action. Prentice-Hall, Englewood Cliffs, NJ

86. Stokols D (1996) Translating social ecological theory into guidelines for community health promotion. Am J Health Promot 10:282–298

87. Richard L, Potvin L, Kishchuk N et al (1996) Assessment of the integration of the ecological approach in health promotion programs. Am J Health Promot 10:318–328

88. Martin PD, Dutton GR, Brantley PJ (2004) Self-Efficacy as a predictor of weight change in African-American Women. Obesity 12:646–651

89. Walcott-Mcquigg J (2000) Psychological factors influencing cardiovascular risk reduction behavior in low and middle income African-American women. Natl Black Nurs Assoc 11:27–35

90. Ainsworth BE, Wilcox S, Thompson WW et al (2003) Personal, social, and physical environmental correlates of physical activity in African-American women in South Carolina. Am J Prev Med 25:23–29

91. Thomas JL, Stewart DW, Lynam IM et al (2009) Support needs of overweight African American women for weight loss (Report). Am J Health Behav 33(4):339

92. Wolfe W (2004) A review: maximizing social support – a neglected strategy for improving weight management with African-American women. Ethn Dis 14:212–218

93. King AC, Jeffery RW, Fridinger F et al (1995) Environmental and policy approaches to cardiovascular disease prevention through physical activity: issues and opportunities. Health Educ Behav 22:499–511

94. Robinson T (2008) Applying the socio-ecological model to improving fruit and vegetable intake among low-income African Americans. J Community Health 33:395–406

95. Kumanyika S (2008) Ethnic minorities and weight control research priorities: where are we now and where do we need to be? Prev Med 47:583–586

96. Huang TT, Drewnowski A, Kumanyika SK et al (2009) A systems-oriented multilevel framework for addressing obesity in the 21st century. Prev Chronic Dis 6:A97

97. Fallon EA, Wilcox S, Ainsworth BE (2005) Correlates of self-efficacy for physical activity in African American women. Women Health 41:47–62

98. Linn BS, Linn MW, Gurel L (1968) Cumulative illness rating scale. J Am Geriatr Soc 16:622–626

99. Miller MD, Paradis CF, Houck PR et al (1992) Rating chronic medical illness burden in geropsychiatric practice and research: application of the cumulative illness rating scale. Psychiatry Res 41:237–248

100. Sallis JF, Pinski RB, Grossman RM et al (1988) The development of self-efficacy scales for health - related diet and exercise behaviors. Health Educ Res 3:283–292

101. Sallis JF, Grossman RM, Pinski RB et al (1987) The development of scales to measure social support for diet and exercise behaviors. Prev Med 16:825–836

102. Echeverria SE, Diez-Roux AV, Link BG (2004) Reliability of self-reported neighborhood characteristics. J Urban Health 81:682–701

103. Casey AA, Elliott M, Glanz K et al (2008) Impact of the food environment and physical activity environment on behaviors and weight status in rural U.S. communities. Prev Med 47:600–604

104. Buzzard IM (1994) Rationale for an international conference on dietary assessment methods. Am J Clin Nutr 59:143S–145S

105. Beaton GH (1994) Approaches to analysis of dietary data: relationship between planned analyses and choice of methodology. Am J Clin Nutr 59:253S–262S

106. Block G, Hartman AM, Dresser CM et al (1986) A data-based approach to diet questionnaire design and testing. Am J Epidemiol 124:453–469

107. Liu K (1994) Statistical issues related to semiquantitative food-frequency questionnaires. Am J Clin Nutr 59:262S–265S

108. Block G, Hartman AM, Naughton D (1990) A reduced dietary questionnaire: development and validation. Epidemiology 1:58–64

109. Norris J, Harnack L, Carmichael S et al (1997) U.S. trends in nutrient intake: the 1987 and 1992 National Health Interview Surveys. Am J Public Health 87:740–746

110. Hartman AM, Block G, Chan W et al (1996) Reproducibility of a self-administered diet history questionnaire administered three times over three different seasons. Nutr Cancer 25:305–315

111. Kriska AM, Caspersen CJ (1997) Introduction to a collection of physical activity questionnaires. Med Sci Sports Exerc 29:5

112. Irwin ML, Crumley D, McTiernan A et al (2003) Physical activity levels before and after a diagnosis of breast carcinoma. Cancer 97:1746–1757

113. Ainsworth BE, Sternfeld B, Slattery ML et al (1998) Physical activity and breast cancer: evaluation of physical activity assessment methods. Cancer 83:611–620

114. Troiano RP, Berrigan D, Dodd KW et al (2008) Physical activity in the United States measured by accelerometer. Med Sci Sports Exerc 40:181–188

115. Carmichael AR (2006) Obesity and prognosis of breast cancer. Obes Rev 7:333–340

116. Ware JE Jr, Sherbourne CD (1992) The MOS 36-item short-form health survey (SF-36). I. Conceptual framework and item selection. Med Care 30:473–483

117. McHorney CA, Ware JE, Lu JFR et al (1994) The MOS 36-Item Short Form Health Survey (SF-36): III. Tests of data quality, scaling assumptions and reliability across diverse patient groups. Med Care 32:40–66
118. Bowen D, Alfano C, McGregor B et al (2007) Possible socioeconomic and ethnic disparities in quality of life in a cohort of breast cancer survivors. Breast Cancer Res Treat 106:85–95
119. Bower JE, Ganz PA, Desmond KA et al (2006) Fatigue in long-term breast carcinoma survivors. Cancer 106:751–758
120. Fallowfield LJ, Leaity SK, Howell A et al (1999) Assessment of quality of life in women undergoing hormonal therapy for breast cancer: validation of an endocrine symptom subscale for the FACT-B. Breast Cancer Res Treat 55:189–199
121. Cella D, Eton DT, Lai J-S et al (2002) Combining anchor and distribution-based methods to derive minimal clinically important differences on the functional assessment of cancer therapy (FACT) anemia and fatigue scales. J Pain Symptom Manage 24:547–561

Chapter 11
The Role of Policy in Reducing Inflammation

Deborah J. Bowen and Stacey Zawacki

Abstract Obesity continues to be a major public health problem affecting approximately 33 % of adults and approximately 20 % of children in the USA (Centers for Disease Control and Prevention (CDC). http://www.cdc.gov/obesity/. Accessed on 31 Oct 2013). The obesity prevalence within minority subgroups of the population is significantly higher (CDC. http://www.cdc.gov/obesity/. Accessed on 31 Oct 2013); differences in dietary and physical activity behaviors likely underpin these disparities (Larson and Story, Ann Behav Med 38:S56–S73, 2009; Story et al., Annu Rev Public Health 29:253–272, 2008). Inflammation occurs because of some combination of genetics and behavior and is often linked with obesity. To address this problem, focused prevention efforts among adults and children is a key strategy (Committee on Prevention of Obesity in Children and Youth, Food and Nutrition Board. Washington, DC: The National Academies Press, 2004; Larson and Story, Ann Behav Med 38:S56–S73, 2009; Story et al., Annu Rev Public Health 29:253–272, 2008). Yet, solely relying on individual-level strategies to change inflammation, obesity, and associated behaviors is not sufficient; creating supportive environments for behavior change is also needed (Affenito et al., J Obes 2012:150732, 2012; Hafekost et al., BMC Med 11:41, 2013; Kahn et al., Am J Prev Med 22 (4 Suppl):73–107, 2002; Larson and Story, Ann Behav Med 38:S56–S73, 2009; Story et al., Annu Rev Public Health 29:253–272, 2008; Mitchell et al., Psychiatr Clin North Am 34(4):717–732, 2011).

D.J. Bowen (✉)
Department of Bioethics and Humanities, University of Washington, 1107 NE 45th Street, Seattle, WA 98105, USA
e-mail: dbowen@uw.edu

S. Zawacki
Department of Nutritional Sciences, Boston University, 635 Commonwealth Ave, Boston, MA 02215, USA
e-mail: szawacki@bu.edu

D.J. Bowen et al. (eds.), *Impact of Energy Balance on Cancer Disparities*, 259
Energy Balance and Cancer 9, DOI 10.1007/978-3-319-06103-0_11,
© Springer International Publishing Switzerland 2014

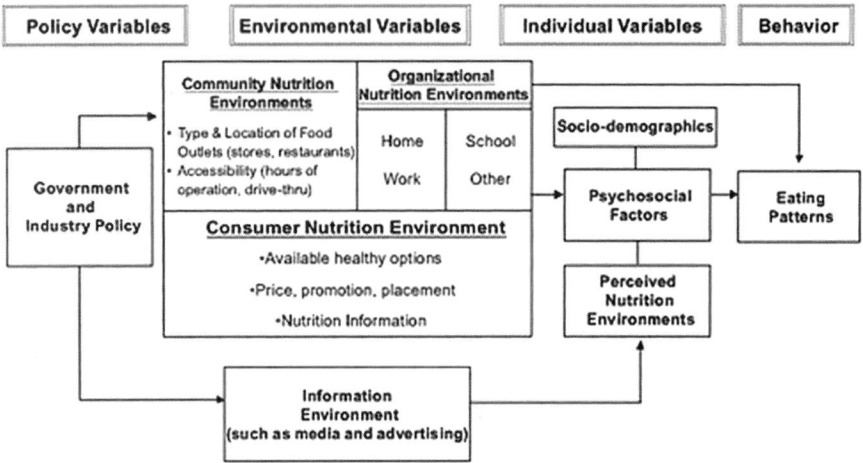

Fig. 11.1 Elements of the nutrition environment (reprinted with permission from American Journal of Health Promotion)

Keywords Inflammatory environment • Obesogenic environment • Nutrition environment • Community effects • Organizational effects • Government effects • Consumer effects • Special Supplemental Nutrition Assistance Program (SNAP) • Soda tax • Menu labeling • Food availability and purchase policy

Why Use Environmental Strategies to Reduce Inflammation

The adjective "obesogenic" has been used to describe environments that promote obesity either through increased energy intake or decreased energy expenditure [119]. For the purpose of this review, we will focus our attention on environments that affect energy intake (i.e., nutrition environments). A socio-ecological framework tailored to characterize nutrition environments [53, 115, 32, 2, 31] has guided much of the research to date. Suggested conceptualization of the nutrition environment centers on individual perceptions of food and activity options in the neighborhood and radiates outwardly to other higher order domains including consumer level (i.e., neighborhood food and activity marketing), community level (i.e., food and activity outlets in the neighborhood), organizational level (i.e., inflammatory opportunity in systems, institutions, or workplaces), informational level (i.e., media and advertising), and policy level (i.e., behaviors affected by governmental decisions) [53] (Fig. 11.1).

Evidence suggests that obesogenic and potentially inflammatory environments are spatially patterned such that they co-occur in areas with larger proportions of low-income and minority populations [115, 78, 126, 66] and may thus contribute to socioeconomic and racial disparities in obesity [115, 78]. Evaluating this potential mechanism has begun by assessing observational associations between mainly availability, accessibility, and affordability dimensions of inflammation, and associated behaviors within domains of the nutrition environment and behavioral indices of high-energy intake seem reasonable [30, 28, 29, 55, 49, 22, 110, 78, 115].

While earlier reviews noted many positive associations between the environment and intake or activity [115, 78], subsequent reviews have concluded that the evidence supporting a causal relationship is moderate to mixed—likely due to conceptual differences and a lack of standardization of measures used to assess the nutrition environment [36, 90, 115, 30, 55, 49, 22, 110]. More in-depth description of measures used to describe the qualities of the environment and methodological considerations are presented elsewhere [55, 74, 32, 84]. In addition to standardizing measures of the environment, research recommendations include conducting studies that (1) employ more rigorous study design, (2) employ a multidimensional approach to characterizing the environment, (3) evaluate a broader array of obesogenic behaviors, (4) address the co-occurrence of obesogenic nutrition and physical activity environments, and (5) measure inflammation as an outcome [30, 23, 55, 49, 22, 110].

Government and Industry Policy Environment

Description

Governmental policy has been used to shape the accessibility of food within organizations, communities, and retailers. Likely the most far-reaching policy is the federal Farm Bill which oversees the largest of the 15 federal nutrition assistance programs—the Special Supplemental Nutrition Assistance Program (SNAP), formerly known as the Food Stamp Program (FSP) [115, 81]. SNAP has significant influence on the accessibility of food overall for low-income Americans [115]; an average of 47 million persons participated in SNAP in 2012 which roughly translates to about 1 in 7 Americans [123]. Eligibility in the program is determined by having a household income ≤130 % of the federal poverty level and <$2,000 in countable assets [81]. By participating in SNAP, income-eligible adults are provided vouchers to purchase foods from program-approved retailers; the current average monthly benefit provided by the program is $133 per person [123].

Other policies implemented at regional and local levels which affect the food environment include food and beverage taxes, menu labeling, commercial zoning policies, as well as licensing and permitting requirements for food outlets [73].

Supporting Evidence

Participation in SNAP has been associated with stocking a wider variety of fruits and vegetables among small retailers as well as increased purchasing of fruits and vegetables among residents in low-income communities [86]. Despite its emphasis on promoting the availability of nutritious foods among low-income Americans,

SNAP has been criticized because benefits can be used to purchase unhealthy food (e.g., baked goods, sweet and salty snacks, and sugar-sweetened beverages) [81]. Studies have identified that SNAP participants do not adhere to the 2005 dietary guidelines for Americans [81, 10]. For example, using NHANES data between 1999 and 2008, SNAP participants reported eating 39 % fewer whole grains, 56 % more potatoes, 46 % more red meat, and 61 % more sugar-sweetened beverages (women only) compared to non-SNAP participants [81]. Similarly, sugar-sweetened beverage purchases were found to account for a greater proportion of all beverage purchases made by SNAP households compared to non-SNAP-eligible households [9, 19]. A recent Institute of Medicine (IOM)/National Research Council (NRC) committee was tasked to ascertain whether the adequacy of the SNAP allotment could be evaluated and concluded that additional factors, *including those related to the nutrition environment*, would need to be considered in the evaluative process [68, 56].

Broader level food and beverage taxes and subsidies have also been proposed to facilitate healthy eating [6, 104, 51, 121, 78, 115], thereby potentially reducing inflammation. Studies have looked at price elasticity, a dimensionless construct representing the percent change in sales resulting from a 1 % price manipulation, to determine the price sensitivity of specific food items among consumers [104, 51]. Fruits and vegetables and fast food tended to have smaller absolute price elasticity values meaning that consumers were generally unwilling to pay higher prices for these items [104]. In a recent review of price elasticity studies, Powell and colleagues found that, in general, higher fast-food prices and lower fruit and vegetable prices were associated with lower body weight outcomes for both children and adults [104]. This consumer purchasing behavior has been suggested to be true for soda as well [8], although perhaps to a lesser degree [104]. It has been estimated that a penny per ounce soda tax could reduce per capita caloric intake by 50 cal, translating to about 3.8 lb/year for adults, if individuals do not substitute another caloric beverage and could also generate significant tax revenue [7, 100]. Among a population-based survey, one-third of respondents indicated that they would cut back on their consumption of sugar-sweetened beverages in response to an added 20 % tax [108].

Menu labeling has also been a proposed strategy to facilitate healthy food choice among consumers. Food away from home now accounts for about one-third of total calories consumed in the USA [71]. Yet, the content of the food provided within restaurant venues is of concern; a review of Web-based nutrition information found that only 4 % of main entrees at major chain restaurants fell within one-third of the recommended daily intake using USDA guidelines for dietary intake of energy, sodium, fat, and saturated fat [133]. While the energy content and diet quality of fast-food meals have not appreciably changed over the past decade, sodium content of meals has increased significantly [14, 64, 112]. Making decisions for healthful eating is difficult in restaurant environments that do not provide point-of-purchase nutritional information and promote energy-dense foods and large portion sizes [115, 78].

Zoning and business permitting or licensure requirements are additional ways cities may impact the nutrition environment, although it is less studied in the public health literature [11]. These policies have traditionally fallen under the purview of urban planning, and use of these strategies to achieve public health objectives is emerging [117, 11]. Cities have begun exploring ways to facilitate a healthy nutrition environment by limiting the number of fast-food restaurants, increasing healthy foods sold within corner stores, and support and space for farmer's markets and community gardens through amending zoning, permitting, and licensure requirements [117, 11]. Impacts of food access within neighborhoods on dietary intake of residents are discussed in further detail in subsequent sections (i.e., "Community Nutrition Environment" and "Consumer Nutrition Environment").

Intervention Results

In 2009, changes, including the offering of additional fresh fruits and vegetables, whole grains, brown rice, soy products, and 2 % milk (instead of whole milk), were made to the SNAP food package to facilitate the meeting of 2005 USDA dietary guidelines among SNAP participants [65]. As a result of this policy change, availability of certified foods increased among samples of supermarket, grocery, corner, and convenience stores [134, 65, 63, 10, 12]. While studies have found that purchasing of whole grains and brown rice increased among SNAP participants after the SNAP benefit change [9, 131, 101], one study found that increased purchasing of whole grains by SNAP participants was not significantly correlated with the actual intake [101]. Additional evaluation of the changes to the SNAP allotment and individual purchasing and consumption of certified foods among SNAP participants is needed. Continued evaluation of the adequacy of the SNAP allotment over time will be informative.

Excluding or limiting sugar-sweetened beverages or providing subsidies for healthy food purchases within SNAP has also been proposed to encourage healthy eating among low-income populations. More studies among SNAP participants across the country are needed to test these relationships among "real" people as well as explore the roles of environmental variables which may differ regionally and locally. Further evaluation of changes to agriculture-led food policies in the USA is needed [37]. Two earlier reviews of menu labeling intervention findings have noted a lack of consistent associations between menu labeling and foods purchased or [118, 61]. However, given the recent federal mandate for menu labeling as a part of the Affordable Care Act in 2010, continued review of this literature is needed as more evidence is published. Ten additional studies have been published since the last published review in 2011 [59, 133, 96, 38, 76, 94, 25, 120, 124]. Again, findings are mixed, although differences could be explained by the study design employed (i.e., experimental versus quasi-experimental). Findings from two studies employing a stronger experimental design were more positive [59, 94]. Typical "treatment" groups included (1) menu label including caloric information only, (2) menu label including a picture cue such as a traffic light in addition to caloric information, and (3) no menu label. Studies also used a quasi-experimental design (i.e., a natural experiment) to evaluate either pre- and post-consumption patterns [96, 38, 76, 120, 124] or changes to restaurant foods offered [133, 96, 25] in response to enacted menu labeling policies.

Overall, findings from these studies do not provide evidence supporting menu labeling as an effective strategy to change purchasing patterns [38]. Analysis by subgroups does provide additional information. One study found that calories purchased decreased among patrons of coffee shops compared to patrons of restaurants [76], while another study found that calories purchased decreased among those who used the menu label in their food purchasing decision compared to those who saw and did not use or did not see the menu label [124]. Restaurants may be responsive to modifying menus in response to consumer demand for more information. Some studies have found that healthy menu options increased after menu labeling enactment [96, 25], although none noted a difference in the overall menu caloric content [133, 96, 25]. Beyond labeling, implementation of nutritional standards for restaurants has also been proposed [35].

Simulation studies on the price elasticity of soda have demonstrated minimal impact on weight-related outcomes [104]. A study using a quasi-experiment design to model changes in taxes on soda within states in relation to population levels of body mass index and obesity over time found evidence of a small positive effect [44] but no data on inflammation or other biological changes. No studies were found that evaluated the effect of a "fat tax" on outcomes within the USA, although evaluation of this strategy has begun in Europe [97] and Great Britain [85]. More studies that evaluate the impact of implemented taxes on actual dietary intake are needed.

Evaluation of interventions employing more local public policy including zoning, permitting, and licensure on obesity-related or dietary outcomes is needed. South Los Angeles implemented a ban on incoming fast-food outlets, which did not appear to change obesity rates at the census tract level [117], but associations with dietary outcomes have yet to be assessed. Also, the Pennsylvania Fresh Food Financing Initiative is a flagship statewide financing program to increase supermarket development in underserved areas [48] which has yet to be evaluated with respect to dietary outcomes. Similar policy development in other regions and cities is emerging [122, 67], and evaluation of these programs will be informative.

Implementation Issues

Acceptability of changes to the SNAP benefit allotment appears high among participants with some notable exceptions [17]. The substitution of whole milk with 2 % milk was not endorsed by Hispanic women, and soy products were not endorsed by the majority of participants [17]. Targeted provision of health information may increase the acceptability of these foods among select populations. Process information is also needed to ensure that changes to the SNAP benefit allotment maximize dimensions of healthy food access and translate to the actual intake. Interestingly, in a cost-effectiveness simulation study on SNAP nutritional changes and chronic disease outcomes, it was noted that banning foods from SNAP allotments may be less desirable as SNAP participants could opt to purchase these items with disposable income, resulting in increased food insecurity; subsidizing healthy foods was found not to affect food security [13]. Instituting a subsidy for healthy foods, therefore, has become a more popular recommendation for revamping SNAP allotments [106], although studies with respect to dietary outcomes are needed. Evaluations of the 2009 change in foods subsidized by SNAP also appear to support this conclusion.

The differences in association between menu labeling and purchasing by subgroup may provide additional process information for this research. Specifically, this strategy as implemented appears to be more effective for more individuals who are more educated or familiar with nutrition labeling in general [111]. Process information in minority and low-income samples indicates, however, that menu labeling implementation can have high fidelity in these communities [80] and that caloric information is salient to purchasing decisions [98]. This nutritional information, however, does not necessarily translate to decreases in consumption [39] which may be related to how and what nutrition information is provided. For example, difficulties in calculating calories per serving arise when ranges of calories are presented for a menu item that comes in multiple flavors or when calories are presented for a menu item that contains multiple servings [35].

The soda tax has not been effective in altering weight-related outcomes, and this may be because the amount of the tax has been too small [34, 104]. Based on price elasticity data, a tax which raises the price of soda by an estimated 10–20 % is needed to significantly impact consumption [104, 34, 8]. Simulation studies and the quasi-experimental study using the 4 % average state soda tax found small effects on obesity [104, 44] which may also provide evidence to raise soda taxes. Barriers to a more rigorous implementation of soda taxes include opposition from the beverage industry as well as established state sales tax exemptions for food and beverages [34, 104]. An excise tax (i.e., a tax on the manufacturer) may be more attractive to manufacturers than a sales tax and is administratively easier for governing bodies to implement and enforce [103]. An excise tax may also be more effective in altering consumer purchasing as it would raise the shelf price of the product which may be a more salient point for purchasing decisions as opposed to during checkout [103, 34]. However, consumers do not appear to support implementation of soda taxes overall, although greater support has been demonstrated among groups of younger age and higher socioeconomic status [108].

Future Directions

Studies are needed to evaluate how changes in the SNAP benefit allotment affect purchasing or dietary intake of participants across the country. Identifying additional strategies to enhance menu-labeling practices is needed as nutrition information alone has not been found to change dietary behavior overall [91, 95]. This may be partly attributed to a lack of understanding of portion sizes, especially in venues outside of the home [43]. Larger portion sizes have been associated with eating meals away from home as well as obesity [40]. In concert with providing more healthy options and nutrition information, limiting portion sizes within restaurants may also be a useful strategy to promote healthy food environments. Further evaluation of menu labeling at the population level is also warranted given the federal enactment of this strategy as a part of the Affordable Care Act [102, 109, 133]. Finally, the need to raise soda taxes may be suggested given that currently enacted tax levels among 28 selected states were not found to be sufficient to impact population-level dietary behaviors or weight-related outcomes.

Organizational Nutritional Environment

Description

This environment is a fertile one for intervention, since strategies can make use of existing structures and of the policies and programs that can be offered within them to influence healthy eating and activity. Also, inflammatory markers could be measured with relative ease in an organizational setting. Therefore, these organizations have great potential for inflammation-related policy. Structures discussed here are workplaces, health care systems, and religious organizations.

Supporting Evidence

Worksites provide access to almost three-quarters of the adult population [24], which makes them ideal settings to implement strategies for changing eating behaviors. A variety of foods and many meals are available and eaten at workplaces, making access to healthy foods at work important, but observational studies of workplace nutrition environments in relation to healthy eating behaviors appear not to have been conducted. Workplaces serve both as existing channels of communication and as social support networks, both important to adult behavior changes from a theoretical perspective. Finally, many opportunities exist in workplaces for environmental and policy change to foster healthy dietary practices. The costs of unhealthy eating are born by employers as well as employees, providing motivation for both groups to push for change [27].

Intervention Results

In contrast to the lack of observational studies of workplaces, intervention studies are plentiful. Workplaces are perhaps one of the most studied of the organizational areas of intervention. There have been Cochrane reviews and other systematic reviews of workplace interventions to improve eating behaviors in adult populations [129]. The Community Guide to Preventive Services Task Force recommends multicomponent interventions that include nutrition and physical activity (including strategies such as providing nutrition education or dietary prescription, physical activity prescription or group activity, and behavioral skill development and training) to control overweight and obesity among adults in worksite settings [5]. In addition, Cochrane reviews [129] found consistent changes as a result of workplace interventions, although most of the focus of workplace interventions has been on physical activity as an outcome. Many workplace interventions in workplaces have included multiple components (materials, classes, access changes in food supplies, and changes in social norms) and have also focused on physical activity to reduce obesity, making it difficult to specify which components are more prominent [5]. Incentive programs at workplaces, providing rewards for healthy eating, physical activity, and/or weight loss, have shown to be of generally positive benefit, although these were not typically used in conjunction with more programmatic or policy interventions.

Health clinics and hospitals are special cases of workplaces, but any policy changes have the potential to reach beyond employees to the patients whom they serve. They have been the subject of multiple intervention projects and reviews. Other organizations, such as public housing developments, look promising [15, 16].

Religious organizations have been used as a setting for policy and programmatic change, both multidenominational settings [21] and primarily Black churches [4] and Hispanic churches. In general these studies have found that a multicomponent intervention that includes policy change, promotional materials, religious organizational involvement, and community health workers has resulted in improvements in eating behaviors. The sustainability and reach of these interventions were tested in follow-up studies, with positive outcomes in organizational changes, indicating high potential for dissemination to occur [3, 60].

Implementation Issues

The field of organizational nutrition environment change is mature enough that there are multiple efficacy studies that target organizations for change and there is a growing literature on dissemination of these types of interventions to change population health. A recent task force on worksite health promotion intervention discussed issues of dissemination and implementation research and recommended several strategies [113]. Issues discussed in this review include the mechanisms by which workplace interventions work, the diversity of workplaces and the individual and unique responses to intervention, and the contributions of individual and environmental multicomponent interventions to changes in workplace behaviors [114]. We are learning about the extent of reach and sustainability of these long-term studies that attempt to put into place policy and programs in organizations that are expected to lead to health promotion changes, such as reductions in obesity, in very large numbers of working adults.

Future Directions

One understudied direction is in research to change opportunity for healthy eating in places that encourage or engage people in inflammation-reducing activities, like healthy eating or activity. Organizations have a remarkable ability to control the offerings that occur in their presence, as exemplified by the use of tobacco sales policies in workplaces and other organizations. A similar set of strategies could be enacted for food or activity availability, increasing the access that individuals have to healthy options. Indeed, policy that improves the availability of healthy foods in workplaces and other organizations and structures seem like a key element of future interventions.

Another area that is also understudied in the literature is the use of multiple strategies or combinations of strategies in multiple channels and organizations. This means that dissemination of policies and programs that have changed small groups now needs to happen at the larger population level. For example, organizational statewide policies that encourage healthy eating need to be evaluated and tested. Again, using tobacco control as a model might provide a blueprint for action in this setting.

The new technologies used in our future national health care system might be an area of focus for healthy eating. This is an area of innovation, as our changing health care system and inclusion of more patient-centered care might open doors for the testing of new interventions that rely on following participants with screening and assistance no matter where they receive care and no matter what care they receive. Engaging these technologies in support of healthy eating might be a novel approach with potential for high yield.

Resolving the gulf between health promotion and worker health and safety might be another area of consideration. There have been a few projects, notably those of Sorenson and Barbeau [113], that have integrated these two areas of workplace health activity, and this might be a promising area for future focus.

Community Nutrition Environment (Neighborhood Level)

Description

The community environment domain has been defined as the distribution of neighborhood outlets including supercenters, supermarkets, grocery stores, convenience stores, and restaurants (both full and quick service), gyms, sports facilities, and activity opportunities [53, 54, 115, 80]. Two methods have mainly been used to quantify the community environment. The most highly used approach involves enumerating the number of food outlets within a specified area such as a census tract or a predefined circular buffer using geographic information systems (GIS), while a second approach involves on-the-ground audits of stores in a neighborhood [30, 110, 86]. Most studies enumerate food sources around the home, but more studies are now assessing food sources in reference to other places such as work-places and schools. Typically, studies have used GIS technology to quantify density and proximity measures of walkability or food outlets within defined geographical units [30, 88]. Linking extant geospatial data to individual-level observational data is attractive to many investigators as this approach is less time and resource intensive than field enumeration of neighborhood outlets.

Supporting Evidence

Reviews of studies of the community environment have noted moderate-to-mixed findings with respect to many dietary outcomes with fruit and vegetable intake being the most studied. Again, this discrepancy is likely attributable to a lack of standardization in measures, especially those derived from GIS methods, used to quantify the environment. Evaluation of studies by assessment method found that findings were the least consistent if derived from GIS methods, while findings using store audits of the community nutrition environment were only slightly more consistent [30]. For example, several studies have found that people who lived closer to supermarkets have higher diet quality whereas people who lived closer to fast food or convenience stores have poorer diet quality [30, 78] or other measure of dietary behavior [30]. That this relationship holds independent of socioeconomic status may be important [82, 89, 92]. On the other hand, not all studies have confirmed these relationships [83, 130, 132]. In one study in California, neighbor-hood density of fast food, but not grocery, outlets was related to dietary intake among a large sample of adults [62]. In a large national study, no relation was found between fast food outlets in neighborhoods and fast food consumption among young adults [107].

Inconsistent findings could be due to differences in variables used to operationalize the environment either due to a lack of "ground truthing" or validity testing of GIS variables within studies or differences in data sources and units of measure used across studies [20, 30, 26]. Lack of consistency among findings could also arise from variation in area sizes used to define the boundaries of the nutrition environment or specific characteristics of the populations under study [30]. In fact, a study among a diverse group of young adults suggested that commonly used GIS boundary sizes in many studies were too small and did not sufficiently represent the food environment for this group [79]. That is, respondents may also access outlets beyond study-defined boundaries of the neighborhood environment [32]. In addi-tion, evaluation of the community food environment (number or proximity of stores in an area) without *concurrently* accounting for what happens within the food store may contribute to some level of misclassification, possibly biasing toward the null, as many supermarkets also contain unhealthy options [125]. Other accessibility dimensions are likely necessary, yet are only just beginning to be incorporated in GIS-based studies [70].

Intervention Results

Several intervention studies have undertaken environmental strategies to increase the number of food outlets offering fresh produce in neighborhoods [33, 46, 93]. Successful interventions include introduction of additional farmer's market days and of community gardens [105, 42, 46, 87]. The Veggie Project was a multicomponent intervention which brought farmer's markets to four Boys' and Girls' Club sites situated within low-income, minority, urban communities in Nashville, Tennessee [46]. The intervention included a discount voucher program to offset the cost of healthy foods for participating families and was associated with a significant increase in purchase of fruits and vegetables, as evaluated by a before–after quasi-experimental design [46].

Three separate ongoing community-level obesity prevention initiatives to change the community environment in California used community-based participatory approaches to identify ways of providing additional healthy outlets into the neighborhood [33]. Selected strategies varied by community and examples included establishing an organic farmer's market, delivering boxes of fruits and vegetables to low-income neighborhood residents, setting up a low-cost fruit and vegetable stand at an elementary school, and transforming a convenience store into a produce outlet [33]. These efforts also proposed to incorporate national low-income voucher programs (e.g., Women, Infants, and Children) to address the economic barriers of people in those communities [33]. Evaluation of these efforts has yet to be completed.

Implementation Issues

Due to the novelty of this intervention research, the main barrier to the implementation of environmental and policy interventions by public health practitioners is a lack of evidence-based model strategies and demonstrated protocols about how to create change in communities [47]. The structural changes to the neighborhood required to implement these environmental approaches can be difficult to achieve in the short term, let alone ensure sustainability, within the timing of grant funding cycles [47]. Other barriers to success include failure to identify intervention partners with all of the expertise needed—interventions may engage community stakeholders who have limited experience with advocating for or implementing structural- or policy-level changes [47]. For example, Morland found an overall lack of business expertise among key personnel [93] when creating a community-based cooperative (co-op) market. Nonetheless, reviewing this type of formative research may be invaluable for the planning and execution of similar strategies. Changelab Solutions (formerly Public Health Law & Policy) provides an online collection of reports which address these issues, although evaluation of these recommendations is needed (http://changelabsolutions.org/. Accessed October 10, 2013). Lessons learned from smoking cessation policy and business change implementation over the last decade may also prove informative, given the possible consideration of unhealthy eating choices as an addictive behavior [75].

Future Directions

Interventions within more rigorous study designs that include a nonintervention control group and that target the community nutrition environment still need to be implemented and evaluated. To more directly attribute behavior change to the structural changes effected, designs should be set up to allow parsing out these effects from other individual-level intervention elements. Using a randomized controlled design, comparison of multiple arms (e.g., environmental change alone, environmental change plus other intervention elements, and control) would be ideal to ascertain whether modifications to the nutrition environment are effective in increasing healthy dietary behaviors. Given that the cost of this study design may not be affordable, another approach could be to estimate the role of the environment as an effect modifier of traditional individual-level intervention approaches on the corresponding outcomes [57]. Evaluation of the impact of environmental strategies on intermediate outcomes such as psychosocial correlates of healthy eating (e.g., social support and self-efficacy) may also provide valuable insights.

Finally, long-term evaluation of any of these strategies is entirely lacking to date, since those studies that are of interventions focused on environmental or policy-level changes to the community environment are still in progress or have only short-term or only behavioral outcomes. Threats to including long-term outcomes include the high mobility of low-income populations. These groups are more likely to be the focus of these interventions but may be difficult to follow for a long term [87].

Consumer Nutrition Environment (Retail Level)

Description

This domain of the nutrition environment includes measures of food availability *within* neighborhood food outlets. Strategies to influence product stocking, pricing, and display are used to affect what is available and promoted to the consumer within a store or a restaurant. Policies to change these are needed as well as evaluation of these policies and evaluation of the link between the policies and inflammation.

Supporting Evidence

The number of studies exploring the consumer environment continues to increase, especially among urban populations [55]. These studies have typically used some version of a store audit and have found more consistent relationships with behaviors than measures of the environment at the community level [30, 55, 88]. Caspi and colleagues noted that 8 of 15 studies using store audit methods found at least one positive association between various measures of neighborhood healthy food access and dietary behavior [30]. In general, greater availability of healthy food/fruit and vegetables in the neighborhood may be associated with higher diet quality/fruit and vegetable consumption among urban neighborhood residents [55]. For example, living near a store that stocked at least five varieties of dark-green or orange vegetables was associated with higher fruit and vegetable consumption [69] while greater store shelf space devoted to fruits and vegetables was associated with greater increases in fruit and vegetable consumption at 1-year follow-up [28]. Lower availability of healthy foods within the nearest store has also been associated with lower diet quality [45]. These findings, however, were not replicated in a study in rural communities in North Carolina [57].

In addition to potentially increased study power by characterizing healthy foods available in the neighborhood environment, additional domains of availability including affordability can also be more elaborately assessed. The pricing of foods offered within stores may also play a significant role in consumer demand [8, 104], yet findings vary by methods used to measure price [41]. Contrary to price elasticity studies, reviewed studies measuring food price via store audit methods in relation to dietary behavior or studies of price changes on purchasing behavior have been null [30] or mixed [41].

Intervention Results

Some intervention studies have been designed to test whether modifying components of the consumer nutrition environment (e.g., availability, price, and acceptability of healthy food) within the retail environment of existing food outlets is associated with increasing healthy dietary choices. Consumer-level strategies share a common goal of increasing the accessibility of healthy food in neighborhood stores and may work through influencing the behaviors of store retailers as well as of consumers.

Targeting store retailers is important as they decide what products to stock, how to display them, and how much variety to offer [51]. Strategies to change retailer food stocking and display practices include offering incentives to provide more healthy foods and produce [33, 50, 102], improving mechanisms and offering incentives for purchasing food from local farms [102], monitoring or purchasing refrigeration units for the storage of perishable produce items [50], placement of healthy foods within the store [1, 33, 50, 51], as well as point-of-purchase promotion of healthy foods and produce [1, 50]. Implementation of these strategies may also require operation within higher policy-level domains.

Similar approaches to increase the availability of healthy foods have also been used within restaurant settings including nutrition training for chefs on how to prepare dishes lower in fat, changing catering policies to require healthy food choices, and point-of-purchase promotion and information such as table tents and menu-labeling policies [52, 76]. Some success has been demonstrated for interventions among small grocery retailers using a combination of product placement and promotion strategies with reported small increases in sales of fruits and vegetables among specific customers [1, 50].

Other environmental level strategies focus on making healthy foods attractive to consumers in store environments. The most popular strategy evidenced in the literature was the implementation of price discounts for healthy food [46, 50, 99, 116, 127], while a few studies also attempted in-store sampling and cooking demonstrations of healthy food items [1, 50]. Although there currently is no evaluation data available for in-store taste demonstration strategies, the use of price discounts has been consistently associated with changing food purchasing in several studies [46, 99, 116, 127]. Larger increases in sales of fruits and vegetables were demonstrated among participants given price discounts within the previously described Veggie Project [46]. A similar increase in healthy food sales was noted at 6 and 12 months post-intervention among those assigned to receiving price discounts in randomized controlled trials [18, 99, 128]. A combination of discount plus nutrition education strategies resulted in an even greater increase in fruit and vegetable purchases, and the number of participants who consumed recommended amounts of fruits and vegetables increased from 42 % at baseline to over 61 % for both discount groups [127]. In an observation study in South Africa, participation in healthy food discount programs was associated with reported consumption of more fruits and vegetables and wholegrain foods as well as decreased consumption of "fast" and fried foods and foods high in sugar or salt [6]. These findings were replicated using grocery receipt scanner data to measure food purchasing [116].

Implementation Issues

Fidelity of intervention strategies targeting small store owners may be a concern that could limit the potential impact to customer food choice via these approaches [1]. Reviews of small store interventions have noted differences in intervention fidelity [50, 1], although these differences could be due to location of small stores studied (i.e., the US versus the UK implementation). Although most small store owners in the UK were supportive of intervention goals to improve health among members of the neighborhood, the inability to compete with larger supermarket pricing of fruits and vegetables was a common barrier to maintaining produce availability [1]. Similar to community-level strategies used in the USA, a lack of business expertise and clear definition of roles and responsibilities was also cited as a significant barrier by store owners for implementing healthy food stocking strategies [1]. Barriers to stocking healthy foods identified by the US small store retailers included a lack of consumer demand, refrigerator or freezer space, and profitability [12] in addition to neighborhood crime [77].

With respect to pricing interventions, the effects of these programs have been demonstrated in mostly white populations, but effects may vary by ethnic group [18]. In a randomized controlled trial in New Zealand, for example, increases in healthy food purchasing tied to the price discount intervention were seen in European and Asian, but not Maori, groups [18]. Selection or cultural adaptation of strategies to address group food preferences may be warranted.

Future Directions

Continued study is needed to determine whether increased availability of healthy foods in retail stores impacts shopping and dietary behaviors of neighborhood customers [51]. Facilitating low-pricing or promoting alternative strategies for small store owners as well as recruiting intervention workers who have both health promotion and retail experience may be helpful. Evaluation of programs using retailer incentives to promote healthy food availability may provide evidence of a solution to address pricing barriers faced by small stores. Exploration of the impact of placement and promotion of healthy foods is also needed as well as further development of tools to assess these components [51]. These may be more attractive alternative strategies for small store owners who are not able to provide discounted pricing on healthy foods. In addition, interventions evaluating price discounting among at-risk groups may be warranted given the noted differences in response in groups differing by indicators of socioeconomic status [18] and to ensure that efforts do not contribute to greater disparities in healthy dietary behaviors. Ensuring that price discounts include culturally appropriate foods is a must.

Conclusions and Future Directions

We have identified recent evidence concerning effective strategies for changing the environment through policy to reduce inflammation. These include instituting a tax on sodas, using worksite environments to support healthful changes in the obesogenic environment, increasing the numbers of farmers' market days, and discount pricing of healthy foods within stores, but we found the results not always to be consistent. Research into the efficacy and effectiveness of intervention modalities incorporating environmental strategies is falling behind associational research. To drive these efforts forward, successful models and demonstrated protocols for community-level change are needed. Incorporation of a nonintervention control group in study designs is ideal for assessing effects directly attributable to changes in the environment. Yet, not everything can be evaluated in the context of a randomized controlled trial. Taking advantage of natural experiments may be one key way to evaluate environmental strategies and be efficient in the use of resources for research. Evaluations of community environment as an effect modifier of other interventions on obesity risk may hold promise. For example, considering the home or the workplace environment when delivering an individually based intervention might help explain findings that vary by a higher group level or might identify future research opportunities.

More studies incorporating both objective and perceived measures of the environment are also needed to further understand relationships with behaviors. Also desperately needed is the use of inflammatory markers as outcomes, alongside obesity and behaviors. Policy changes are clearly needed and so is evaluation of such policies. Given the lack of success demonstrated by informational strategies alone (e.g., menu labeling), perhaps we should take advantage of lessons learned from smoking cessation policy to invoke desired behavioral change. There is now a rigorously conducted set of studies that supports the use of tobacco control policies of many types implemented concurrently to help smokers quit and to prevent youngsters from beginning to smoke. Using this evidence and applying it to eating behavior change may yield novel and more effective approaches to using the information and policy environment to effect dietary behavior change. For example, behavioral change as a result of currently enacted soda taxes has not been demonstrated. Increasing the tax further, as was done for tobacco products, for sugar-sweetened beverages may be called for. To be commensurate with taxes placed on cigarettes, for example, soda tax would need to be an added 58 % which would translate to a population-level shift in BMI of 0.6 kg/m^2.

Policy and evaluation of policy to control marketing and promotion of sales of food and activity opportunity is a key element of the promotion of healthy behaviors. Given the outlay of money for advertising, these efficacious messaging systems do change people's behaviors and need to be considered and curtailed to reduce future increases in inflammation. These types of policies are under-evaluated and often misunderstood, as in the case of the recent law to limit sales of large soda containers in New York City [135]. This law is currently under review by a higher court, but this attempt to limit soda manufacturers from selling oversized soda containers in New York City brought on a campaign by the beverage distributors to label the law as anti-freedom and misguided in its attempts to control soda sales [95]. Evaluating the effects of these types of policies on both individual behavior and on the behaviors of food companies will be enlightening.

Counteradvertising such as is being done with tobacco might be an additional answer. Counteradvertising to provide truthful messages to the public about tobacco products has been shown to be efficacious in changing individual smoking behavior. How it will affect healthy behaviors is currently unknown. In addition, how to engage counteradvertising without increasing stigma for overweight people is also unknown and should be the focus of research.

A larger focus on the use of multiple organizational units (worksites, hospitals, schools, communities) within natural policy units such as cities, counties, or states is a promising direction. The use of theory to guide these interventions is often quite poor, and the role of theory in this area of research is not well established. Some interventions might be more powerful or more reaching and sustainable with the use of theory. This area needs research attention.

References

1. Adams J, Halligan J et al (2012) The change4life convenience store programme to increase retail access to fresh fruit and vegetables: a mixed methods process evaluation. PLoS One 7 (6):e39431
2. Affenito SG, Franko DL et al (2012) Behavioral determinants of obesity: research findings and policy implications. J Obes 2012:150732
3. Allicock M, Campbell MK, Valle CG, Carr C, Resnicow K, Gizlice Z (2012) Evaluating the dissemination of Body & Soul, an evidence-based fruit and vegetable intake intervention: challenges for dissemination and implementation research. J Nutr Educ Behav 44(6):530–538
4. Allicock M, Johnson L-S, Leone L, Carr C, Walsh J, Ni A, Resnicow K, Pignone M, Campbell M (2013) Promoting fruit and vegetable consumption among members of black churches, Michigan and North Carolina, 2008-2010. Prev Chronic Dis 10:E33
5. Anderson LM, Quinn TA, Glanz K, Ramirez G, Kahwati LC, Johnson DB, Buchanan LR et al (2009) The effectiveness of worksite nutrition and physical activity interventions for controlling employee overweight and obesity: a systematic review. Am J Prev Med 37 (4):340–357
6. An R, Patel D et al (2013) Eating better for less: a national discount program for healthy food purchases in South Africa. Am J Health Behav 37(1):56–61
7. Andreyeva T, Chaloupka FJ et al (2011) Estimating the potential of taxes on sugar-sweetened beverages to reduce consumption and generate revenue. Prev Med 52(6):413–416
8. Andreyeva T, Long MW et al (2010) The impact of food prices on consumption: a systematic review of research on the price elasticity of demand for food. Am J Public Health 100 (2):216–222
9. Andreyeva T, Luedicke J (2013) Federal food package revisions: effects on purchases of whole-grain products. Am J Prev Med 45(4):422–429
10. Andreyeva T, Luedicke J et al (2012) Positive influence of the revised Special Supplemental Nutrition Program for women, infants, and children food packages on access to healthy foods. J Acad Nutr Diet 112(6):850–858
11. Ashe M, Feldstein LM et al (2007) Local venues for change: legal strategies for healthy environments. J Law Med Ethics 35(1):138–147
12. Ayala GX, Laska MN et al (2012) Stocking characteristics and perceived increases in sales among small food store managers/owners associated with the introduction of new food products approved by the Special Supplemental Nutrition Program for Women, Infants, and Children. Public Health Nutr 15(9):1771–1779
13. Basu S, Seligman H et al (2013) Nutritional policy changes in the supplemental nutrition assistance program: a microsimulation and cost-effectiveness analysis. Med Decis Making 33 (7):937–948
14. Bauer KW, Hearst MO et al (2012) Energy content of U.S. fast-food restaurant offerings: 14-year trends. Am J Prev Med 43(5):490–497
15. Bennett GG, McNeill LH, Wolin KY, Duncan DT, Puleo E, Emmons KM (2007) Safe to walk? Neighborhood safety and physical activity among public housing residents. PLoS Med 4(10):e306
16. Bennett GG, Wolin KY, Puleo E, Emmons KM (2006) Pedometer-determined physical activity among multiethnic low-income housing residents. Med Sci Sports Exerc 38 (4):768–773

17. Black MM, Hurley KM et al (2009) Participants' comments on changes in the revised special supplemental nutrition program for women, infants, and children food packages: the Maryland food preference study. J Am Diet Assoc 109(1):116–123

18. Blakely T, Ni Mhurchu C et al (2011) Do effects of price discounts and nutrition education on food purchases vary by ethnicity, income and education? Results from a randomised, controlled trial. J Epidemiol Community Health 65(10):902–908

19. Bleich SN, Vine S et al (2013) American adults eligible for the Supplemental Nutritional Assistance Program consume more sugary beverages than ineligible adults. Prev Med 57 (6):894–899

20. Blitstein JL, Snider J et al (2012) Perceptions of the food shopping environment are associated with greater consumption of fruits and vegetables. Public Health Nutr 15 (6):1124–1129

21. Bowen DJ, Beresford SAA, Christensen CL, Kuniyuki AA, McLerran D, Feng Z, Hart A Jr, Tinker L, Campbell M, Satia J (2009) Effects of a multilevel dietary intervention in religious organizations. Am J Health Promot 24(1):15–22

22. Bodor JN, Rice JC et al (2010) The association between obesity and urban food environments. J Urban Health 87(5):771–781

23. Brownson RC, Haire-Joshu D, Luke DA (2006) Shaping the context of health: a review of environmental and policy approaches in the prevention of chronic diseases. Annu Rev Public Health 27:341–370

24. Brownson RC, Hopkins DP, Wakefield MA (2002) Effects of smoking restrictions in the workplace. Annu Rev Public Health 23:333–348

25. Bruemmer B, Krieger J et al (2012) Energy, saturated fat, and sodium were lower in entrees at chain restaurants at 18 months compared with 6 months following the implementation of mandatory menu labeling regulation in King County, Washington. J Acad Nutr Diet 112 (8):1169–1176

26. Burgoine T, Alvanides S et al (2013) Creating 'obesogenic realities'; do our methodological choices make a difference when measuring the food environment? Int J Health Geogr 12:33

27. Burns DM, Shanks TG, Major JM, Gower KB (2000) Restrictions on smoking in the workplace. In: Population-based smoking cessation: proceedings of a conference on What works to influence cessation in the general population, vol 12, NCI tobacco control monograph. US DHHS, Bethesda, MD, pp 99–128

28. Caldwell EM, Miller Kobayashi M et al (2009) Perceived access to fruits and vegetables associated with increased consumption. Public Health Nutr 12(10):1743–1750

29. Caspi CE, Kawachi I et al (2012) The relationship between diet and perceived and objective access to supermarkets among low-income housing residents. Soc Sci Med 75(7):1254–1262

30. Caspi CE, Sorensen G et al (2012) The local food environment and diet: a systematic review. Health Place 18(5):1172–1187

31. Centers for Disease Control and Prevention (CDC) (2013) "Overweight and obesity." http://www.cdc.gov/obesity/. Accessed on 31 Oct 2013

32. Charreire H, Casey R et al (2010) Measuring the food environment using geographical information systems: a methodological review. Public Health Nutr 13(11):1773–1785

33. Cheadle A, Samuels SE et al (2010) Approaches to measuring the extent and impact of environmental change in three California community-level obesity prevention initiatives. Am J Public Health 100(11):2129–2136

34. Chriqui JF, Chaloupka FJ et al (2013) A typology of beverage taxation: multiple approaches for obesity prevention and obesity prevention-related revenue generation. J Public Health Policy 34(3):p403

35. Cohn EG, Larson EL et al (2012) Calorie postings in chain restaurants in a low-income urban neighborhood: measuring practical utility and policy compliance. J Urban Health 89 (4):587–597

36. Committee on Prevention of Obesity in Children and Youth, Food and Nutrition Board (2004) Preventing childhood obesity: health in the balance. The National Academies Press, Washington, DC

37. Dangour AD, Hawkesworth S et al (2013) Can nutrition be promoted through agriculture-led food price policies? A systematic review. BMJ Open 3(6):e002937
38. Downs JS, Wisdom J et al (2013) Supplementing menu labeling with calorie recommendations to test for facilitation effects. Am J Public Health 103(9):1604–1609
39. Elbel B, Kersh R et al (2009) Calorie labeling and food choices: a first look at the effects on low-income people in New York City. Health Aff (Millwood) 28(6):w1110–w1121
40. Ello-Martin JA, Ledikwe JH et al (2005) The influence of food portion size and energy density on energy intake: implications for weight management. Am J Clin Nutr 82 (1 Suppl):236S–241S
41. Epstein LH, Jankowiak N et al (2012) Experimental research on the relation between food price changes and food-purchasing patterns: a targeted review. Am J Clin Nutr 95 (4):789–809
42. Evans AE, Jennings R et al (2012) Introduction of farm stands in low-income communities increases fruit and vegetable among community residents. Health Place 18(5):1137–1143
43. Faulkner GP, Pourshahidi LK et al (2012) Serving size guidance for consumers: is it effective? Proc Nutr Soc 71(4):610–621
44. Fletcher JM, Frisvold D et al (2010) Can soft drink taxes reduce population weight? Contemp Econ Policy 28(1):23–35
45. Franco M, Diez-Roux AV et al (2009) Availability of healthy foods and dietary patterns: the Multi-Ethnic Study of Atherosclerosis. Am J Clin Nutr 89(3):897–904
46. Freedman DA, Bell BA et al (2011) The veggie project: a case study of a multi-component farmers' market intervention. Journal of Primary Prevention 32(3–4):213–224
47. Gantner LA, Olson CM (2012) Evaluation of public health professionals' capacity to implement environmental changes supportive of healthy weight. Eval Program Plann 35 (3):407–416
48. Giang T, Karpyn A et al (2008) Closing the grocery gap in underserved communities: the creation of the Pennsylvania Fresh Food Financing Initiative. J Public Health Manag Pract 14 (3):272–279
49. Giskes K, van Lenthe F et al (2011) A systematic review of environmental factors and obesogenic dietary intakes among adults: are we getting closer to understanding obesogenic environments? Obes Rev 12(5):e95–e106
50. Gittelsohn J, Rowan M et al (2012) Interventions in small food stores to change the food environment, improve diet, and reduce risk of chronic disease. Prev Chronic Dis 9(2):E59
51. Glanz K, Bader MD et al (2012) Retail grocery store marketing strategies and obesity: an integrative review. Am J Prev Med 42(5):503–512
52. Glanz K, Hoelscher D (2004) Increasing fruit and vegetable intake by changing environments, policy and pricing: restaurant-based research, strategies, and recommendations. Prev Med 39(Suppl 2):S88–S93
53. Glanz K, Sallis JF et al (2005) Healthy nutrition environments: concepts and measures. Am J Health Promot 19(5):330–333, ii
54. Gustafson A, Christian JW et al (2013) Food venue choice, consumer food environment, but not food venue availability within daily travel patterns are associated with dietary intake among adults, Lexington Kentucky 2011. Nutr J 12:17
55. Gustafson A, Hankins S et al (2012) Measures of the consumer food store environment: a systematic review of the evidence 2000-2011. J Community Health 37(4):897–911
56. Gustafson A, Lewis S et al (2013) Neighbourhood and consumer food environment is associated with dietary intake among Supplemental Nutrition Assistance Program (SNAP) participants in Fayette County, Kentucky. Public Health Nutr 16(7):1229–1237
57. Gustafson AA, Sharkey J et al (2011) Perceived and objective measures of the food store environment and the association with weight and diet among low-income women in North Carolina. Public Health Nutr 14(6):1032–1038
58. Hafekost K, Lawrence D et al (2013) Tackling overweight and obesity: does the public health message match the science? BMC Med 11:41

59. Hammond D, Goodman S et al (2013) A randomized trial of calorie labeling on menus. Prev Med 57(6):860–866
60. Hannon PA, Bowen DJ, Christensen CL, Kuniyuki A (2008) Disseminating a successful dietary intervention to faith communities: feasibility of using staff contact and encouragement to increase uptake. J Nutr Educ Behav 40(3):175–180
61. Harnack LJ, French SA (2008) Effect of point-of-purchase calorie labeling on restaurant and cafeteria food choices: a review of the literature. Int J Behav Nutr Phys Act 5:51
62. Hattori A, An R et al (2013) Neighborhood food outlets, diet, and obesity among California adults, 2007 and 2009. Prev Chronic Dis 10:E35
63. Havens EK, Martin KS et al (2012) Federal nutrition program changes and healthy food availability. Am J Prev Med 43(4):419–422
64. Hearst MO, Harnack LJ et al (2013) Nutritional quality at eight U.S. fast-food chains: 14-year trends. Am J Prev Med 44(6):589–594
65. Hillier A, McLaughlin J et al (2012) The impact of WIC food package changes on access to healthful food in 2 low-income urban neighborhoods. J Nutr Educ Behav 44(3):210–216
66. Hilmers A, Hilmers DC et al (2012) Neighborhood disparities in access to healthy foods and their effects on environmental justice. Am J Public Health 102(9):1644–1654
67. Hood C, Martinez-Donate A et al (2012) Promoting healthy food consumption: a review of state-level policies to improve access to fruits and vegetables. WMJ 111(6):283–288
68. IOM (Institute of Medicine) and NRC (National Research Council) (2013) Supplemental Nutrition Assistance Program: examining the evidence to define benefit adequacy. The National Academies Press, Washington, DC
69. Izumi BT, Zenk SN et al (2011) Associations between neighborhood availability and individual consumption of dark-green and orange vegetables among ethnically diverse adults in Detroit. J Am Diet Assoc 111(2):274–279
70. Jiao J, Moudon AV et al (2012) How to identify food deserts: measuring physical and economic access to supermarkets in King County, Washington. Am J Public Health 102 (10):e32–e39
71. Johnson DB, Payne EC et al (2012) Menu-labeling policy in King County, Washington. Am J Prev Med 43(3 Suppl 2):S130–S135
72. Kahn EB, Ramsey LT, Brownson RC, Heath GW, Howze EH, Powell KE, Stone EJ, Rajab MW, Corso P (2002) The effectiveness of interventions to increase physical activity. A systematic review. Am J Prev Med 22(4 Suppl):73–107
73. Kaiser HJ (2004) The role of media in childhood obesity: issue brief. Henry J. Kaiser Foundation, Menlo Park, CA
74. Kelly B, Flood VM et al (2011) Measuring local food environments: an overview of available methods and measures. Health Place 17(6):1284–1293
75. Kenny PJ (2011) Reward mechanisms in obesity: new insights and future directions. Neuron 69(4):664–679
76. Krieger JW, Chan NL et al (2013) Menu labeling regulations and calories purchased at chain restaurants. Am J Prev Med 44(6):595–604
77. Larson C, Haushalter A et al (2013) Development of a community-sensitive strategy to increase availability of fresh fruits and vegetables in Nashville's urban food deserts, 2010-2012. Prev Chronic Dis 10:E125
78. Larson N, Story M (2009) A review of environmental influences on food choices. Ann Behav Med 38:S56–S73
79. Laska MN, Graham DJ et al (2010) Young adult eating and food-purchasing patterns food store location and residential proximity. Am J Prev Med 39(5):464–467
80. Lee-Kwan SH, Goedkoop S et al (2013) Development and implementation of the Baltimore healthy carry-outs feasibility trial: process evaluation results. BMC Public Health 13:638
81. Leung CW, Ding EL et al (2012) Dietary intake and dietary quality of low-income adults in the Supplemental Nutrition Assistance Program. Am J Clin Nutr 96(5):977–988

82. Longacre MR, Drake KM et al (2012) Fast-food environments and family fast-food intake in nonmetropolitan areas. Am J Prev Med 42(6):579–587
83. Lucan SC, Mitra N (2012) Perceptions of the food environment are associated with fast-food (not fruit-and-vegetable) consumption: findings from multi-level models. Int J Public Health 57(3):599–608
84. Lytle LA (2009) Measuring the food environment: state of the science. Am J Prev Med 36 (4 Suppl):S134–S144
85. Madden D (2013) The poverty effects of a 'Fat-Tax' in Ireland. Health Econ
86. Martin KS, Havens E et al (2012) If you stock it, will they buy it? Healthy food availability and customer purchasing behaviour within corner stores in Hartford, CT, USA. Public Health Nutr 15(10):1973–1978
87. McCormack LA, Laska MN et al (2010) Review of the nutritional implications of farmers' markets and community gardens: a call for evaluation and research efforts. J Am Diet Assoc 110(3):399–408
88. McKinnon RA, Reedy J et al (2009) Measures of the food environment a compilation of the literature, 1990-2007. Am J Prev Med 36(4):S124–S133
89. Mercille G, Richard L et al (2012) Associations between residential food environment and dietary patterns in urban-dwelling older adults: results from the VoisiNuAge study. Public Health Nutr 15(11):2026–2039
90. Mitchell NS, Catenacci VA et al (2011) Obesity: overview of an epidemic. Psychiatr Clin North Am 34(4):717–732
91. Moore LV, Diez Roux AV et al (2008) Comparing perception-based and geographic information system (GIS)-based characterizations of the local food environment. J Urban Health 85(2):206–216
92. Moore LV, Diez Roux AV et al (2009) Fast-food consumption, diet quality, and neighborhood exposure to fast food: the multi-ethnic study of atherosclerosis. Am J Epidemiol 170 (1):29–36
93. Morland KB (2010) An evaluation of a neighborhood-level intervention to a local food environment. Am J Prev Med 39(6):e31–e38
94. Morley B, Scully M et al (2013) What types of nutrition menu labelling lead consumers to select less energy-dense fast food? An experimental study. Appetite 67:8–15
95. Nalebuff B (2012) Ban calories, not ounces, to regulate sugary beverages. http://www.bloomberg.com/news/2012-09-10/ban-calories-not-ounces-to-regulate-sugary-beverages.html
96. Namba A, Auchincloss A et al (2013) Exploratory analysis of fast-food chain restaurant menus before and after implementation of local calorie-labeling policies, 2005-2011. Prev Chronic Dis 10:E101
97. Nederkoorn C, Havermans RC et al (2011) High tax on high energy dense foods and its effects on the purchase of calories in a supermarket. An experiment. Appetite 56(3):760–765
98. Nevarez CR, Lafleur MS et al (2013) Salud Tiene Sabor: a model for healthier restaurants in a Latino community. Am J Prev Med 44(3 Suppl 3):S186–S192
99. Ni Mhurchu C, Blakely T et al (2010) Effects of price discounts and tailored nutrition education on supermarket purchases: a randomized controlled trial. Am J Clin Nutr 91 (3):736–747
100. Novak NL, Brownell KD (2011) Taxation as prevention and as a treatment for obesity: the case of sugar-sweetened beverages. Curr Pharm Des 17(12):1218–1222
101. Odoms-Young AM, Kong A et al (2013) Evaluating the initial impact of the revised Special Supplemental Nutrition Program for Women, Infants, and Children (WIC) food packages on dietary intake and home food availability in African-American and Hispanic families. Public Health Nutr 17(1):83–93
102. Ohri-Vachaspati P, Leviton L et al (2012) Strategies proposed by healthy kids, healthy communities partnerships to prevent childhood obesity. Prev Chronic Dis 9:E11
103. Pomeranz JL (2012) Advanced policy options to regulate sugar-sweetened beverages to support public health. J Public Health Policy 33(1):75–88

104. Powell LM, Chriqui JF et al (2013) Assessing the potential effectiveness of food and beverage taxes and subsidies for improving public health: a systematic review of prices, demand and body weight outcomes. Obes Rev 14(2):110–128
105. Quandt SA, Dupuis J et al (2013) Feasibility of using a community-supported agriculture program to improve fruit and vegetable inventories and consumption in an underresourced urban community. Prev Chronic Dis 10:E136
106. Richards MR, Sindelar JL (2013) Rewarding healthy food choices in SNAP: behavioral economic applications. Milbank Q 91(2):395–412
107. Richardson AS, Boone-Heinonen J et al (2011) Neighborhood fast food restaurants and fast food consumption: a national study. BMC Public Health 11:543
108. Rivard C, Smith D et al (2012) Taxing sugar-sweetened beverages: a survey of knowledge, attitudes and behaviours. Public Health Nutr 15(8):1355–1361
109. Robert Wood Johnson (RWJ) Foundation (2013) Healthy eating research. http://www.healthyeatingresearch.org/. Accessed on 31 Oct 2013
110. Rose D, Bodor JN et al (2010) The importance of a multi-dimensional approach for studying the links between food access and consumption. J Nutr 140(6):1170–1174
111. Roseman MG, Mathe-Soulek K et al (2013) Relationships among grocery nutrition label users and consumers' attitudes and behavior toward restaurant menu labeling. Appetite 71C:274–278
112. Rudelt A, French S et al (2013) Fourteen-year trends in sodium content of menu offerings at eight leading fast-food restaurants in the USA. Public Health Nutr 10:1–7
113. Sorensen G, Barbeau E (2012) Steps to a healthier US workforce: integrating occupational health and safety and worksite health promotion: state of the science. US Department of Health and Human Services, Public Health Service, Centers for Disease Control and Prevention, National Institute for Occupational Safety and Health, Washington, DC
114. Sorensen G, Landsbergis P, Hammer L, Amick BC 3rd, Linnan L, Yancey A, Welch LS et al (2011) Preventing chronic disease in the workplace: a workshop report and recommendations. Am J Public Health 101(Suppl 1):S196–S207
115. Story M, Kaphingst KM et al (2008) Creating healthy food and eating environments: policy and environmental approaches. Annu Rev Public Health 29:253–272
116. Sturm R, An R et al (2013) A cash-back rebate program for healthy food purchases in South Africa: results from scanner data. Am J Prev Med 44(6):567–572
117. Sturm R, Cohen DA (2009) Zoning for health? The year-old ban on new fast-food restaurants in South LA. Health Aff (Millwood) 28(6):w1088–w1097
118. Swartz JJ, Braxton D et al (2011) Calorie menu labeling on quick-service restaurant menus: an updated systematic review of the literature. Int J Behav Nutr Phys Act 8:135
119. Swinburn B, Egger G et al (1999) Dissecting obesogenic environments: the development and application of a framework for identifying and prioritizing environmental interventions for obesity. Prev Med 29(6 Pt 1):563–570
120. Tandon PS, Zhou C et al (2011) The impact of menu labeling on fast-food purchases for children and parents. Am J Prev Med 41(4):434–438
121. Tiffin R, Arnoult M (2011) The public health impacts of a fat tax. Eur J Clin Nutr 65 (4):427–433
122. Ulmer VM, Rathert AR et al (2012) Understanding policy enactment: the New Orleans Fresh Food Retailer Initiative. Am J Prev Med 43(3 Suppl 2):S116–S122
123. USDA Food and Nutrition Service Programs (2013). Summary of annual data, FY 2008-2012. http://www.fns.usda.gov/pd/Overview.htm. Accessed on 28 Oct 2013
124. Vadiveloo MK, Dixon LB et al (2011) Consumer purchasing patterns in response to calorie labeling legislation in New York City. Int J Behav Nutr Phys Act 8:51
125. Vernez Moudon A, Drewnowski A et al (2013) Characterizing the food environment: pitfalls and future directions. Public Health Nutr 16:1238–1243
126. Walker RE, Keane CR et al (2010) Disparities and access to healthy food in the United States: a review of food deserts literature. Health Place 16(5):876–884

127. Waterlander WE, de Boer MR et al (2013) Price discounts significantly enhance fruit and vegetable purchases when combined with nutrition education: a randomized controlled supermarket trial. Am J Clin Nutr 97(4):886–895

128. Waterlander WE, Steenhuis IH et al (2012) The effects of a 25% discount on fruits and vegetables: results of a randomized trial in a three-dimensional web-based supermarket. Int J Behav Nutr Phys Act 9:11

129. Waters E, de Silva-Sanigorski A, Hall BJ, Brown T, Campbell KJ, Gao Y, Armstrong R, Prosser L, Summerbell CD (2011) Interventions for preventing obesity in children. In: The Cochrane Collaboration, Waters E (eds) Cochrane database of systematic reviews. John Wiley & Sons, Ltd., Chichester

130. Webber CB, Sobal J et al (2010) Shopping for fruits and vegetables. Food and retail qualities of importance to low-income households at the grocery store. Appetite 54(2):297–303

131. Whaley SE, Ritchie LD et al (2012) Revised WIC food package improves diets of WIC families. J Nutr Educ Behav 44(3):204–209

132. Wieland LS, Falzon L, Sciamanna CN, Trudeau KJ, Brodney S, Schwartz JE, Davidson KW (2012) Interactive computer-based interventions for weight loss or weight maintenance in overweight or obese people. Cochrane Database Syst Rev 8:CD007675

133. Wu HW, Sturm R (2013) What's on the menu? A review of the energy and nutritional content of US chain restaurant menus. Public Health Nutr 16(1):87–96

134. Zenk SN, Odoms-Young A et al (2012) Fruit and vegetable availability and selection: federal food package revisions, 2009. Am J Prev Med 43(4):423–428

135. 2012. A soda ban too far. The New York Times, May 31, Opinion

Chapter 12
Cancer Prevention Through Policy Interventions That Alter Childhood Disparities in Energy Balance

Debra Haire-Joshu

Abstract Dramatic increases in obesity and cancer risk are more pronounced across various socially disadvantaged populations (Dixon et al., Adv Nutr 3(1): 73–82, 2012). These disparities have their beginnings in early childhood with devastating effects that track into adulthood. The purpose of this chapter is to describe the (1) prevalence of obesity disparities in youth, (2) social determinants and dimensions of obesity disparities, (3) influences on stages of obesity development, and (4) role and types of policies designed to prevent obesity disparities in young children. The need to behaviorally disrupt the intergenerational cycle of obesity that begins in early life is discussed. How socioeconomic, sociocultural, living and working conditions, and life course exposures influence this cycle is addressed. The role of evidence-based policies, and their impact across target environments where children spend time, is presented.

Keywords Obesity prevention • Early childhood • Social determinants • Target environments • Evidence-based policy

Introduction

Almost 69 % of adults and 32 % of children are overweight and obese, the result of energy imbalance between calorie intake and expenditure [1, 2]. Obesity is associated with increased risk of several cancer types including colon, breast, endometrium, liver, kidney, esophagus, gastric, pancreatic, gallbladder, and leukemia and can lead to poorer treatment and increased cancer-related mortality [3–5]. In 2007 about 34,000 new cases of cancer in men (4 %) and 50,500 in women (7 %) were due to obesity [6]. Cancer attributed to obesity was as high as 40 % for

D. Haire-Joshu (✉)
The Brown School, Public Health, Washington University, Campus Box 1196, One Brookings Drive, St. Louis, MO 63130-4899, USA
e-mail: djoshu@wustl.edu

D.J. Bowen et al. (eds.), *Impact of Energy Balance on Cancer Disparities*,   283
Energy Balance and Cancer 9, DOI 10.1007/978-3-319-06103-0_12,
© Springer International Publishing Switzerland 2014

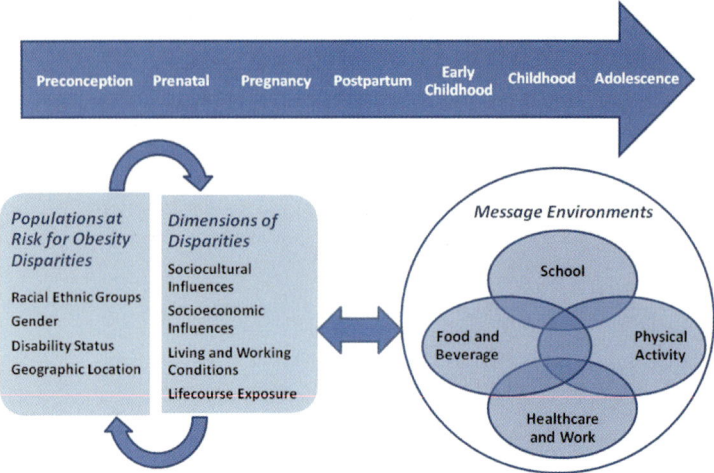

Fig. 12.1 An organizational model for addressing disparities and policy interventions to prevent obesity. Adopted from Institute of Medicine. *Evaluating Obesity Prevention Efforts: A Plan for Measuring Progress.* Washington, DC: The National Academies Press, 2013

endometrial cancer and esophageal adenocarcinoma [6]. The link between obesity and cancer underscores the importance of maintaining a healthy body weight throughout life as one of the most important ways to protect against cancer. This is in addition to other sequelae associated with obesity including metabolic syndrome, diabetes, cardiovascular disease, hypertension, and other chronic diseases [7].

Dramatic increases in obesity and cancer risk have been more pronounced across various racial ethnic groups and socially disadvantaged populations [8]. These disparities have their beginnings in early childhood with devastating effects that track into adulthood. Additionally, the onset of the obesity epidemic in the 1980s has yielded a generation of adult mothers who now parent at-risk children [9].

Eating and physical activity behaviors, learned and reinforced by parents and families, directly impact energy balance [10, 11]. However, social determinants influence the quality of the environments that impact these same behaviors [12, 13]. Exposure to environments of negative quality can impact the onset of obesity at the earliest stages of life. Methods for addressing and halting the intergenerational obesity epidemic are needed to prevent an epidemic of cancer manifested among our youth, further contributing to health disparities. Policy interventions promote opportunities supporting children in attaining healthy behaviors across the multiple environments where they spend time [14–16]. Evidence-based policies help to assure the health quality of those environments.

The purpose of this chapter is to further describe (1) the prevalence of obesity disparities in youth, (2) social determinants and dimensions of obesity disparities, (3) the influences on stages of obesity development, and (4) the role and types of policies designed to prevent obesity disparities in young children. Figure 12.1 present an organizing model for the chapter, which promotes the understanding of how populations at risk are impacted by dimensions of disparities and how evidence-based policies can reduce disparities by promoting the quality of target environments across key stages of obesity development.

Obesity Disparities in Youth

Cancer disparities associated with obesity in adulthood have their beginnings in early childhood. Approximately 17 % (or 12.5 million) of children and adolescents aged 2–19 years are obese [2]. One-third of babies are overweight or obese by age 9 months [1]. Rapid weight gain contributes to 34 % of overweight or obese children by age 2 years [17]. By age 5, 21 % of children are overweight or obese, a number that has doubled in the past 30 years [18]. Older children and adolescents had higher obesity prevalence than preschool children [19]. In addition to the prevalence of obesity, the type of obesity has shifted. Specifically, youth are becoming more centrally obese as defined by higher waist circumference in contrast to body mass index (BMI) with several minority groups exhibiting increases in overall fat in contrast to white counterparts [20].

Obesity disparities are pronounced across various racial ethnic groups including African Americans, Hispanics, American Indian and Alaskan Natives, Asian Americans, Hawaiian, and Pacific Islanders [21, 22]. Wang [19] analyzed national survey data concluding that American Indians had the highest obesity prevalence while blacks and Mexican Americans had rates higher than those in whites and Asians. Numerous studies support these conclusions noting Hispanic (21 %) and non-Hispanic black (24 %) youth have higher rates of obesity than non-Hispanic white youth (14 %) [23]. Obesity is most prevalent in children aged 2–4 years who are American Indian or Alaska Natives (21.1 %) and Hispanic (17.6 %); Hawaiian–Pacific Islanders (Samoan children) report 17.5–27 % prevalence among 1–4-year-olds [24]. The early onset and cumulative effects of obesity increase the risk for cancer and poor health outcomes that track into adulthood with Native Hawaiians/Pacific Islanders 70 % more likely to be obese than non-Hispanic whites [23, 25] and non-Hispanic adult blacks with the highest age-adjusted rates of obesity (49.5 %) compared with Mexican Americans (40.4 %), all Hispanics (39.1 %), AI/AN (39.9 %), and whites (34.3 %) [23, 25].

Gender disparities are also evident at a young age. Among boys, Mexican Americans report the highest combined prevalence (40.5 % vs. 34.5 % in whites and 32.1 % in blacks) [19]. In girls, blacks have the highest prevalence (44.5 % vs. 31.7 % in whites and 37.1 % in Mexican Americans). In all age groups, between 1988–1994 and 2007–2008, the estimated average annual increase in obesity was ~0.6 percentage points. Among children, boys had a faster increase in obesity than girls (0.7 vs. 0.5 percentage points), although the increases were similar in adolescents (~0.5 %) [19]. These obesity disparities place children from high-risk racial ethnic groups at greater risk chronic diseases including cancer [9].

Disparities not only are present by racial ethnic groups but also exist among children with disabilities. For example, adolescents with autism and Down syndrome were two to three times more likely to be obese than adolescents in the general population; obesity among adolescents with physical and cognitive disabilities (17.5 %) is significantly higher than among adolescents without disabilities (13.0 %) [26, 27].

Finally, obesity also varies by urban versus rural geographic location. YRBSS data have shown considerable disparities in obesity rates across the covered states and cities in the USA [28]; nine of the ten states with the highest rates of obese children are in the South with evidence that rural children are more likely than urban children to be obese [29, 30]. More than one-third of children in both large (34.6 %) and small rural areas (35.2 %) had a BMI at or above the 85th percentile for their age and sex, compared to 30.9 % of urban children, a finding further enhanced by children living in poverty [31]. In 2007, across the 39 included states, obesity prevalence ranged from 8.7 to 17.9 % and was 20.4 to 35.8 % for the combined prevalence [32]. Utah (8.7 %) had the lowest and Mississippi (17.9 %) had the highest obesity prevalence [19]. In children aged 6–9 years, the combined prevalence was higher in urban areas (26.1 % vs. 22.8 %), but in adolescents, it was slightly higher in rural areas (27.2 % vs. 24.4 %) [19].

The prevalence of obesity among disadvantaged youth, and resulting health risks for cancer and related chronic diseases, identifies the critical need to better define and understand factors that influence or cause these disparities.

Dimension of Disparities and Childhood Obesity

Health disparities have been defined by Braveman et al. [33, 34] as a "difference in health that can be shaped by policies, in which disadvantaged social groups systematically experience worse health or more health risks than do more advantaged social groups." Determinants of disparities in obesity are complex driven by interacting factors across multiple levels of influence. These "upstream" determinants have been defined by socioeconomic, sociocultural, living conditions, and life course influences that, in turn, impact "downstream" outcomes defined by individual behavior among populations at risk for obesity disparities.

The health behavior of a child cannot be divorced from *socioeconomic influences* that impact healthy choices (e.g., wealth vs. poverty, food security, quality of education) [35, 36]. Several studies show that education or occupation, as indicators of economic resources and social class, can influence the risk for obesity [37–39]. In 2008, over 14 million children less than 18 years of age lived in families with incomes at or below the federal poverty level while almost 30 million children lived in low-income families [40–43]. Children less than 6 years of age accounted for 44 % of those living in low-income families. African American, Hispanic, or American Indians were twice as likely as other racial-ethnic groups to be living in economic deprivation [9, 44]. Children living in poverty are at higher risk for obesity and its negative health outcomes including cancer, diabetes, and early cardiovascular disease [12, 45, 46]. For example, poverty impacts food security, a driver of obesity and cancer risk [47]. Food security exists when people have physical, social, and economic access to sufficient and nutritious food that meets their dietary needs [48]; food insecurity describes an environment which lacks access to sufficient quality and quantity of foods with periods of hunger common [49–53]. In 2008, nearly 23 % of children (16.7 million) lived in food-insecure households defined by interrupted food intake.

Sociocultural influences, such as racism and discrimination, are associated with poorer living conditions due to residential segregation and chronic stress related to racial/ethnic bias [37, 54, 55]. Children exposed to discrimination and limited educational or occupational benefits are at risk for the development of early obesity [56, 57]. Discrimination contributes to poor *living and working conditions* described by the quality of the physical or the built environment including housing conditions, access to healthy food outlets, and street connectivity or density [13, 35, 58, 59]. Appropriate and stable housing, safe living conditions, and access to healthy food outlets can have a direct impact on physical activity and eating behaviors associated with the prevention and control of childhood obesity [60]. Poor-quality living conditions limit access to other services that impact young children. For example, children of the working poor are at high risk for poor-quality childcare services further exacerbating the lack of quality nutrition during the early years [61–63]. Approximately 60 % of young children are cared for in out-of-home childcare settings [64, 65]. Lack of ability to pay for quality childcare services places children at risk for obesity due to poorer access to quality nutrition and physical activity [64, 66].

Finally, *life course exposure* to childhood disadvantage has been associated with increased risk of obesity into adulthood; those with the longest exposure to disadvantaged circumstances across the life course are at highest risk for negative health outcomes [67]. The pervasive role of stress associated with disadvantaged environments influences neuroendocrine, inflammatory, or immune outcomes that lead to cancer, diabetes, and heart disease [68]. The cumulative burden of adverse dimensions of disparities significantly impacts the health of minority populations in contrast to short-term exposure [69, 70].

Evidence-Based Policy and Target Environments

Over the past decade, the role of policy as a means of altering upstream factors to positively influence eating and activity behaviors has become central to obesity prevention efforts. Evidence-based policies provide a mechanism for creating an environment that eliminates barriers and provides support for healthy behavioral choices. Policies can be initiated at the local, organizational, state, or federal level. They hold promise as a strategy that will "level the playing field" by assuring that all populations have access to, and are supported by, resources and environments that promote healthy outcomes and reduce cancer risk. Policies can promote health and eliminate disparities associated with economic, cultural, or physical factors.

Table 12.1 Examples of upstream obesity policies by target environment

Environment	Sample policies	Applied examples
Food and beverage	Policies related to agriculture, food availability, and access	Adopt nutrition standards for federal child nutrition programs that are aligned with guidance on optimal nutrition ensure food literacy
		Government agencies should maximize participation in federal nutrition assistance programs and increase access to healthy foods at the community level
Physical activity	Policies related to urban planning, housing and the built environment, land use and transportation, recreational facilities	Increase access to safe places for physical activity, joint use agreements between parks and schools
School–childcare	Policies related to quality childcare, education, and services	Increase the number of states with licensing regulation for physical activity in childcare that require the number of minutes of physical activity per day or by length of time
		Require schools to implement policies to assure that students meet national guidelines for physical activity and nutrition; implement curriculum designed to support healthy lifestyle behaviors
Health Care–worksite	Policies related to businesses and work environments	Promote support for breastfeeding at the workplace
	Policies on health care infrastructure and financing, public health systems, and access to and delivery of care	Institute baby-friendly hospital practices; educate community on importance of prepregnancy weight control and prenatal care
Message	Policies on media, marketing, and information	Implement social marketing campaign to promote consistent messaging of healthy foods and physical activity, reduce marketing of poor-quality foods to children

Modified from Institute of Medicine Reports on Childhood Obesity [71–73]

They can target societal or social factors that can create support for obesity prevention behaviors.

The Institute of Medicine (IOM) identified evidence-based policy as a mechanism for influencing five environments or settings targeted as critical to eliminating childhood obesity depicted previously in Fig. 12.1, with examples of upstream policies provided in Table 12.1 [71]. The food and beverage environment is defined by policies promoting access and availability to healthy, quality foods [74, 75]. Policies targeting the physical activity environment address the distribution of resources that encourage physical activity or active play for children [76, 77]. School and early childcare environments are focused on assuring that the youngest of our children have access to healthy food and physical activity delivered by qualified caregivers trained in practices designed to prevent the onset of obesity [78]. Health care and work environments are focused on assuring access to quality health care and supportive food and physical activity across environments [70, 71]. Finally, the message environment addresses marketing strategies that promote consumption of healthy food or physical activity [66–68].

Stages of Obesity Development

Limiting exposure to negative influences and stressors is critical to impacting the life course of obesity development and halting the intergenerational obesity and cancer risk. Policies can serve as an upstream intervention with promise to support healthy, individual behaviors of children yielding optimal health outcomes [34, 79]. Policy can impact the development of obesity in children across the life course defined for purposes of this chapter as pregnancy and the intrauterine environment, early childhood (0–5 years), and school-aged youth (6–18 years) [79].

Pregnancy and the intrauterine environment. The earliest risk for the onset of childhood obesity occurs during pregnancy. In the USA, more than one-half of pregnant women are overweight or obese putting them at a greater risk of pregnancy complications [80]. The prevalence of racial-ethnic disparities among women in general is also present among pregnant women [81–83]. The pathways that link the mother's prenatal status to her offspring include metabolic effects of obesity on the intrauterine environment of the child, maternal behaviors that impact gestational weight gain (GWG), and a genetic predisposition transmitted to the child [83, 84]. Maternal obesity before and during pregnancy and excess GWG increase both maternal and neonatal morbidity and mortality [84]. Obesity is associated with adverse pregnancy outcomes including stillbirth, neonatal death, miscarriage, congenital malformations, preeclampsia, gestational diabetes mellitus, preterm birth, macrosomia, and cesarean delivery [15, 85]. It disrupts glucose homeostasis, insulin sensitivity, amino acid synthesis, and fat metabolism, directly increasing the risk for obesity in the child [86–88]. Additionally, there is evidence that the food choices a mother makes during her pregnancy may set the stage for an infant's later acceptance of solid foods, influencing childhood obesity development past pregnancy [89]. The experiences of taste and smell function during fetal life provide a "flavor bridge" that plays a key role in the acquisition of food and flavor preferences of the child [90].

An important step in preventing obesity is to design evidence-based policies that interrupt the pathways contributing to obesity that begin in early life and are prevalent in populations at risk for disparities [72]. Interventions that encourage ideal preconception weight, appropriate GWG, and reduction in postpartum weight retention (PPWR) can interrupt the pathway that links maternal and child obesity. Table 12.2 gives examples of these interventions, introduced next.

Table 12.2 Sample policy and practice recommendations

Sample policy or practice recommendations	Application to practice
Prenatal, pregnancy, and postpartum	
Promote early prenatal, postnatal, and interconceptual care	Health care providers should incorporate most recent IOM/ACOG guidelines in prenatal/interconceptual care
Achieve health gestational weight gain	Health care providers should offer consistent health recommendations regarding breastfeeding and infant nutrition across prenatal and postnatal health care providers
Encourage postpartum return to healthy weight	
Early childhood (0–5 years)	
Assess, monitor, and track growth from birth to age 5	Health care providers should track growth from birth to 5 using standardized approaches and WHO or CDC growth charts AND assess parent BMI in routine assessment of child risk
Promote the consumption of a variety of nutrition foods and encourage and support breastfeeding during infancy	Implement baby-friendly initiatives in hospitals
Create a healthful eating environment responsive to children's hunger and fullness cues	Encourage employers to implement policies that support breastfeeding mothers after they return to work
	State child care regulatory agencies should require childcare providers to practice responsive feeding in accordance with CACFP guidelines
Increase physical activity and decrease sedentary activity in young children	Childcare regulatory agencies should require childcare providers to provide opportunities for the infant to move freely and explore their environments and for toddlers to engage in physical activity 15 min per hour
Limit young children's screen time and exposure to food and beverage marketing	Childcare settings should limit screen time
	Health care providers should counsel parents to limit screen time
Promote age-appropriate sleep durations among children	Childcare agencies should adopt practices that promote age-appropriate sleep durations for children
	Childcare regulatory agencies should require childcare agencies to environments, behaviors, and practices that promote restful sleep
Childhood to adolescence (6–18 years)	
Require quality physical education and opportunities for physical activity in schools	Increase the numbers of school districts that require elementary schools recess for an appropriate amount of time
Ensure strong nutritional standards for all foods and beverages sold or provided through schools	Increase the availability of fruits and vegetables with foods offered in school
	Assure that meals meet 2012 federal nutrition standards for school meals
Promote food literacy in schools	Schools should require curriculum that meets National Health Education Standards

Modified from Institute of Medicine [71, 91] and Nader et al. [73]

There is a strong link between maternal *preconception weight* and the likelihood that these women will give birth to infants and children who will develop obesity [92]. There appears to be a dose–response association with the magnitude of maternal obesity related to that of her child [93, 94]. Strategies for improving women's preconception health include greater individual responsibility across the life-span, preventive visits and intervening for identified risks, interconception care, prepregnancy checkup, public health programs and strategies, and monitoring improvements [94]. Policies that encourage ideal maternal weight prior to pregnancy are likely to impact the prevalence of early childhood obesity.

Increased *GWG* elevates the risk for excessive weight in the offspring of the mother [91]. Recent studies found a preponderance of evidence to support the association between GWG and childhood weight as a continuous measure and risk of overweight and obesity [86, 87, 95]. In 2009, the IOM published revised GWG guidelines that are based on prepregnancy BMI ranges for underweight, normal-weight, overweight, and obese women recommended by the World Health Organization [95]. Recent studies underscore the importance of dietary therapy, and encouraging an increase in exercise, for controlling GWG among obese women [3, 96–98]. Women who routinely participate in exercise during pregnancy reduce their incidence of gestational diabetes, have less GWG, have less PPWR, and give birth to smaller but normal weight range infants than those who do not exercise [49, 99, 100]. In general these studies identify strategies that promote ideal nutrition and physical activity during pregnancy, coupled with close monitoring of weight, which are protective against excessive GWG [15, 85, 101–103]. These findings suggest the importance of encouraging weight control within the IOM-ACOG guidelines to promote the optimal uterine environment and infant outcomes.

Maternal obesity is associated with *PPWR* that is exacerbated by obesity and multiple pregnancies [95, 104, 105]. PPWR is associated with preconception weight and gestational weight gain [106]. Nehring et al. found that women who gained above the IOM recommendations retained an additional 3.06 kg after 3 years and 4.72 kg on average after ≥ 15 years postpartum [107]. Several provider-based interventions have been designed with a goal to reduce PPWR [108, 109]. For example, the Fit for Pregnancy study included one face-to-face visit; weekly mailed materials that promoted an appropriate weight gain, healthy eating, and exercise; individual graphs of weight gain; and telephone-based feedback. This intervention reduced excessive GWG and prevented PPWR in normal-weight, overweight, and obese women [96]. Policies that encourage provider-based interventions can impact the risk for postpartum weight gain and obesity onset and reduce the risk for cancer and related disease for a long term.

Early Childhood (0–5 Years)

Offspring born to obese mothers have a greater risk of neurodevelopmental delay, atypical neurodevelopment, becoming obese, and other sequelae when compared to those born to healthy, lean women [5]. These offspring are also at risk for the onset of rapid weight gain during infancy that is associated with the onset of early-childhood obesity [17].

Policy interventions hold promise for preventing obesity and related illnesses, such as cancer, in very young children from disadvantaged populations. The IOM targeted several key areas in which policies can support behavioral and practice changes that impact young children [91]. These include growth monitoring, breastfeeding, healthy eating, physical activity, screen time and marketing, and adequate sleep.

Growth monitoring, defined by routine tracking of height and weight from birth to age 5, is needed to assure optimal child growth and development and identify early-onset obesity. Parental perceptions about weight can influence their behaviors in child feeding [110]. Recent studies suggest that parents are unlikely to recognize overweight in their young child [111]. Additionally, overweight mothers tend to underestimate the weight of their child, a finding associated with disparities of less income and education [112]. Continual monitoring of infant and child weight through well child visits affords health care providers the opportunity to encourage appropriate child feeding and activity [91]. This also allows for early intervention at the first sign of excessive weight gain.

Breastfeeding is recommended as the optimal feeding method for the first 6 months of life, followed by the introduction of solids and continued breastfeeding for a minimum of 1 year [113]. Breast milk supports normal growth, has immuno-logical properties that provide some early protection from infection, and may protect against obesity [114, 115]. Breastfeeding may also impact food acceptance and encourage the infant to self-regulate or control energy intake [90, 116]. For example, Mennella and colleagues found that experience with different flavors in breast milk facilitated the infant's acceptance of foods of the adult diet, especially those foods consumed by the mother during lactation [89]. In contrast to bottle-fed infants, breastfeeding encourages infants to self-regulate by adjusting the volume of milk consumed [113, 117–120]. Despite these advantages, disparities in the breastfeeding practices of the US women are quite evident [121]. Breastfeeding initiation rates are markedly lower among black women (60 %) compared with other ethnic groups. Also of concern, Hispanic and black women have the highest rates of formula supplementation of breast-fed infants [121, 122]. The low rates and rapid attrition of breastfeeding by mothers in general, and those from minority groups in particular, suggest a need for strategies to promote the consumption of a variety of nutritious foods and encourage and support breastfeeding during infancy [123, 124]. Policies are needed that adequately prepare a mother to breastfeed through educational initiatives and to encourage adults who care for infants to support breastfeeding [74, 75]. Examples of policies are provided in Table 12.1 [115, 121, 123, 125] designed to promote breastfeeding.

In addition to support for breastfeeding, strategies also need to create a *healthful eating environment* responsive to children's hunger and fullness cues. The majority of young children in this country are not consuming healthy diets [120]. Parental understanding of the normal growth and development of their child, and the negative sequelae of excessive weight gain, is critical as they promote the food environment of the infant [126, 127]. The feeding practices of parents influence the development of food preference beginning in infancy [90]. Infants respond to differences in energy density early in life, self-regulating their intake to accommodate nutritional needs [128, 129]. Parental feeding practices focused on type, amount, and patterns of child feeding can influence the child's ability to self-regulate energy intake, a finding that has been associated with differences in weight status [130–133]. Parents should be encouraged to practice responsive feeding practices, including age-appropriate foods and portion sizes, provided through daily routines for timing of snacks and meals [90, 132, 133]. Policies are needed to maximize access to healthy foods, critical to eliminating obesity disparities associated with poverty and low-income areas [134–137]. Strategies that maximize participation in federal nutrition assistance programs are one means of increasing access to healthy foods for the youngest children.

Another goal for young children is to *increase physical activity and decrease sedentary activity*. This goal is based on the extensive evidence base suggesting that physical activity benefits older children. Safety concerns often confine the young child to cribs, car seats, and strollers for extended periods of time [91]. Infants and young children need opportunities to engage in unrestricted play to encourage physical activity, prevent excessive weight gain, and maximize the developmental potential of the child [138, 139]. A recent report by the IOM identified policy initiatives to promote physical activity in the very young, primarily focused on educational interventions for parents and caregivers, as well as regulatory guidance to childcare facilities [91]. Recommendations include the provision of opportunities for safe physical activity with guidance suggesting that infants should have access to "tummy time"; toddlers and preschoolers should be active at least 15 min per hour of physical activity [91]. Table 12.2 provides additional examples of recommendations.

Marketing strategies encourage excess consumption of food or discourage physical activity and contribute to obesity disparities in children [140–142]. The use of media to entertain infants, by television and other screen time equipment, has also become common [143]. This suggests that adults and caregivers of the very young should *limit their exposure to screen time and marketing*. There is substantial research to suggest that marketing works to influence food preferences, purchases, and immediate consumption of older children and adults [144–146]. Exposure to TV and advertising is also associated with inactivity and increased snacking and overweight–obesity [143, 147, 148]. Additionally, the American Academy of Pediatrics statement recommending no television viewing for children under 2 years of age is based on concerns that significant brain development that occurs during the first 18–24 months of life can be impacted by exposure to media [149]. The evidence supports strategies to eliminate screen time exposure of children from birth to age 2 and limit exposure to less than 2 h per day for children aged 3–5 [91]. Health care providers should communicate these guidelines to parents and caregivers.

Finally, it is important to promote age-appropriate *sleep duration* among children [8]. Sleep deprivation and shorter durations of sleep are a risk factor for obesity and related disease [150–154]. Less sleep or irregular sleeping habits are inversely associated with elevated BMI among young children 4 years of age [155]. This association between sleep deprivation and obesity appears to track into adulthood [156]. Insufficient childhood sleep may yield metabolic dysfunction [157–159]. Poor sleep habits might also be associated with environmental distractions such as televisions in the bedroom [117, 160–162]. Support for healthy sleep patterns, without food soothing, is critical.

Childhood to Adolescence (6–18 Years)

Over the past three decades, obesity has more than doubled for adolescents 12–19 years and tripled for children 1–6 years of age [163]. The prevalence of obesity is even higher among youth from diverse racial ethnic backgrounds [23]. Approximately 60 % of children and adolescents consume high-fat diets, 79 % do not eat the recommended amount of fruits and vegetables, and 40 % drink soda at least once per day. Children also report watching television for 3 or more hours on an average school day, and 65 % do not meet the recommended levels of moderate and vigorous exercise [164]. Minority youth have higher use of media per day compared to white youth, up to 8 h per day [140, 165–167].

The focus on intervention across the target environments, as noted in early childhood, is relevant for school-aged and adolescent youth as well. Strategies to promote the physical activity environment, food and beverage environments, and message environments parallel those of early childhood. The influence of health care and work environments is also relevant, particularly as adolescent youth enter the workforce or, in some cases, become parents. Of particular relevance to youth is the role of the school environment. Children spend more time in schools than in any other environment away from home. More than 48 million students attend 94,000 public elementary, middle, and secondary schools each day, and an additional 5.3 million students attend 30,000 private schools [168]. More than 95 % of American youth aged 5–17 are enrolled in school, and no other institution has as much continuous and intensive contact and influence on children during their first two decades of life. Health and education success are intertwined: schools cannot achieve their primary mission of education if students are not healthy and fit [169–171]. While the schools alone cannot solve the childhood obesity epidemic, it also is unlikely that childhood obesity rates can be reversed without strong school-based policies and programs to support healthy eating and physical activity.

The IOM identified several policy strategies relevant to the role of schools in preventing obesity and risk for chronic disease and cancer in children and youth [71]. First, the report urged communities to make schools a focal point of obesity prevention efforts. The school environment is interrelated with all aspects of the community and can have a powerful and influential role in structuring healthy environments for children. The 2005 IOM report Preventing Childhood Obesity: Health in the Balance recommends that children and adolescents participate in a minimum of 30 min of physical activity during the school day [172]. Actions that reflect this include requiring quality physical education and opportunities for physical activity in schools [71]. There is substantial evidence documenting the value of recess and activity in schools [173–175]. Despite this, there is also a wide variation in number, timing, and quality of physical activity breaks and recess afforded in schools. In 2006, only 4 % of elementary schools and 2 % of high schools required daily physical education. Among all school districts across the country, only 39 % required 30 min or more of recess per day. While this is an improvement over the past decade, additional work is needed to enable students to achieve the goal of 60 min of activity per day.

The role of nutrition in schools is critical to obesity prevention among youth. The goal of school nutrition programs is to assure that children are nourished effectively and educated as to healthy dietary patterns. Youth with access to healthy foods are more likely to consume those foods [176–180]. Knowledge about healthy choices is also critical to long-term nutrition and health. The Healthy Hunger Free Kids Act of 2010 gave USDA the mandate to develop regulations governing all foods sold and served on school campuses. Children participating in government nutrition programs (e.g., school lunch), many of who are minority children living in poverty, receive nearly half their daily calories from foods received in these programs. Schools provide a mechanism for encouraging appropriate eating patterns, portion sizes, and foods that meet the 2010 dietary guidelines (e.g., fruits and vegetables, low-fat dairy products). Policy strategies to encourage schools and communities to implement these approaches to a broad and comprehensive extent can have immediate and direct impacts on the local youth.

Finally, schools have an important role in assuring the food literacy of children and adolescents [181, 182]. Nutrition education can have a positive impact on a child's food and nutrition intake by changing knowledge and attitudes [183]. While several studies note the importance and impact of nutrition education, more instructional time is needed. Current data suggests that students are exposed to 1–2 min per day of information on nutrition [71]. The success of several school-based programs in preventing obesity, including the CATCH trial and others, suggests the value of nutrition education within a multicomponent intervention [184–186].

Conclusion

The national obesity epidemic is even more devastating among disadvantaged populations. The prevalence of obesity disparities in youth, and the subsequent risk for developing cancer and related chronic disease, suggests the need to behaviorally disrupt the intergenerational cycle of obesity that begins in early life. Dimensions of disparities as defined by socioeconomic, sociocultural, living and working conditions, and life course exposures influence obesity onset. Evidence-based policies can influence the qualities of the environments where children spend time and interrupt the pathways that link behaviors to obesity, reducing the risk for cancer development.

References

1. Moss BG, Yeaton WH (2011) Young children's weight trajectories and associated risk factors: results from the Early Childhood Longitudinal Study-Birth Cohort. Am J Health Promot 25(3):190–198
2. Dalenius K, Borland E, Smith B, Polhamus B, Grummer-Strawn L, Pediatric Nutrition Surveillance 2010 Report 2012, U.S. Department of Health and Human Services, Centers for Disease Control and Prevention: Atlanta, GA
3. Olson CM (2007) A call for intervention in pregnancy to prevent maternal and child obesity. Am J Prev Med 33(5):435–436
4. Ishihara T et al (2003) Relationships between infant lifestyle and adolescent obesity. The Enzan maternal-and-child health longitudinal study [Nihon koshu eisei zasshi]. Jpn J Pub Health 50(2):106–117
5. Rasmussen KM, Kjolhede CL (2008) Maternal obesity: a problem for both mother and child. Obesity 16(5):929–931
6. Obesity and cancer risk. Fact sheet. http://www.cancer.gov/cancertopics/factsheet/Risk/obesity
7. Costa-Font J, Gil J (2013) Intergenerational and socioeconomic gradients of child obesity. Soc Sci Med 93:29–37
8. Dixon B, Pena MM, Taveras EM (2012) Life course approach to racial/ethnic disparities in childhood obesity. Adv Nutr 3(1):73–82
9. Muhlhausler BS et al (2013) Nutritional approaches to breaking the intergenerational cycle of obesity. Can J Physiol Pharmacol 91(6):421–428
10. Sutherland G, Brown S, Yelland J (2013) Applying a social disparities lens to obesity in pregnancy to inform efforts to intervene. Midwifery 29(4):338–343
11. Cole TJ, Power C, Moore GE (2008) Intergenerational obesity involves both the father and the mother. Am J Clin Nutr 87(5):1535–1536, author reply 1536–1537
12. Braveman PA et al (2011) Health disparities and health equity: the issue is justice. Am J Public Health 101(Suppl 1):S149–S155
13. Woolf SH, Braveman P (2011) Where health disparities begin: the role of social and economic determinants–and why current policies may make matters worse. Health Aff 30(10):1852–1859
14. Hendler I et al (2006) Association of obesity with pulmonary and nonpulmonary complications of pregnancy in asthmatic women. Obstet Gynecol 108(1):77–82
15. Dietl J (2005) Maternal obesity and complications during pregnancy. J Perinat Med 33(2):100–105

16. Haire-Joshu D et al (2011) The quality of school wellness policies and energy-balance behaviors of adolescent mothers. Prev Chronic Dis 8(2):A34
17. Goodell LS, Wakefield DB, Ferris AM (2009) Rapid weight gain during the first year of life predicts obesity in 2-3 year olds from a low-income, minority population. J Community Health 34(5):370–375
18. Ogden C, Carroll M (2010) Prevalence of obesity among children and adolescents: United States, Trends 1963–1965 Through 2007–2008. NCHS Health E-Stat 2010; http://www.cdc.gov/nchs/data/hestat/obesity_child_07_08/obesity_child_07_08.htm
19. Wang Y (2011) Disparities in pediatric obesity in the United States. Adv Nutr 2(1):23–31
20. Beydoun MA, Wang Y (2009) Gender-ethnic disparity in BMI and waist circumference distribution shifts in US adults. Obesity 17(1):169–176
21. Bellamy GR, Bolin JN, Gamm LD (2011) Rural healthy people 2010, 2020, and beyond: the need goes on. Fam Community Health 34(2):182–188
22. Brown DW (2009) The dawn of healthy people 2020: a brief look back at its beginnings. Prev Med 48(1):94–95
23. Ogden CL et al (2012) Prevalence of obesity and trends in body mass index among US children and adolescents, 1999-2010. JAMA 307(5):483–490
24. Baruffi G et al (2004) Ethnic differences in the prevalence of overweight among young children in Hawaii. J Am Diet Assoc 104(11):1701–1707
25. Flegal KM et al (2012) Prevalence of obesity and trends in the distribution of body mass index among US adults, 1999-2010. JAMA 307(5):491–497
26. Yamaki K et al (2011) Prevalence of obesity-related chronic health conditions in overweight adolescents with disabilities. Res Dev Disabil 32(1):280–288
27. Rimmer JH et al (2011) Obesity and overweight prevalence among adolescents with disabilities. Prev Chronic Dis 8(2):A41
28. Eaton DK et al (2012) Youth risk behavior surveillance - United States, 2011. MMWR Morb Mortal Wkly 61(4):1–162
29. Levi J, Segal L, St. Laurent R, Lang A, Rayburn J (2012) F as in fat: how obesity threatens America's future 2012. Trust for America's Health, Washington, DC
30. Bethell C et al (2009) Consistently inconsistent: a snapshot of across- and within-state disparities in the prevalence of childhood overweight and obesity. Pediatrics 123(Suppl 5): S277–S286
31. National Survey of Children's Health (2011) The health and well-being of children in rural areas: a portrait of the nation 2007. U.S. Department of Health and Human Services, Health Resources and Services Administration, Maternal and Child Health Bureau, Rockville, MD
32. Brener ND et al (2013) Behaviors related to physical activity and nutrition among U.S. High school students. J Adolesc Health 53(4):539–546
33. Braveman P (2006) Health disparities and health equity: concepts and measurement. Annu Rev Public Health 27:167–194
34. Braveman P (2009) A health disparities perspective on obesity research. Prev Chronic Dis 6(3):A91
35. Braveman P, Egerter S, Williams DR (2011) The social determinants of health: coming of age. Annu Rev Public Health 32:381–398
36. Braveman P, Gruskin S (2003) Poverty, equity, human rights and health. Bull World Health Organ 81(7):539–545
37. Williams DR, Sternthal M (2010) Understanding racial-ethnic disparities in health: sociological contributions. J Health Soc Behav 51(Suppl):S15–S27
38. Dehlendorf C et al (2010) Sociocultural determinants of teenage childbearing among Latinas in California. Matern Child Health J 14(2):194–201
39. Kawachi I, Daniels N, Robinson DE (2005) Health disparities by race and class: why both matter. Health Aff 24(2):343–352
40. Braveman P et al (2001) Measuring socioeconomic status/position in studies of racial/ethnic disparities: maternal and infant health. Publ Health Rep 116(5):449–463

41. Braveman P, Gruskin S (2003) Defining equity in health. J Epidemiol Community Health 57(4):254–258

42. Braveman P et al (2010) Poverty, near-poverty, and hardship around the time of pregnancy. Matern Child Health J 14(1):20–35

43. Keene DE, Geronimus AT (2011) "Weathering" HOPE VI: the importance of evaluating the population health impact of public housing demolition and displacement. J Urb Health Bull N Y Acad Med 88(3):417–435

44. Evans-Campbell T (2008) Historical trauma in American Indian/Native Alaska communities: a multilevel framework for exploring impacts on individuals, families, and communities. J Interpers Violence 23(3):316–338

45. Braveman P, Krieger N, Lynch J (2000) Health inequalities and social inequalities in health. Bull World Health Organ 78(2):232–234, discussion 234-5

46. Braveman PA (2003) Monitoring equity in health and healthcare: a conceptual framework. J Health Popul Nutr 21(3):181–192

47. Morris PM, Neuhauser L, Campbell C (1992) Food security in rural America - a study of the availability and costs of food. J Nutr Educ 24(1):S52–S58

48. Artal R et al (2007) A lifestyle intervention of weight-gain restriction: diet and exercise in obese women with gestational diabetes mellitus. Appl Physiol Nutr Metab 32(3):596–601

49. Tovar A et al (2010) Knowledge, attitudes, and beliefs regarding weight gain during pregnancy among Hispanic women. Matern Child Health J 14(6):938–949

50. Hay J et al (2012) Physical activity intensity and cardiometabolic risk in youth. Arch Pediatr Adolesc Med 166(11):1022–1029

51. Downs SM et al (2012) Geography influences dietary intake, physical activity and weight status of adolescents. J Nutr Metabol 2012:816834

52. Al-Bahry SN et al (2013) Escherichia coli tetracycline efflux determinants in relation to tetracycline residues in chicken. Asian Pac J Trop Med 6(9):718–722

53. Murthy KR et al (2013) Comparison of profile of retinopathy of prematurity in semiurban/rural and urban NICUs in Karnataka, India. Br J Ophthalmol 97(6):687–689

54. Chakraborty BM, Chakraborty R (2010) Concept, measurement and use of acculturation in health and disease risk studies. Colleg Antropolog 34(4):1179–1191

55. Flaskerud JH, DeLilly CR (2012) Social determinants of health status. Iss Mental Health Nurs 33(7):494–497

56. Nadimpalli SB, Hutchinson MK (2012) An integrative review of relationships between discrimination and Asian American health. J Nurs Scholar 44(2):127–135

57. Braveman P (2010) Social conditions, health equity, and human rights. Health Hum Right 12(2):31–48

58. Lovasi GS et al (2009) Built environments and obesity in disadvantaged populations. Epidemiol Rev 31:7–20

59. Braveman PA, Egerter SA, Mockenhaupt RE (2011) Broadening the focus: the need to address the social determinants of health. Am J Prev Med 40(1 Suppl 1):S4–S18

60. Brownson RC et al (2009) Measuring the built environment for physical activity: state of the science. Am J Prev Med 36(4 Suppl):S99–123.e12

61. Hearst MO et al (2011) The co-occurrence of obesity, elevated blood pressure, and acanthosis nigricans among American Indian school children: identifying individual heritage and environment-level correlates. Am J Hum Biol 23(3):346–352

62. Pwint MK et al (2013) Prevalence of overweight and obesity in Chinese preschoolers in Singapore. Ann Acad Med Singapore 42(2):66–72

63. Scott F, Rhodes RE, Downs DS (2009) Does physical activity intensity moderate social cognition and behavior relationships? J Am Col Health 58(3):213–222

64. Ozkok E et al (2008) Combined impact of matrix metalloproteinase-3 and paraoxonase 1 55/192 gene variants on coronary artery disease in Turkish patients. Med Sci Monit 14(10):CR536–CR542

65. Misawa Y et al (2009) Vitamin D(3) induces expression of human cathelicidin antimicrobial peptide 18 in newborns. Int J Hematol 90(5):561–570
66. Ahmad M et al (2011) Ultrastructural and histochemical evaluation of appositional mineralization of circumpulpal dentin at the crown- and root-analog portions of rat incisors. J Electr Microsc 60(1):79–87
67. Coogan PE et al (2012) Life course educational status in relation to weight gain in African American women. Ethnic Dis 22(2):198–206
68. Golden SH et al (2012) Health disparities in endocrine disorders: biological, clinical, and nonclinical factors – an Endocrine Society scientific statement. J Clin Endocrinol Metab 97(9):E1579–E1639
69. Love C et al (2010) Exploring weathering: effects of lifelong economic environment and maternal age on low birth weight, small for gestational age, and preterm birth in African-American and white women. Am J Epidemiol 172(2):127–134
70. Das A (2013) How does race get "under the skin"?: inflammation, weathering, and metabolic problems in late life. Soc Sci Med 77:75–83
71. Glickman D, Parker L, Sim LJ (2012) Accelerating progress in obesity prevention: solving the weight of the nation. The National Academy of Science, Washington, DC
72. C.o.E.P.o.O.P.E.I.o. Medicine (2013) Evaluating obesity prevention efforts: a plan for measuring progress. National Academy of Science, Washington, DC
73. Nader PR et al (2012) Next steps in obesity prevention: altering early life systems to support healthy parents, infants, and toddlers. Child Obes 8(3):195–204
74. Glanz K et al (2005) Healthy nutrition environments: concepts and measures. Am J Health Promot 19(5):330–333, ii
75. Glanz K et al (2007) Nutrition Environment Measures Survey in stores (NEMS-S): development and evaluation. Am J Prev Med 32(4):282–289
76. Gordon-Larsen P et al (2006) Inequality in the built environment underlies key health disparities in physical activity and obesity. Pediatrics 117(2):417–424
77. Jones-Smith JC et al (2011) Cross-national comparisons of time trends in overweight inequality by socioeconomic status among women using repeated cross-sectional surveys from 37 developing countries, 1989-2007. Am J Epidemiol 173(6):667–675
78. Gittelsohn J, Rowan M (2011) Preventing diabetes and obesity in American Indian communities: the potential of environmental interventions. Am J Clin Nutr 93(5):1179S–1183S
79. Braveman P, Barclay C (2009) Health disparities beginning in childhood: a life-course perspective. Pediatrics 124(Suppl 3):S163–S175
80. American College of, O. and Gynecologists (2013) ACOG Committee opinion no. 548: weight gain during pregnancy. Obstet Gynecol 121(1):210–212
81. Cubbin C et al (2002) Socioeconomic and racial/ethnic disparities in unintended pregnancy among postpartum women in California. Matern Child Health J 6(4):237–246
82. Kramer MS et al (2000) Socio-economic disparities in pregnancy outcome: why do the poor fare so poorly? Paediat Perinat Epidemiol 14(3):194–210
83. Alexander GR, Cornely DA (1987) Racial disparities in pregnancy outcomes: the role of prenatal care utilization and maternal risk status. Am J Prev Med 3(5):254–261
84. Rooney BL, Schauberger CW, Mathiason MA (2005) Impact of perinatal weight change on long-term obesity and obesity-related illnesses. Obstet Gynecol 106(6):1349–1356
85. Emerson RG (1962) Obesity and its association with the complications of pregnancy. Br Med J 2(5303):516–518
86. Guelinckx I et al (2008) Maternal obesity: pregnancy complications, gestational weight gain and nutrition. Obes Rev 9(2):140–150
87. Magann EF et al (2011) Pregnancy, obesity, gestational weight gain, and parity as predictors of peripartum complications. Arch Gynecol Obstet 284(4):827–836
88. Marshall NE, Spong CY (2012) Obesity, pregnancy complications, and birth outcomes. Sem Reprod Med 30(6):465–471

89. Mennella JA, Jagnow CP, Beauchamp GK (2001) Prenatal and postnatal flavor learning by human infants. Pediatrics 107(6):E88
90. Savage JS, Fisher JO, Birch LL (2007) Parental influence on eating behavior: conception to adolescence. J Law Med Ethics 35(1):22–34
91. Birch LL, Parker L, Annina B (eds) (2011) Early childhood obesity prevention policies. The National Academies Press, Washington, DC
92. Zhang S et al (2013) Racial disparities in economic and clinical outcomes of pregnancy among medicaid recipients. Matern Child Health J 17(8):1518–1525
93. Herring SJ, Oken E (2011) Obesity and diabetes in mothers and their children: can we stop the intergenerational cycle? Curr Diabet Rep 11(1):20–27
94. Johnson K et al (2006) Recommendations to improve preconception health and health care – United States. A report of the CDC/ATSDR Preconception Care Work Group and the Select Panel on Preconception Care. MMWR Morb Mortal Wkly 55(6):1–23
95. Siega-Riz AM et al (2009) A systematic review of outcomes of maternal weight gain according to the Institute of Medicine recommendations: birth weight, fetal growth, and postpartum weight retention. Am J Obstet Gynecol 201(4):339.e1–14
96. Phelan S et al (2011) Randomized trial of a behavioral intervention to prevent excessive gestational weight gain: the Fit for Delivery Study. Am J Clin Nutr 93(4):772–779
97. Sagedal LR et al (2013) Study protocol: fit for delivery - can a lifestyle intervention in pregnancy result in measurable health benefits for mothers and newborns? A randomized controlled trial. BMC Publ Health 13:132
98. Skouteris H et al (2012) Protocol for a randomized controlled trial of a specialized health coaching intervention to prevent excessive gestational weight gain and postpartum weight retention in women: the HIPP study. BMC Publ Health 12:78
99. Chasan-Taber L et al (2013) Physical activity and gestational weight gain in Hispanic women. Obesity 22(3):909–918
100. Tovar A et al (2012) Acculturation and gestational weight gain in a predominantly Puerto Rican population. BMC Pregnan Childbirth 12:133
101. Davenport MH et al (2013) Timing of excessive pregnancy-related weight gain and offspring adiposity at birth. Obstet Gynecol 122(2, PART 1):255–261
102. Davenport RK Jr, Menzel EW Jr, Rogers CM (1961) Maternal care during infancy: its effect on weight gain and mortality in the chimpanzee. Am J Orthopsychiat 31:803–809
103. Galtier-Dereure F, Boegner C, Bringer J (2000) Obesity and pregnancy: complications and cost. Am J Clin Nutr 71(5 Suppl):1242S–1248S
104. Begum F et al (2012) Gestational weight gain and early postpartum weight retention in a prospective cohort of Alberta women. J Obstetr Gynaecol Can 34(7):637–647
105. Vesco KK et al (2009) Excessive gestational weight gain and postpartum weight retention among obese women. Obstet Gynecol 114(5):1069–1075
106. Mannan M, Doi SA, Mamun AA (2013) Association between weight gain during pregnancy and postpartum weight retention and obesity: a bias-adjusted meta-analysis. Nutr Rev 71(6):343–352
107. Nehring I et al (2011) Gestational weight gain and long-term postpartum weight retention: a meta-analysis. Am J Clin Nutr 94(5):1225–1231
108. Mottola MF et al (2010) Nutrition and exercise prevent excess weight gain in overweight pregnant women. Med Sci Sport Exerc 42(2):265–272
109. Ruchat SM et al (2012) Nutrition and exercise reduce excessive weight gain in normal-weight pregnant women. Med Sci Sport Exerc 44(8):1419–1426
110. Berge JM et al (2011) Are parents of young children practicing healthy nutrition and physical activity behaviors? Pediatrics 127(5):881–887
111. Larson N et al (2011) What role can child-care settings play in obesity prevention? A review of the evidence and call for research efforts. J Am Diet Assoc 111(9):1343–1362
112. Rietmeijer-Mentink M et al (2013) Difference between parental perception and actual weight status of children: a systematic review. Mat Child Nutr 9(1):3–22

113. Gartner LM et al (2005) Breastfeeding and the use of human milk. Pediatrics 115(2):496–506
114. Dewey KG (2003) Is breastfeeding protective against child obesity? J Hum Lact 19(1):9–18
115. Kramer MS, Kakuma R (2004) The optimal duration of exclusive breastfeeding: a systematic review. Adv Exp Med Biol 554:63–77
116. Fisher JO et al (2000) Breast-feeding through the first year predicts maternal control in feeding and subsequent toddler energy intakes. J Am Diet Assoc 100(6):641–646
117. Monasta L et al (2010) Early-life determinants of overweight and obesity: a review of systematic reviews. Obes Rev 11(10):695–708
118. Moss BG, Yeaton WH (2014) Early childhood healthy and obese weight status: potentially protective benefits of breastfeeding and delaying solid foods. Matern Child Health J 18 (5):1224–1232. doi:10.1007/s10995-013-1357-z, Published online September 22, 2013, Springer publishing
119. Owen CG et al (2005) Effect of infant feeding on the risk of obesity across the life course: a quantitative review of published evidence. Pediatrics 115(5):1367–1377
120. Saavedra JM et al (2013) Lessons from the feeding infants and toddlers study in north America: what children eat, and implications for obesity prevention. Ann Nutr Metab 62(Suppl 3):27–36
121. Chapman DJ, Perez-Escamilla R (2012) Breastfeeding among minority women: moving from risk factors to interventions. Adv Nutr 3(1):95–104
122. Chapman DJ (2010) Exploring breastfeeding ambivalence among low-income, minority women. J Hum Lact 26(1):82–83
123. Dowling S, Brown A (2013) An exploration of the experiences of mothers who breastfeed long-term: what are the issues and why does it matter? Breastfeed Med 8(1):45–52
124. Hundalani SG et al (2013) Breastfeeding among inner-city women: from intention before delivery to breastfeeding at hospital discharge. Breastfeed Med 8(1):68–72
125. Phares TM et al (2004) Surveillance for disparities in maternal health-related behaviors–selected states, Pregnancy Risk Assessment Monitoring System (PRAMS), 2000–2001. MMWR Morb Mortal Wkly 53(4):1–13
126. Anzman SL, Rollins BY, Birch LL (2010) Parental influence on children's early eating environments and obesity risk: implications for prevention. Int J Obes 34(7):1116–1124
127. Birch LL, Anzman-Frasca S (2011) Learning to prefer the familiar in obesogenic environments. Nestle Nutr Workshop Ser Pediatr Program 68:187–196, discussion 196-9
128. Faith MS et al (2004) Parent-child feeding strategies and their relationships to child eating and weight status. Obes Res 12(11):1711–1722
129. Birch LL, Ventura AK (2009) Preventing childhood obesity: what works? Int J Obes 33(Suppl 1):S74–S81
130. Cutting TM et al (1999) Like mother, like daughter: familial patterns of overweight are mediated by mothers' dietary disinhibition. Am J Clin Nutr 69(4):608–613
131. Fisher JO, Birch LL (1999) Restricting access to palatable foods affects children's behavioral response, food selection, and intake. Am J Clin Nutr 69(6):1264–1272
132. Savage JS et al (2012) Serving smaller age-appropriate entree portions to children aged 3-5 y increases fruit and vegetable intake and reduces energy density and energy intake at lunch. Am J Clin Nutr 95(2):335–341
133. Savage JS et al (2012) Do children eat less at meals when allowed to serve themselves? Am J Clin Nutr 96(1):36–43
134. Morland K et al (2002) Neighborhood characteristics associated with the location of food stores and food service places. Am J Prev Med 22(1):23–29
135. Cullen KW et al (2003) Availability, accessibility, and preferences for fruit, 100% fruit juice, and vegetables influence children's dietary behavior. Health Educ Behav 30(5):615–626
136. Helling A, Sawicki DS (2003) Race and residential accessibility to shopping and services. Hous Policy Debate 14(1–2):69–101
137. Kelly B, Flood VM, Yeatman H (2011) Measuring local food environments: an overview of available methods and measures. Health Place 17(6):1284–1293

138. Lambourne K, Donnelly JE (2011) The role of physical activity in pediatric obesity. Pediatr Clin North Am 58(6):1481–1491, xi–xii
139. Loprinzi PD et al (2012) Benefits and environmental determinants of physical activity in children and adolescents. Obes Facts 5(4):597–610
140. Boyland EJ, Halford JC (2013) Television advertising and branding. Effects on eating behaviour and food preferences in children. Appetite 62:236–241
141. Cairns G et al (2013) Systematic reviews of the evidence on the nature, extent and effects of food marketing to children. A retrospective summary. Appetite 62:209–215
142. Pomeranz JL (2012) Advanced policy options to regulate sugar-sweetened beverages to support public health. J Public Health Policy 33(1):75–88
143. Duch H et al (2013) Screen time use in children under 3 years old: a systematic review of correlates. Int J Behav Nutr Phys Act 10(1):102
144. Marsh S, Ni Mhurchu C, Maddison R (2013) The non-advertising effects of screen-based sedentary activities on acute eating behaviours in children, adolescents, and young adults. A systematic review. Appetite 71C:259–273
145. Montoye AH et al (2013) Junk food consumption and screen time: association with childhood adiposity. Am J Health Behav 37(3):395–403
146. O'Connor TM et al (2013) Physical activity and screen-media-related parenting practices have different associations with children's objectively measured physical activity. Child Obes 9(5):446–453
147. Lampard AM, Jurkowski JM, Davison KK (2013) The family context of low-income parents who restrict child screen time. Child Obes 9(5):386–392
148. Steffen LM et al (2013) Relation of adiposity, television and screen time in offspring to their parents. BMC Pediatr 13(1):133
149. A.A.o. Pediatrics (2009) Media and children. AAP, Elk Grove Village, IL
150. Gunderson EP et al (2008) Association of fewer hours of sleep at 6 months postpartum with substantial weight retention at 1 year postpartum. Am J Epidemiol 167(2):178–187
151. Haines J et al (2013) Healthy habits, happy homes: randomized trial to improve household routines for obesity prevention among preschool-aged children. JAMA Pediatr 167(11): 1072–1079
152. Nevarez MD et al (2010) Associations of early life risk factors with infant sleep duration. Academic Pediatr 10(3):187–193
153. Schmidt ME et al (2009) Television viewing in infancy and child cognition at 3 years of age in a US cohort. Pediatrics 123(3):e370–e375
154. Taveras EM et al (2010) Racial/ethnic differences in early-life risk factors for childhood obesity. Pediatrics 125(4):686–695
155. Anderson SE, Whitaker RC (2010) Household routines and obesity in US preschool-aged children. Pediatrics 125(3):420–428
156. Al Mamun A et al (2007) Do childhood sleeping problems predict obesity in young adulthood? Evidence from a prospective birth cohort study. Am J Epidemiol 166(12):1368–1373
157. Knutson KL (2013) Sociodemographic and cultural determinants of sleep deficiency: implications for cardiometabolic disease risk. Soc Sci Med 79:7–15
158. Reiter RJ et al (2012) Obesity and metabolic syndrome: association with chronodisruption, sleep deprivation, and melatonin suppression. Ann Med 44(6):564–577
159. Weiss R, Bremer AA, Lustig RH (2013) What is metabolic syndrome, and why are children getting it? Ann N Y Acad Sci 1281:123–140
160. Jolin EM, Weller RA (2011) Television viewing and its impact on childhood behaviors. Curr Psychiatry Rep 13(2):122–128
161. Kuhl ES, Clifford LM, Stark LJ (2012) Obesity in preschoolers: behavioral correlates and directions for treatment. Obesity 20(1):3–29
162. Must A, Parisi SM (2009) Sedentary behavior and sleep: paradoxical effects in association with childhood obesity. Int J Obes 33(Suppl 1):S82–S86

163. Iannotti RJ, Wang J (2013) Trends in physical activity, sedentary behavior, diet, and BMI Among US adolescents, 2001-2009. Pediatrics 132(4):606–614
164. Eaton DK et al (2008) Youth risk behavior surveillance – United States, 2007. MMWR Morb Mortal Wkly 57(4):1–131
165. Demissie Z et al (2013) Electronic media and beverage intake among United States high school students-2010. J Nutr Educ Behav 45(6):756–760
166. Carter MA et al (2012) Availability and marketing of food and beverages to children through sports settings: a systematic review. Publ Health Nutr 15(8):1373–1379
167. Chandon P, Wansink B (2012) Does food marketing need to make us fat? A review and solutions. Nutr Rev 70(10):571–593
168. Frumkin H (2006) The measure of place. Am J Prev Med 31(6):530–532
169. Branscum P, Sharma M (2012) After-school based obesity prevention interventions: a comprehensive review of the literature. Int J Environ Res Publ Health 9(4):1438–1457
170. Marcus MD et al (2013) Lessons learned from the HEALTHY primary prevention trial of risk factors for type 2 diabetes in middle school youth. Curr Diabet Rep 13(1):63–71
171. Silveira JA et al (2013) The effect of participation in school-based nutrition education interventions on body mass index: a meta-analysis of randomized controlled community trials. Prev Med 56(3–4):237–243
172. Koplan JP, Liverman CT, Kraak VA (2005) Preventing childhood obesity: health in the balance. The National Academies Press, Washington, DC
173. Chin JJ, Ludwig D (2013) Increasing children's physical activity during school recess periods. Am J Public Health 103(7):1229–1234
174. Ickes MJ, Erwin H, Beighle A (2012) Systematic review of recess interventions to increase physical activity. J Phys Activ Health 10(6):910–926
175. Turner L, Chriqui JF, Chaloupka FJ (2013) Withholding recess from elementary school students: policies matter. J Sch Health 83(8):533–541
176. Larson N, Neumark-Sztainer D (2009) Adolescent nutrition. Pediatr Rev 30(12):494–496
177. Larson N, Story M (2010) Are 'competitive foods' sold at school making our children fat? Health Aff 29(3):430–435
178. Nelson Laska M et al (2010) Dietary patterns and home food availability during emerging adulthood: do they differ by living situation? Publ Health Nutr 13(2):222–228
179. Nelson MC et al (2009) Disparities in dietary intake, meal patterning, and home food environments among young adult nonstudents and 2- and 4-year college students. Am J Public Health 99(7):1216–1219
180. Robinson-O'Brien R et al (2009) Characteristics and dietary patterns of adolescents who value eating locally grown, organic, nongenetically engineered, and nonprocessed food. J Nutr Educ Behav 41(1):11–18
181. Cluss PA et al (2013) Nutrition knowledge of low-income parents of obese children. Transl Behav Med 3(2):218–225
182. Deng F, Zhang A, Chan CB (2013) Acculturation, dietary acceptability, and diabetes management among Chinese in North America. Front Endocrinol (Lausanne) 4:108
183. Doak CM et al (2006) The prevention of overweight and obesity in children and adolescents: a review of interventions and programmes. Obes Rev 7(1):111–136
184. Hoelscher DM et al (2004) School-based health education programs can be maintained over time: results from the CATCH Institutionalization study. Prev Med 38(5):594–606
185. Hoelscher DM et al (2003) How the CATCH eat smart program helps implement the USDA regulations in school cafeterias. Health Educ Behav 30(4):434–446
186. Hoelscher DM et al (2010) Reductions in child obesity among disadvantaged school children with community involvement: the Travis County CATCH Trial. Obesity 18(Suppl 1): S36–S44

Index

A

Abdel-Rahman, M.A., 23
Acheson Report, 20
Adipose tissue biology, 178–179
Adverse treatment effects, 67–70, 75–77
Affordable Care Act (ACA), 81, 82, 263
African American women, 38, 72–75, 80, 124,
126, 134, 136, 167, 180–182, 234, 235,
237, 239, 241, 243, 245, 251
Aging, 70, 91–107, 169, 217
Agurs-Collins, T., 63–83
Appalachia, 1–15

B

Barbeau, E., 268
Beydoun, M.A., 206
Black Report, 20, 25–26
Body fat distribution, 40, 80, 96, 133, 169,
173–178, 219
Bohlke, K., 119–126
Bowen, D.J., 259–276
Bradley, C.J., 70
Brancati, F.L., 134
Braveman, P., 286
Breast cancer, 6, 23, 41, 66, 101, 123, 167, 206,
233
Breast cancer survival, 72, 79

C

Cancer incidence, 5–8, 26, 38, 39, 43, 104, 107,
122, 123, 133, 167, 206, 208, 219
Cancer related fatigue (CRF), 69, 76–77, 251

Cancer survivorship, 63–83
Carson, K., 119–126
Carstairs deprivation index, 22
Chang, S.-H., 119–126
Cheng, I., 139
Cohen, S.S., 201–219
Colditz, G.A., 119–126
Coleman, M.P., 22, 23, 26
Colorectal cancer, 6–8, 14, 30, 40, 41, 46, 66,
67, 69, 73–75, 131–145, 167, 206
Colorectal cancer survival, 69
Community based interventions, 241–245
Community effects, 275
Conroy, S.M., 79, 80, 90
Consumer effects, 260
Cook, L., 37–53
Cortisol induced obesity, 171
Crosby, B., 47

D

Dai, J., 138
De Luis, D.A., 140
Dein, S., 29
Denis, G.V., 91–107
Dennis-Parker, E.A., 235
Diet, 2, 11–15, 25, 28, 41, 42, 52, 66, 80, 105–
107, 124, 132–135, 140, 141, 145, 173,
175, 181, 204, 207, 208, 234, 235, 238–
241, 244–246, 249–251, 262, 269, 272,
292
Dietary fat, 40–42, 141, 234, 235, 238, 239,
241, 244
Djuric, Z., 240

D.J. Bowen et al. (eds.), *Impact of Energy Balance on Cancer Disparities*,
Energy Balance and Cancer 9, DOI 10.1007/978-3-319-06103-0,
© Springer International Publishing Switzerland 2014

E
Early childhood, 284, 285, 289–294
Economically stressed neighborhoods, 237
Ectopic fat, 97, 175, 178, 181, 183
Endometrial cancer survival, 69, 73
Evidence based policy, 285, 287–289
Exercise, 9, 10, 52, 97, 98, 105–107, 133, 136,
        141, 145, 181, 235–240, 242–249, 291,
        294
Exercise barriers and facilities, 235, 238

F
Finnie, R., 29
Fish consumption, 41
Fisher, J.L., 1–15
Fontaine-Bisson, B., 140
Food availability and purchase policy, 262, 263
Freddy Wen, C.K., 37–53
Friedlander, Y., 139
Friedman, G.D., 125
Fruit and vegetable consumption, 11, 235, 244,
        272
Fu, O.S., 76

G
Garcia, R., 37–53
Gehlert, S., 63–83
Genetic susceptibility studies, 132, 144–145
Genome wide association studies (GWAS),
        137, 138, 145
Gewandter, J.S., 77
Goodin, 47
Goran, M.I., 165–184
Government effects, 260
GWAS. *See* Genome wide association studies
        (GWAS)

H
Haire-Joshu, D., 283–296
Harsh, J., 47
Hart, J.T., 31
Hasan, S., 77
Hasson, R.E., 165–184
Health
    behaviors, 34, 52, 53, 76, 235–239, 242,
        244, 286
    inequalities, 19, 20, 25, 26, 28, 29, 33
Heath, F., 33
Henderson, S.O., 80
Hepatic fat, 173, 176–178, 181, 183

Herrinton, L.J., 125
Hicken, M.T., 47
Hourdequin, K.C., 68
Hsu, Y.-W., 37–53
Huffman, D.M., 91–107
Hulver, M.W., 214
Hyperinsulinemia, 96, 134, 167–174, 177, 183,
        184, 218, 241, 250

I
IGF. *See* Insulin-like growth factor (IGF)
Inflammation, 2, 67, 92, 96–103, 132, 133,
        135–137, 141, 143, 145, 171, 175, 176,
        179–181, 184, 209, 212–213, 218, 219,
        234, 241, 250, 259–275
Inflammatory environment, 67, 260
Insulin, 6, 40, 67, 96, 132, 165, 209,
        234, 289
Insulin and inflammatory cytokines, 132
Insulin-like growth factor (IGF), 96, 133, 134,
        137, 172, 179, 209, 210, 234, 299
Insulin resistance, 96, 97, 99, 100, 132–135,
        138–141, 145, 165–184, 209–211, 214,
        215, 218, 219, 234
Intramyocellular lipid, 175–176, 181
Inverse Care Law, 31

K
Kang, M., 131–145
Kawachi, I., 26
Keku, T.O., 131–145, 141, 142
Kroenke, C., 26
Kwan, M.L., 80, 81

L
LeBourgeois, M.K., 47
LeBrasseur, N.K., 91–107
Lewis, J.M., 31
Lipworth, L., 201–219
Liu, L., 139
Lyratzopoulos, G., 23

M
Macintyre, S., 25
Macleod, U., 19–33
Marmot Review, 20
Marshall, S.J., 204
McGuire, L., 47
McLaren, L., 205

Mennella, J.A., 292
Menu labeling, 261–266, 273, 275
Monoclonal gammopathy of unknown
    significance (MGUS), 121–123, 126
Moore, L.L., 96
Morimoto, L.M., 139
Morland, K.B., 270
Moser, K., 29
Moss, M., 29
Moving Forward, 235, 241–251
M protein, 121, 126
Multiple myeloma, 70, 119–126

N
Nader, P.R., 290
National Health Service Cancer Plans, 20
Nehring, I., 291
Neuhouser, M.L., 63–83
Nguyen-Rodriguez, S.T., 37–53
Nicolson, D., 19–33
Non-esterified fatty acids (NEFA), 181–182
Nutrition environment, 260–263, 266–274

O
Obesity, 2, 28, 38, 64, 92, 122, 132, 166, 201,
    234, 260, 283
Obesity prevention, 270, 284, 287, 288, 295
Obesogenic environment, 274
Organizational effects, 267
Oxidative stress, 104, 182–184

P
Palmer, N.R., 76
Pancreatic fat, 177–178, 183
Paskett, E.D., 1–15
Patel, N.P., 46
Patient Protection and Affordable Health Care
    Act, 66, 71
Pechlivanis, S., 139
Physical activity, 2, 28, 39, 66, 92, 122, 132,
    169, 204, 234, 261, 294
Pickett, K., 33
Plascak, J.J., 1–15
Pollack, L., 63–83
Poverty, 4, 13, 46, 81, 82, 103, 261, 286, 293,
    295
Powe, B.D., 29
Prevention and Public Health Fund, 81, 82
Prostate cancer survival, 69, 76
Psychological stress, 103, 171

Q
Quality of life, 32, 64, 68, 69, 75–76, 93, 234–
    237, 242, 246, 251

R
Rachet, B., 22
Racial differences
    adiponectin, 214–215
    inflammation, 212–213
    insulin-IGF-1, 209–211
    leptin, 216–217
    sex steroid hormones, 211–212
Racial disparities in obesity, 203, 260
Racial/ethnic disparities, 50, 52, 53
Raine, R., 32
Rasmussen, S.K., 140
Red meat consumption, 41
Reilly, G.A.O., 37–53
Robb, K., 30
Rural health, 11

S
Schmitz, K.H., 63–83
Schutte, A.E., 135
Self-efficacy, 243–245, 248, 271
Senescence, 95, 96, 99
SES. *See* Socio economic status (SES)
Single nucleotide polymorphism (SNP), 131–
    145
Sleep, 39, 45–48, 52, 53, 76, 290, 292, 294
Smith, M.T., 47
Smoking, 6, 25, 28, 29, 38, 39, 48–53, 79, 80,
    93, 132, 140, 144, 145, 169, 181, 271,
    275
Smoldering multiple myeloma, 121
SNP influence on adipokines, 132, 138–140
Social cognitive therapy, 244
Social determinants, 81, 92, 103, 284
Socio-ecological model (SEM), 244
Socioeconomic disparities, 38–43, 45–49, 103
Socio economic status (SES), 26, 28–32,
    42–49, 71, 73–76, 81, 205, 206, 234
Soda tax, 262, 265, 266, 275
Sontag, S., 29
Sorenson, G., 268
Soto-Salgado, M., 73
Special Supplemental Nutrition Assistance
    Program (SNAP), 261–265
Spruijt-Metz, D., 37–53
Sridhar, G., 74
Sternthal, M.J., 53

Stolley, M., 233–251
Sugar consumption, 40

T
Takatsuno, Y., 138
Target environments, 285, 287–289, 294
Tomfohr, L., 47
Troiano, R.P., 250
Tsilidis, K.K., 143

U
Unger, J.B., 37–53
United Kingdom, 19–33, 273

V
Vasku, A., 139
Vitamin D, 106, 126, 132, 136, 139

W
Waldenström Macroglobulinemia, 121
Wang, Y., 206, 285

Wassel Fyr, C., 214
Weier, R.C., 1–15
Weight loss, 48, 69, 92, 100, 105, 106, 133,
          135, 136, 174, 213, 234, 235, 238–248,
          251, 267
Weight loss barriers and facilitators, 235, 237
Weller, D., 19–33
White, C.A., 32
Wilkinson, R., 33
Wong, H.L., 139
Woo, J.G., 140

X
Xing, L.L., 144

Y
Yabroff, K.R., 70

Z
Zanetti, K.A., 144
Zawacki, S., 259–276

Printed by Printforce, the Netherlands